Studies in Computational Intelligence

Volume 1121

Series Editor

Janusz Kacprzyk, Polish Academy of Sciences, Warsaw, Poland

The series "Studies in Computational Intelligence" (SCI) publishes new developments and advances in the various areas of computational intelligence—quickly and with a high quality. The intent is to cover the theory, applications, and design methods of computational intelligence, as embedded in the fields of engineering, computer science, physics and life sciences, as well as the methodologies behind them. The series contains monographs, lecture notes and edited volumes in computational intelligence spanning the areas of neural networks, connectionist systems, genetic algorithms, evolutionary computation, artificial intelligence, cellular automata, self-organizing systems, soft computing, fuzzy systems, and hybrid intelligent systems. Of particular value to both the contributors and the readership are the short publication timeframe and the world-wide distribution, which enable both wide and rapid dissemination of research output.

Indexed by SCOPUS, DBLP, WTI Frankfurt eG, zbMATH, SCImago.

All books published in the series are submitted for consideration in Web of Science.

Farhad Hosseinzadeh Lotfi · Tofigh Allahviranloo ·
Witold Pedrycz · Mohammadreza Shahriari ·
Hamid Sharafi · Somayeh Razipour GhalehJough

Fuzzy Decision Analysis: Multi Attribute Decision Making Approach

Springer

Farhad Hosseinzadeh Lotfi
Department of Mathematics,
Science and Research Branch
Islamic Azad University
Tehran, Iran

Witold Pedrycz
Department of Electrical and Computer Engineering
University of Alberta
Edmonton, AB, Canada

Hamid Sharafi
Department of Mathematics,
Science and Research Branch
Islamic Azad University
Tehran, Iran

Tofigh Allahviranloo
Faculty of Engineering & Natural Sciences
Istinye University
Istanbul, Turkey

Mohammadreza Shahriari
Department of Industrial Management,
South Tehran Branch
Islamic Azad University
Tehran, Iran

Somayeh Razipour GhalehJough
Department of Mathematics,
Science and Research Branch
Islamic Azad University
Tehran, Iran

ISSN 1860-949X ISSN 1860-9503 (electronic)
Studies in Computational Intelligence
ISBN 978-3-031-44741-9 ISBN 978-3-031-44742-6 (eBook)
https://doi.org/10.1007/978-3-031-44742-6

This Springer imprint is published by the registered company Springer Nature Switzerland AG
The registered company address is: Gewerbestrasse 11, 6330 Cham, Switzerland

Paper in this product is recyclable.

Preface

In the digital age, our daily lives are constantly evolving due to rapid technological advancements. This transformation has had a profound impact on businesses, increasing their reliance on technology and necessitating a greater level of information, knowledge, and expertise. Concepts such as digital transformation, digital economy, and artificial intelligence have reshaped the competitive landscape, making survival and success more challenging for organizations compared to the past.

Today, businesses cannot thrive without recognizing the significance of competition in the contemporary sense. The advent of digital transformation has given rise to numerous start-up ventures that have achieved multi-billion-dollar valuations, while simultaneously leading to the downfall of several established companies. This reality necessitates a departure from traditional economic, industrial, and business approaches, as they may no longer be effective in the face of digital disruption.

To remain competitive, organizations must embrace digitalization and leverage the opportunities provided by technological advancements. This entails adapting to changing market dynamics, implementing innovative strategies, and incorporating digital tools and solutions into their operations. It requires a shift in mindset and a willingness to embrace new approaches that align with the demands of the digital economy.

In this context, businesses must continuously evolve, learn, and acquire the necessary skills to navigate the complexities of the digital age. They need to embrace agility and flexibility, continuously innovate, and stay abreast of emerging technologies. Only by embracing these changes can businesses effectively compete and thrive in today's dynamic and highly competitive landscape.

Throughout history, the goals of industrial revolutions have evolved from a focus on increasing production to the optimization of operations and decision-making. The first, second, and third industrial revolutions witnessed a shift from manual labor to machine-based production, revolutionizing industries such as agriculture and transportation. However, as time progressed, the sole emphasis on production quantity and cost-effectiveness became insufficient.

The concept of productivity became intricately linked with optimization techniques and research science in operations, leading to the development of various models and approaches across different industries. These advancements symbolized the onset of the fourth and subsequent industrial revolutions. Marketing also became an integral part of this paradigm, as the competitiveness of businesses relied heavily on effective decision-making.

In the era of digital transformation, characterized by technological breakthroughs like artificial intelligence, robotics, cryptocurrency, blockchain, cloud computing, and the Internet of Things, decision-making has gained unparalleled significance. It is not only crucial for maintaining current achievements, but also acts as the foundation for future advancements.

In today's complex and uncertain world, the ability to make accurate decisions has become the key to success and fulfillment. This process requires a sophisticated combination of science, rationality, art, logic, introspection, and intuition. Decision-making serves as the central axis that shapes the path toward progress and prosperity. It is a dynamic interplay between various factors and is vital in navigating the complexities of the digital age and unlocking its transformative potential.

This book delves into the psychological foundations that influence our decision-making processes. By drawing upon insights from cognitive science, behavioral economics, and psychology, we aim to explore the intricacies of human thinking and the biases that often hinder our judgment. Through this exploration, we hope to shed light on the fascinating interplay between conscious and unconscious thought processes, equipping readers with heightened awareness and a stronger foundation for making decisions based on quantitative and mathematical approaches.

While the introductory section provides an overview of decision theory fundamentals, the focus of this book revolves around multi-criteria decision-making (MADM) models. These models offer a framework and methods for evaluating and comparing options based on diverse criteria. In this book, we extensively analyze and examine this specific branch, discussing its applications and implications in both precise and imprecise data environments. We acknowledge the reality of imprecise data, and the inherent uncertainties and ambiguities often present in real-world decision-making scenarios.

In such environments, logical and fuzzy approaches, as well as interval calculus, take center stage. These methodologies enable decision-makers to navigate the complexities of qualitative and imprecise information. We delve into the intricacies of this approach, providing insights into how to implement logical processes when employing multi-criteria decision models in the face of ambiguous or uncertain data.

Furthermore, we highlight that the science of decision-making is now considered an interdisciplinary field, finding applications in various domains such as applied mathematics, economics, management, engineering, healthcare, accounting, meteorology, environmental studies, and related industries and businesses. The utility and versatility of decision-making principles are widespread across these fields.

Throughout this book, we aim to equip readers with practical knowledge and tools to enhance their decision-making capabilities. By combining theoretical foundations with real-world applications, we strive to provide a comprehensive resource that can benefit individuals across diverse disciplines and industries.

Tehran, Iran	Farhad Hosseinzadeh Lotfi
Istanbul, Turkey	Tofigh Allahviranloo
Edmonton, Canada	Witold Pedrycz
Tehran, Iran	Mohammadreza Shahriari
Tehran, Iran	Hamid Sharafi
Tehran, Iran	Somayeh Razipour GhalehJough

Acknowledgements

We would like to take this moment to express our heartfelt gratitude to the remarkable individuals whose unwavering support and contributions have played an invaluable role in the creation of this book.

First and foremost, we extend our deepest appreciation to esteemed researchers, university professors, and authors of articles and books in the field of decision-making. Your extensive knowledge, groundbreaking research, and insightful perspectives have been instrumental in shaping the content of this book.

A special thank you goes to the Iranian DEA society for their unwavering spiritual support and consensus in the writing of this book. Your guidance, encouragement, and firm belief in the importance of decision-making have provided us with a solid framework to explore and delve into the depths of this subject. Your invaluable support has been truly remarkable, and we are deeply grateful for the opportunity to collaborate with such esteemed professionals.

To our family and friends, we cannot express enough gratitude for your unwavering support and encouragement. Your patience during our writing endeavors and your constant motivation have been the driving force behind our efforts.

I want to express my sincere appreciation for the dedicated efforts and persistent follow-up Ms. Saranya Kalidoss has invested in the development of the book titled *Fuzzy Decision Analysis: Multi-Attribute Decision-Making Approach.* Her commitment to this project has not gone unnoticed.

I would also like to extend our heartfelt gratitude to all the team members at Springer Publisher for their invaluable contributions. The collaborative spirit exhibited throughout the entire process has undoubtedly played a significant role in enhancing the quality and success of the book. Their collective dedication and expertise have truly made a difference, and we are genuinely appreciative of their efforts.

Once again, we extend our heartfelt appreciation to all those who have contributed to this book in various ways. Your support, expertise, and guidance have played an instrumental role in its creation, and we are truly grateful for your invaluable contributions.

Farhad Hosseinzadeh Lotfi
Tofigh Allahviranloo
Witold Pedrycz
Mohammadreza Shahriari
Hamid Sharafi
Somayeh Razipour GhalehJough

Introduction

Decision-making is the compass that guides us through the labyrinth of possibilities, empowering us to transform uncertainty into opportunity and shape our own destiny.

Henry Mintzberg, a prominent Professor, and strategist in the field of management studies and decision-making behavioral obedience, introduces the most important role of a manager. With the consensus of famous researchers in the field of management, decision-making is both science and art. This statement captures the complex and multifaceted nature of the process that governs our choices and actions. In this introduction, we begin the exploration of decision-making as a discipline that combines scientific principles and artistic sensibilities. At its core, decision-making is a science—an intellectual pursuit guided by rigorous analysis, systematic methodology, and empirical evidence, and drawing on diverse scientific fields such as psychology, economics, management, mathematics, and computer science to Reveal the underlying mechanisms and patterns that shape our choices.

The scientific side of decision-making provides us with valuable frameworks, models, and tools for navigating the complexities of a choice. From precise and deterministic decision-making models to imprecise and qualitative approaches, all these scientific foundations equip us with a systematic approach to evaluation.

However, apart from being a science, decision-making is also an art. A complex dance of intuition, creativity, and judgment that is deeply influenced by our values, experiences, and feelings, and our inner intuition and perception. The human element in decision-making introduces nuances that cannot be fully captured by scientific formulas or algorithms. Therefore, it requires us to consider subjective factors such as intuition and perception, social dynamics, and ethical considerations in the decision-making process.

The artistic dimension of decision-making is reflected in our ability to synthesize information, imagine possibilities, and make choices that align with our desires and values. It requires intuition, empathy, and the ability to navigate ambiguity and uncertainty. The art of decision-making lies in the ability to balance logical analysis with emotional intelligence, combining logic with intuition to make choices that go beyond data-driven calculations.

In this essay on the realms of decision-making, we will explore the synergistic relationship between the science and art of decision-making, addressing the theories, models, and techniques that form the foundation of decision science. At the same time, we will understand the importance of judgment, creativity, and human intuition in shaping our decisions and embracing the artistic aspects of this process.

Additionally, expanding on the previous explanations and in accordance with the book's title, we will delve into the domain of decision-making science and examine the topic of Multi-Criteria Decision-Making.

In the dynamic and rapidly changing world of today, leaders encounter progressively intricate and demanding circumstances that necessitate thoughtful evaluation of numerous attributes or criteria. Multiple-Attribute Decision-Making (MADM) has emerged as a crucial area of research, providing valuable perspectives and resources to assist decision-makers in their pursuit of optimal resolutions.

This book takes a comprehensive and in-depth approach to MADM, delving into various facets of the discipline to equip readers with a solid understanding of its principles and applications. It explores the intricate nuances of decision-making problems in uncertain and fuzzy environments, where traditional decision-making techniques may fall short.

One fundamental aspect covered in this book is the identification and definition of attributes and criteria. It provides guidance on how to select relevant factors that capture the essence of the decision problem and reflect the decision-maker's preferences. By carefully crafting these attributes, decision-makers can gain a holistic view of the problem at hand and consider multiple dimensions simultaneously.

Furthermore, this book goes beyond the conventional deterministic approaches and explores the challenges posed by uncertainty and fuzziness. It investigates various methodologies and techniques that address these complexities, enabling decision-makers to make informed choices in situations where information is incomplete, imprecise, or subjective.

Among the methods discussed in this book, readers will find valuable insights into fuzzy logic, which offers a flexible framework for handling imprecise information. Fuzzy sets and interval calculus provide decision-makers with the tools to model and reason about uncertainty, allowing them to assess the consequences of their decisions under varying conditions.

Now, we will explore the interpretation of the chapter titles and expected learning objective in the book:

Chapter 1: "Foundations of Decision"

Learning Objective: By the end of this chapter, readers will have a solid understanding of the foundational elements of decision-making. They will be able to explain the key concepts and principles involved in the decision-making process, comprehend the role of decision theory and decision science, and analyze the historical context of decision-making. Additionally, readers will gain insights into the reputable domains and applications of decision-making and understand the significance of scale measurements in data analysis.

Chapter 2: "Fuzzy Introductory Concepts"

Learning Objective: Upon completing this chapter, readers will be able to grasp the fundamental principles of fuzzy mathematics and its application in decision-making. They will develop an understanding of fuzzy set theory, fuzzy numbers, and fuzzy logic, and how these concepts help handle uncertainty and vagueness. Additionally, readers will learn about fuzzy ranking and its role in Multiple-Attribute Decision-Making (MADM), enabling them to effectively consider imprecise and uncertain information in decision-making processes.

Chapter 3: "Weight Determination Methods in Fuzzy Environment"

Learning Objective: At the conclusion of this chapter, readers will have a comprehensive understanding of various methods for determining attribute weights in fuzzy decision-making contexts. They will be able to apply techniques such as fuzzy analytic hierarchy process (AHP), fuzzy entropy, and fuzzy preference programming to assign appropriate weights to attributes. This chapter will equip readers with the skills necessary to handle imprecision and uncertainty when determining the importance of attributes in decision-making processes.

Chapter 4: "Non-Compensatory Methods in Uncertainty Environment"

Learning Objective: By the end of this chapter, readers will be able to analyze non-compensatory decision-making methods and their application in uncertain decision scenarios. They will understand the concept of non-compensatory relationships among attributes and how methods like lexicographic ordering and satisficing approaches effectively handle uncertainty. Readers will develop the ability to make decisions that consider non-compensatory relationships and effectively address uncertainty in decision-making processes.

Chapter 5: "Simple Additive Weighting (SAW) Method in Fuzzy Environment"

Learning Objective: Upon completing this chapter, readers will have a comprehensive understanding of the Simple Additive Weighting (SAW) method and its application in Multiple-Attribute Decision-Making (MADM). They will be able to assign weights to attributes, compute overall performance scores for alternatives, and make informed decisions based on the SAW method. Readers will gain insights into the benefits, limitations, and practical applications of the SAW method in decision-making contexts.

Chapter 6: "Technique for Order Preferences by Similarity to Ideal Solutions (TOPSIS) in Uncertainty Environment"

Learning Objective: At the conclusion of this chapter, readers will be able to apply the Technique for Order of Preference by Similarity to Ideal Solution (TOPSIS) in uncertain decision environments. They will understand how TOPSIS identifies the most favorable alternative by evaluating its proximity to the ideal solution and how it handles uncertainty by considering fuzzy numbers and uncertainty analysis. Readers will gain practical knowledge on implementing TOPSIS effectively in decision-making processes characterized by uncertainty.

Chapter 7: "Elimination Choice Translating Reality (ELECTRE) in Uncertainty Environment"

Learning Objective: By the end of this chapter, readers will have a comprehensive understanding of the ELECTRE method and its application in decision-making problems involving uncertainty. They will learn how ELECTRE utilizes outranking relationships to rank alternatives and how it addresses uncertainty by handling imprecise data and conducting sensitivity analysis. This chapter will equip readers with the skills to effectively utilize ELECTRE in uncertain decision scenarios and analyze uncertain information within the decision-making process.

Chapter 8: "Analytical Hierarchy Process (AHP) in Fuzzy Environment"

Learning Objective: Upon completing this chapter, readers will possess a thorough understanding of the Analytic Hierarchy Process (AHP) and the Analytic Network Process (ANP) as powerful decision-making methods. They will comprehend the principles, steps, and applications of AHP and ANP in handling complex decision problems. Readers will gain insights into how AHP and ANP can address uncertainty and fuzzy information in Multiple-Attribute Decision-Making (MADM). This chapter will provide practical guidance on utilizing AHP and ANP effectively in decision-making processes characterized by uncertainty, offering real-world examples and highlighting their practical applications.

Chapter 9: "VIKOR Method in Uncertainty Environment"

Learning Objective: At the conclusion of this chapter, readers will be proficient in applying the VlseKriterijumska Optimizacija I Kompromisno Resenje (VIKOR) method in decision problems influenced by uncertainty and imprecise data. They will understand how VIKOR, as a compromise-based Multiple-Attribute Decision-Making (MADM) technique, balances the maximum group utility and the minimum individual regret. This chapter will provide insights into effectively handling imprecise data and addressing uncertainty using VIKOR, enabling readers to implement it in uncertain decision environments.

Chapter 10: "The Measuring Attractiveness by a Categorical Based Evaluation Technique (MACBETH) in Uncertainty Environment"

Learning Objective: By the end of this chapter, readers will possess a comprehensive understanding of the Measuring Attractiveness by a Categorical Based Evaluation Technique (MACBETH) and its application in decision-making processes affected by uncertainty. They will learn how MACBETH utilizes pairwise comparisons and interval scales to evaluate alternatives based on multiple criteria. Readers will gain practical knowledge on how MACBETH handles uncertainty and imprecise evaluations within decision-making contexts, enabling them to effectively navigate uncertain environments and implement MACBETH in practice.

Chapter 11: "Multi-Attributive Border Approximation Area Comparison (MABAC) in Uncertainty Environment"

Learning Objective: Upon completing this chapter, readers will be proficient in applying the Multi-Attributive Border Approximation Area Comparison (MABAC) method in decision problems characterized by uncertainty and imprecise data. They will understand how MABAC, utilizing fuzzy sets and ranking methods, assesses alternatives considering both positive and negative ideal solutions. This chapter will provide insights into effectively handling uncertainty and imprecise information using MABAC, enabling readers to make informed decisions in uncertain decision-making processes.

Chapter 12: "The COmplex PRoportional ASsessment (COPRAS) in Uncertainty Environment"

Learning Objective: At the conclusion of this chapter, readers will have a comprehensive understanding of the Complex Proportional Assessment (COPRAS) method and its application in decision-making processes characterized by uncertainty and imprecision. They will comprehend how COPRAS utilizes fuzzy preference relations and decision matrices to rank alternatives across multiple criteria. This chapter will equip readers with practical knowledge on effectively handling uncertainty and imprecision in decision-making processes using COPRAS, enabling them to implement it in uncertain decision environments.

Chapter 13: "The Criteria Importance Through Inter-Criteria Correlation (CRITIC) in Uncertainty Environment"

Learning Objective: By the end of this chapter, readers will be proficient in applying the Criteria Importance Through Inter-Criteria Correlation (CRITIC) method in decision-making processes characterized by uncertainty. They will understand how CRITIC is designed to determine the importance of criteria and their relationships in decision-making. Readers will gain insights into effectively navigating and assessing interdependencies among criteria in uncertain decision scenarios using CRITIC, enabling them to make well-informed decisions under uncertain conditions.

Chapter 14: "The Multi-objective Optimization Ratio Analysis (MOORA) in Uncertainty Environment"

Learning Objective: Upon completing this chapter, readers will possess a comprehensive understanding of the Multi-objective Optimization Ratio Analysis (MOORA) method and its application in decision-making problems involving conflicting objectives. They will learn how MOORA combines ratio analysis and optimization techniques to determine the most favorable alternative in uncertain decision environments. Readers will gain insights into the principles and practical applications of MOORA, including its effectiveness in handling uncertainty and optimizing decision outcomes in complex decision scenarios.

In conclusion, this book provides readers with a comprehensive exploration of multiple attribute decision-making (MADM) methods in uncertain and fuzzy environments. Each chapter focuses on a specific method or concept and offers a separate learning objective to ensure readers' understanding and mastery of the subject matter. By the end of the book, readers will possess the knowledge and skills necessary to apply a range of MADM methods effectively in decision-making processes characterized by uncertainty, vagueness, and imprecise data.

Each chapter of the book contributes to a deeper understanding and application of MADM in complex decision scenarios. By integrating theoretical concepts with practical examples, this book serves as a valuable resource for decision-makers, equipping them with the necessary tools and knowledge to effectively tackle real-world decision problems. It provides a comprehensive overview of various MADM methods and their suitability for uncertain and fuzzy environments, empowering decision-makers to make informed choices and achieve favorable outcomes.

Farhad Hosseinzadeh Lotfi
Tofigh Allahviranloo
Witold Pedrycz
Mohammadreza Shahriari
Hamid Sharafi
Somayeh Razipour GhalehJough

Contents

Chapter 1
Foundations of Decision

1.1 Introduction

Welcome to the opening chapter of this book, where we embark on a captivating exploration of the concepts that lie at the heart of decision theory, decision science, and the art of decision making. In this chapter, we will delve into the depths of these interconnected fields, unraveling their intricacies and uncovering the profound impact they have on our lives.

Decision making is a constant companion on our journey through life, guiding our choices, shaping our paths, and molding our destinies. It is a process that encompasses a myriad of factors, from rational analysis to intuitive leaps, from calculated risks to emotional considerations. By delving into the concepts of decision theory and decision science, we equip ourselves with powerful frameworks and methodologies to navigate this intricate landscape with greater clarity and purpose.

Furthermore, we will shine a light on the field of decision science, which serves as an interdisciplinary bridge, drawing from mathematics, economics, psychology, and other fields to provide us with a holistic understanding of decision making. Through the lens of decision science, we gain insights into the cognitive biases that influence our judgments, the systematic processes that underlie our choices, and the ways in which we can optimize our decision-making strategies.

So, join us in this captivating journey of discovery and enlightenment as we delve into the realms of decision theory and decision science. Together, we will unravel the mysteries, broaden our perspectives, and hone our decision-making prowess, empowering ourselves to navigate the complexities of life with confidence and wisdom [1, 2].

F. Hosseinzadeh Lotfi et al., *Fuzzy Decision Analysis: Multi Attribute Decision Making Approach*, Studies in Computational Intelligence 1121,
https://doi.org/10.1007/978-3-031-44742-6_1

1.2 Decision Theory

Decision Theory is a field of study that explores the process of making choices and decisions in various contexts. It provides a framework for understanding how individuals and organizations assess alternatives, evaluate potential outcomes, and select the best course of action based on rationality, preferences, and available information. The history of decision theory dates to ancient times, where philosophers and thinkers pondered the intricacies of decision-making. However, it wasn't until the mid-twentieth century that decision theory as a formal field of study began to take shape.

At its core, decision theory seeks to answer fundamental questions such as: How do we make decisions? What factors influence our choices? How can we optimize decision-making to achieve desired outcomes? By examining the principles, models, and techniques of decision theory, we can gain insights into the complexities of decision-making and enhance our ability to make informed and effective decisions.

Decision theory is an interdisciplinary field that intersects with various disciplines, including Philosophy, Mathematics, Economics, Engineering Sciences, Management, Psychology, Stock Market, Politics, Negotiation techniques, etc. It is widely recognized as one of the most utilized sciences, with ongoing development and applications across different domains. By drawing upon sciences and subjects such as operations research, probability theory, game theory, utility theory, and behavioral science, decision theory offers a comprehensive understanding of decision-making processes. Whether applied to personal life, business, public policy, manufacturing, performance evaluation, finance and accounting, planning, or any other field, decision theory equips us with valuable tools and frameworks for analyzing and enhancing decision-making practices.

In the continuation of this section, we will present the history and development of decision theory in detail [3–5].

1.3 Existential Philosophy of Decision Theory

The existential philosophy of decision theory is an approach that incorporates existentialist perspectives into the study of decision-making. It examines the existential dimensions of decision theory, focusing on the individual's subjective experience, freedom, responsibility, and the search for meaning in decision-making processes.

It provides a unique perspective that complements traditional decision theory approaches. It offers insights into the subjective and existential aspects of decision-making, emphasizing individual agency, personal responsibility, and the pursuit of meaning and authenticity in decision-making processes [6–8].

Key aspects of the existential philosophy of decision theory include:

- Subjective Experience: It emphasizes the subjective and personal experience of decision-making. The existential perspective recognizes that decision-making is deeply intertwined with an individual's unique thoughts, emotions, desires, and values. It explores how these subjective aspects shape the decision-making process and outcomes.

- Freedom and Authenticity: Existential philosophy emphasizes the freedom of the individual to make choices. In decision theory, this perspective highlights the significance of personal agency and the responsibility that comes with it. It explores how individuals can make authentic decisions that align with their values, beliefs, and sense of self.
- Anxiety and Uncertainty: Existentialism acknowledges the inherent anxiety and uncertainty that accompany decision-making. The existential philosophy of decision theory examines how individuals grapple with the anxiety of choosing and the inherent uncertainty of outcomes. It explores how individuals confront these challenges and make decisions despite the existential anxieties they may experience.
- Meaning and Purpose: Existentialism is concerned with the search for meaning and purpose in life. The existential philosophy of decision theory explores how decision-making plays a role in individuals' quest for meaning. It considers how decisions can be seen as expressions of one's values, desires, and aspirations, and how they contribute to a sense of purpose in life.
- Authenticity and Self-Reflection: Existential philosophy emphasizes self-reflection and self-awareness in decision-making. It encourages individuals to examine their own values, motives, and biases to make choices that align with their authentic selves. The existential philosophy of decision theory explores how self-reflection and introspection can enhance the quality of decisions and promote personal growth.
- Ethical Implications: The existential philosophy of decision theory also considers ethical dimensions. It examines how decisions can be morally significant and how individuals can navigate ethical dilemmas within the context of their personal values and the impact on others [9–11].

The existential philosophy of decision theory offers a distinctive viewpoint that complements conventional approaches in decision theory. By delving into the subjective and existential dimensions of decision-making, it provides valuable insights into the intricacies of human experience when faced with choices. This perspective places emphasis on the individual's sense of agency, emphasizing their capacity to act and shape their own decisions. It also underscores the significance of personal responsibility, as individuals are encouraged to take ownership of their choices and acknowledge the consequences that arise from them [12].

Moreover, the existential philosophy of decision theory recognizes the profound human desire to find meaning and authenticity in decision-making processes. It acknowledges that decisions are not merely analytical calculations or objective assessments, but rather deeply intertwined with one's values, beliefs, and aspirations. By exploring the existential aspects of decision-making, individuals are prompted to reflect on their own values, motives, and biases, enabling them to make choices that align with their authentic selves [13].

This existential perspective adds a crucial layer of understanding to decision theory by highlighting the subjective, existential, and ethical dimensions that influence the decision-making process. It recognizes that decision-making is a deeply human endeavor, where individuals navigate the complexities of uncertainty, anxiety, and the search for personal fulfillment. By embracing this perspective, decision-makers are encouraged to engage in self-reflection, confront existential anxieties, and make choices that contribute to a greater sense of meaning and purpose in their lives.

1.4 Decision Science

In this opening part, we embark on a comprehensive exploration of decision science, aiming to provide readers with a clear understanding of its essence, scope, objectives, and fundamental principles. Decision science is introduced as a dynamic and interdisciplinary field that draws upon a diverse range of disciplines, including mathematics, statistics, economics, psychology, and computer science. It serves as a framework for studying and comprehending the intricacies of decision-making processes.

Decision science, at its core, seeks to unravel the complexities of decision-making by employing a systematic and analytical approach. It involves the application of rigorous methodologies, models, and theories to enhance our understanding of how individuals, groups, and organizations make decisions in various contexts. By integrating insights from multiple disciplines, decision science offers a holistic perspective that encompasses both rational and behavioral aspects of decision-making.

The scope of decision science extends across a broad spectrum of domains, encompassing business, management, finance, healthcare, public policy, environmental management, and more. Its methodologies and principles are applicable to diverse decision-making scenarios, ranging from routine operational choices to high-stakes strategic decisions [12, 14].

Decision science is a multidisciplinary field that utilizes quantitative techniques to aid decision-making, spanning from individual choices to societal contexts. It encompasses various methodologies, including decision analysis, risk assessment, cost–benefit analysis, optimization within constraints, simulation modeling, and behavioral decision theory. Drawing upon diverse disciplines such as Operations Research,[1] microeconomics, statistical inference, management control, cognitive

[1] Operations Research (OR) is a field of study that utilizes advanced analytical methods to facilitate improved decision-making. It draws upon mathematical sciences, such as mathematical modeling, statistical analysis, and mathematical optimization, to find optimal or near-optimal solutions to complex decision problems. With a focus on practical applications and human-technology interaction, operations research intersects with disciplines like industrial engineering, operations management, psychology, and organization science.

The primary objective of operations research is to determine the best possible outcome for a given real-world objective. This could involve maximizing profit, performance, or yield, or minimizing

and social psychology, and computer science, decision science provides a unique framework for understanding and addressing public health issues. By prioritizing the examination of decisions, themselves, decision science facilitates a comprehensive understanding of these challenges and enables the development of effective policies to tackle them. While many fields focus on generating new knowledge, decision science stands out by emphasizing the attainment of optimal choices based on existing information. It aims to uncover the scientific aspects and value judgments that underlie these decisions, while also highlighting the potential trade-offs associated with different courses of action or inaction. Decision science employs a diverse range of tools, including models for decision-making under uncertain conditions, experimental and descriptive studies of decision-making behavior, economic analysis of competitive and strategic choices, and techniques for facilitating group decision-making. Additionally, mathematical modeling techniques play a crucial role. The applications of decision science span various domains such as business, management, law, education, environmental regulation, military science, public health, and public policy [15].

The objectives of decision science are multifaceted. Firstly, it aims to provide decision-makers with analytical tools and frameworks to support informed decision-making. Decision science equips individuals with techniques to assess alternatives, evaluate risks and uncertainties, and optimize outcomes based on specific criteria and constraints.

Secondly, decision science seeks to improve the quality of decisions by incorporating a scientific and evidence-based approach. By leveraging empirical data, statistical analysis, and mathematical models, decision science enables decision-makers to make well-informed choices, minimizing the impact of biases, subjective judgments, and intuitive errors.

Thirdly, decision science explores the underlying cognitive and psychological processes that influence decision-making. It delves into the realm of human behavior, examining the factors that affect choices, including heuristics, biases, emotions, and social influences. By understanding these behavioral aspects, decision science strives to enhance decision-making effectiveness and design interventions to mitigate decision biases.

Fundamental principles underpin decision science, guiding its methodologies and frameworks. These principles include rationality, consistency, utility, risk analysis, and value trade-offs. Rationality assumes that individuals aim to make decisions that

---

loss, risk, or cost. By employing mathematical modeling techniques and sophisticated analysis, operations research provides insights and recommendations that guide decision-makers in making more effective choices. Although operations research initially emerged from military efforts before World War II, its methodologies and techniques have evolved over time and are now applied across a wide range of industries. The versatility of operations research allows it to address complex problems in various fields, including manufacturing, logistics, transportation, healthcare, finance, and telecommunications. By employing operations research, organizations can optimize their processes, allocate resources efficiently, and enhance overall performance. It enables decision-makers to analyze various scenarios, evaluate trade-offs, and make informed choices based on quantitative evidence. Through its interdisciplinary nature and emphasis on practical applications, operations research continues to contribute to the advancement and efficiency of decision-making in diverse industries.

maximize expected outcomes, given the available information. Consistency emphasizes the need for coherence and logical consistency in decision-making processes. Utility theory posits that decision-makers assign values and preferences to different outcomes, reflecting their subjective evaluations. Risk analysis involves assessing and managing uncertainties and their potential consequences. Value trade-offs acknowledge that decision-making often involves considering multiple conflicting objectives and constraints.

In conclusion, the field of decision science emerges as an interdisciplinary endeavor, integrating diverse disciplines to comprehend and improve decision-making processes. By leveraging insights from mathematics, statistics, economics, psychology, and computer science, decision science equips decision-makers with analytical tools, evidence-based approaches, and behavioral insights to make well-informed and effective decisions. Its scope spans various domains, and its fundamental principles provide a solid foundation for studying and advancing decision science.

1.5 The Importance and Applications of Decision Science

Decision science plays a crucial role in numerous domains and holds significant importance in supporting informed decision-making. Its applications span a wide range of fields, providing valuable insights and methodologies for optimizing choices. Here are some key points highlighting the importance and applications of decision science:

- Improved Decision-Making: Decision science offers a systematic approach to decision-making, incorporating quantitative techniques, models, and frameworks. By leveraging data analysis, probabilistic reasoning, and optimization methods, decision science enhances the quality of decisions, leading to more favorable outcomes.
- Risk Management: Decision science provides valuable tools for assessing and managing risks. It helps individuals and organizations evaluate potential hazards, quantify uncertainties, and make informed choices that minimize risks and maximize rewards. This is essential in industries such as finance, insurance, project management, and healthcare.
- Resource Allocation: Decision science aids in efficient resource allocation. It enables organizations to optimize the allocation of limited resources, whether it's budgetary funds, human capital, or time. By using techniques such as mathematical optimization and simulation modeling, decision science helps identify optimal resource allocation strategies for improved productivity and effectiveness.
- Strategic Planning: Decision science supports strategic decision-making. It assists in analyzing complex scenarios, evaluating alternative strategies, and identifying the most favorable course of action. By incorporating decision analysis, game

theory, and scenario planning, decision science helps organizations formulate robust strategies and adapt to changing environments.

- Policy Development: Decision science informs policy-making processes. By considering various stakeholders, evaluating trade-offs, and quantitatively assessing policy impacts, decision science aids in designing effective public policies. It facilitates evidence-based decision-making in areas such as environmental regulations, healthcare policy, transportation planning, and economic development.
- Personal Decision-Making: Decision science offers valuable tools for individuals to make better personal choices. It helps individuals assess risks, evaluate alternatives, and consider trade-offs when making important decisions. Whether it's financial planning, career choices, or personal life decisions, decision science provides frameworks and methodologies to support rational and informed decision-making [12, 13].

Overall, decision science has broad-reaching applications across diverse fields. Its importance lies in providing systematic and quantitative approaches to decision-making, enabling individuals and organizations to make optimal choices, manage risks, allocate resources efficiently, and formulate effective policies. By integrating data, models, and analysis, decision science contributes to improved outcomes and enhances the overall decision-making process.

1.6 The Decision-Making Theories

The concept of decision-making is deeply connected to the future. Our choices and decisions could shape the trajectory of our personal or professional lives, with each carrying its own significance and influence. To reliably accept the consequences and transformations that arise from decision-making and embrace what awaits us, it is essential to integrate our hearts, minds, emotions, intuition, logic, and calculations. By harmonizing these aspects, we can navigate with acceptance and determination, forging a path towards a brighter and more promising tomorrow [16].

Decision-making is a fundamental cognitive process that individuals engage in daily. To comprehend the foundations of decision-making, it is essential to explore the underlying principles, concepts, and theories that shape this process. Let's delve deeper into these foundations:

I. *Rationality and Expected Utility*
Rationality serves as the cornerstone of decision-making. It assumes that individuals strive to maximize their expected outcomes based on available information. Rational decision-making involves systematically evaluating alternatives, considering potential consequences, and selecting the option that offers the highest expected utility. Expected utility theory, developed by economists such as Daniel Bernoulli and Leonard Savage, provides a framework for quantifying preferences and evaluating choices based on their expected outcomes and

associated utilities. According to this theory, decision-makers assign subjective values (utilities) to different outcomes and make decisions to maximize their expected utility.

II. *Decision-Making Under Uncertainty*
Decision-making often occurs in situations where outcomes are uncertain, or probabilities are unknown. To address uncertainty, decision science employs various techniques such as probability theory, statistical analysis, and decision trees. Probability theory enables decision-makers to assign probabilities to different outcomes, facilitating a rational evaluation of alternatives. Decision trees provide a visual representation of decision-making scenarios, capturing possible choices, outcomes, and their associated probabilities. By quantifying uncertainty and analyzing decision tree structures, decision-makers can identify optimal strategies under uncertainty.

III. *Decision-Making Biases and Heuristics*
Behavioral research has uncovered the influence of cognitive biases and heuristics on decision-making. Cognitive biases are systematic errors in thinking that lead to deviations from rational decision-making. Heuristics are mental shortcuts or rules of thumb that individuals use to simplify decision-making processes. Examples of cognitive biases include confirmation bias (favoring information that confirms pre-existing beliefs), availability bias (overestimating the importance of easily recalled information), and framing bias (decision preferences influenced by how options are presented). These biases can lead to suboptimal decisions and deviations from rationality.

IV. *Prospect Theory and Loss Aversion*
Prospect theory, developed by Daniel Kahneman and Amos Tversky, challenges the traditional assumption of rational decision-making by introducing the concept of loss aversion. Prospect theory suggests that individuals evaluate potential gains and losses relative to a reference point and are more sensitive to losses than gains. According to prospect theory, individuals tend to exhibit risk aversion when facing potential gains but become risk-seeking when facing potential losses. This theory highlights the importance of framing decisions and considering the reference point, as it significantly influences decision preferences.

V. *Emotions and Intuition*
Decision-making is not solely a rational process but is also influenced by emotions and intuition. Emotional states can impact decision-making, as individuals' moods and emotional reactions influence their evaluation of options and risk perceptions. Intuition, often referred to as gut feeling or instinct, plays a role in decision-making by relying on implicit knowledge and pattern recognition. It involves quick and automatic processing, which can lead to efficient decision-making in certain contexts [16].
Daniel Kahneman, a renowned psychologist, and Nobel laureate in economics, has contributed significantly to the understanding of decision-making processes, including the role of emotions and intuition. Kahneman's perspective emphasizes the influence of emotions and intuitive thinking on our decision-making.

According to Kahneman, emotions play a crucial role in decision-making and can significantly impact our choices. He argues that emotions, such as fear or excitement, can shape our perceptions of risks and rewards, leading to biased decision-making. Emotions can cloud our judgment and lead to irrational behavior or choices that may not align with logical reasoning.

In addition to emotions, Kahneman also highlights the importance of intuition in decision-making. In his book "Thinking Fast and Slow" distinguishes between two systems of thinking: System 1 and System 2. System 1 thinking is fast, intuitive, and automatic, while System 2 thinking is slow, deliberate, and analytical. Kahneman suggests that much of our decision-making occurs through the intuitive and automatic processes of System 1, which rely on heuristics and mental shortcuts. However, he also notes that intuitive thinking can be prone to biases and errors.

Daniel Kahneman has extensively studied the cognitive biases and traps that can influence decision-making. One of the key concepts he highlights is the "trap of decision-making," which refers to the various biases and errors that can lead individuals to make suboptimal or irrational decisions.

Kahneman argues that decision-making is often affected by cognitive biases, which are systematic errors in thinking that can skew our judgment. These biases can lead to deviations from rational decision-making and impact the quality of our choices. Some examples of cognitive biases include confirmation bias, availability bias, anchoring bias, and overconfidence bias.

Confirmation bias, for instance, refers to the tendency to favor information that confirms our preexisting beliefs or hypotheses while disregarding or downplaying conflicting evidence. Availability bias, on the other hand, occurs when we rely heavily on readily available examples or information, even if they may not accurately represent the overall picture.

Kahneman's research suggests that these biases can lead to flawed decision-making processes and result in suboptimal outcomes. He emphasizes the importance of recognizing and mitigating these biases to make more informed and rational decisions.

Additionally, Kahneman discusses the influence of heuristics, which are mental shortcuts or rules of thumb that individuals often rely on to simplify decision-making. While heuristics can be helpful in many situations, they can also lead to biases and errors. For example, the representativeness heuristic can lead us to make judgments based on stereotypes or superficial similarities, rather than considering the relevant statistical information [12].

Overall, Kahneman's view on the trap of decision-making highlights the presence of cognitive biases and heuristics that can hinder our ability to make optimal choices. By understanding and addressing these biases, individuals can improve their decision-making processes and avoid falling into common traps.

VI. *Normative and Descriptive Decision-Making*

Decision science distinguishes between normative and descriptive approaches to decision-making. Normative decision-making focuses on identifying optimal decisions based on rational principles and maximizing expected utility. It provides guidelines for ideal decision-making, assuming full information, rationality, and logical consistency. Descriptive decision-making, on the other hand, aims to understand and describe how decisions are made, considering the influence of cognitive biases, heuristics, and emotions. Descriptive approaches acknowledge that decision-making is often influenced by bounded rationality, limited information, and cognitive limitations [16].

1.7 The Reputable Domains and Applications of Decision Making

By exploring these foundational aspects of decision-making, researchers and practitioners can gain a deeper understanding of the complexities involved in decision processes. This knowledge helps in developing strategies to improve decision-making, enhance outcomes, and design interventions that account for human biases, uncertainties, and emotional states. The foundations of decision-making provide valuable insights across various domains and applications [12, 14, 17].

1.7.1 Decision Support Systems and Business Intelligence

Decision science has contributed to the development of decision support systems (DSS). These computer-based tools assist decision-makers in analyzing complex problems, exploring alternative solutions, and evaluating potential outcomes. DSS utilizes mathematical models, data analysis techniques, and visualization tools to enhance decision-making processes [18, 19].

Let's explore the key components and benefits of Decision Support Systems (DSS) in more detail:

- Components of Decision Support Systems: a. Data Management: DSS integrates data from diverse sources, both internal and external, including real-time data feeds. Data management involves collecting, storing, retrieving, and preprocessing data to ensure accurate and relevant information is available for decision-making. b. Models and Analytical Techniques: DSS utilizes mathematical and statistical models, algorithms, and analytical techniques to analyze data, simulate scenarios, and generate insights. These may include optimization models, simulation models, forecasting models, and data mining techniques, among others. The selection of appropriate models depends on the specific decision problem and available data. c. User Interface: DSS provides user-friendly interfaces that allow decision-makers to interact with the system. They can input their preferences and constraints, visualize data and results, and explore different decision alternatives. The user interface enables decision-makers to navigate through the decision-making process effectively and interpret the outputs of the DSS. d. Knowledge

Base: DSS may incorporate a knowledge base that stores domain-specific knowledge, rules, best practices, and guidelines. This knowledge base provides decision-makers with access to relevant information and expertise, assisting them in making informed decisions based on established knowledge and experience.

- Benefits of Decision Support Systems: a. Improved Decision-Making: DSS helps decision-makers structure problems, explore alternatives, and evaluate potential outcomes. By providing analytical tools and real-time insights, DSS enhances the quality and accuracy of decision-making. Decision-makers can consider multiple scenarios, perform "what-if" analyses, and assess risks and benefits associated with different options. b. Enhanced Efficiency: DSS automates complex calculations, data analysis, and report generation, reducing the time and effort required for decision-making tasks. Decision-makers can access relevant information and perform analyses more efficiently, enabling quicker responses to changing conditions or opportunities. c. Increased Collaboration: DSS supports collaboration among decision-makers and stakeholders by providing a common platform to share information, discuss alternatives, and coordinate decision-making efforts. DSS facilitates communication, consensus-building, and the integration of diverse perspectives in the decision-making process. d. Agility and Adaptability: DSS enables decision-makers to respond quickly to dynamic environments and evolving conditions. Real-time data integration and analysis capabilities allow decision-makers to monitor key indicators, detect trends, and adjust decisions accordingly. DSS also facilitates scenario planning and sensitivity analysis, enabling decision-makers to assess the potential impacts of different factors and adapt their strategies accordingly. e. Documentation and Auditability: DSS capture and store decision-related data, models, and analyses, providing a comprehensive record of the decision-making process. This documentation facilitates accountability, traceability, and auditability, which are essential in regulated industries or situations where decision justifications need to be reviewed or revisited. f. Continuous Improvement: DSS can be updated and refined over time based on feedback, user experiences, and performance evaluations. Continuous improvement ensures that the DSS remains relevant, accurate, and aligned with the evolving needs of decision-makers and the organization.
- Data Source: Business Intelligence serves as the foundation for Decision Support Systems. BI systems gather and consolidate data from various sources, such as transactional databases, external sources, and operational systems, to create a unified and comprehensive view of the organization's data.
- Data Analysis and Reporting: Business Intelligence systems provide the analytical capabilities needed to process and analyze large volumes of data. They utilize techniques such as data mining, statistical analysis, and visualization to identify patterns, trends, and relationships within the data. These insights are then used by decision-makers within the Decision Support Systems to evaluate alternatives and make informed decisions.
- Decision-Making Support: Decision Support Systems build upon the data and insights provided by Business Intelligence systems. DSS incorporates decision models, algorithms, and analytical tools to facilitate decision-making processes.

It leverages the outputs of BI systems to assist decision-makers in evaluating different scenarios, assessing risks, and selecting the most suitable course of action.

- Interactive and User-Friendly Interface: Both DSS and BI systems aim to provide decision-makers with user-friendly interfaces to interact with the data and analysis. BI systems often include dashboards, reports, and visualizations that allow users to explore data and monitor performance indicators. DSS systems incorporate similar interfaces but focus more on providing decision-specific functionalities, such as sensitivity analysis, goal-seeking, and "what-if" simulations.

In conclusion, Decision Support Systems (DSS) serve as powerful aids for decision-makers, equipping them with essential tools, data, and analytical capabilities. The integration of data, models, and user interfaces within DSS enhances decision-making processes in terms of efficiency, accuracy, collaboration, and adaptability. DSS proves particularly valuable in complex decision scenarios, where it enables decision-makers to evaluate alternatives, assess risks, and optimize outcomes within dynamic environments.

While Business Intelligence (BI) systems focus on the collection, analysis, and presentation of data to generate insights, the true strength of DSS lies in its ability to leverage these insights. DSS bridges the gap between data-driven insights generated by BI systems and decision-making processes by providing decision-makers with the necessary tools and information. The integration of BI and DSS significantly enhances decision-making capabilities by combining data-driven insights with interactive and analytical decision support functionalities.

By utilizing DSS, organizations can empower decision-makers with comprehensive and actionable insights, leading to more informed and effective decision-making. As decision-making continues to grow in complexity and organizations rely increasingly on data-driven insights, the integration of BI and DSS becomes a critical aspect of successful decision-making processes. Ultimately, the combination of these systems serves as a powerful ally, aiding decision-makers in navigating complex scenarios and achieving optimal outcomes.

1.7.2 Strategic Management

Decision science plays a significant role in strategic management by providing valuable insights and tools to aid organizations in making informed strategic choices. In today's dynamic business landscape, organizations face numerous challenges, such as evolving market conditions, intense competition, and shifting industry trends. Decision science enables organizations to assess and analyze these factors to gain a deeper understanding of their implications and make proactive decisions. Through the application of decision analysis techniques, organizations can evaluate various strategic options and their potential outcomes. This involves considering different scenarios, assessing risks and uncertainties, and quantifying the potential impact of each decision.

Additionally, decision science helps in scenario planning, which involves developing alternative future scenarios and exploring their implications. By considering a range of plausible scenarios, organizations can anticipate potential disruptions, identify emerging opportunities, and develop robust strategies to navigate uncertainty. This proactive approach enables organizations to adapt and respond effectively to changing environments, ensuring their long-term success. Moreover, optimization techniques are employed in strategic decision-making to identify the most optimal course of action. These techniques help organizations maximize their objectives, whether it's increasing market share, optimizing resource allocation, or minimizing costs. By leveraging mathematical modeling and algorithms, decision science enables organizations to make data-driven decisions that align with their strategic goals [20].

In summary, decision science plays a crucial role in strategic management by providing organizations with the necessary tools and insights to assess market conditions, analyze competitive landscapes, and identify industry trends. It empowers organizations to make informed strategic choices, develop robust strategies, and adapt to changing environments. By integrating decision science into strategic management processes, organizations can enhance their competitiveness, drive innovation, and achieve long-term success.

1.7.3 Healthcare and Medicine

Decision science plays a crucial role in healthcare and medicine, providing valuable support in clinical decision-making, treatment planning, and resource allocation. The complex and dynamic nature of the healthcare industry requires evidence-based approaches to optimize patient care and efficiently allocate limited resources.

One key application of decision science in healthcare is in clinical decision-making. Healthcare professionals often face complex situations where they need to choose the most appropriate diagnostic tests, treatment options, or interventions for their patients. Decision models and algorithms aid in evaluating various factors such as patient characteristics, medical history, test results, and treatment outcomes to make informed decisions. By considering the available evidence and utilizing decision support tools, healthcare professionals can enhance the accuracy and effectiveness of their clinical decisions, leading to improved patient outcomes.

Another important aspect is treatment planning. Decision science techniques, such as decision trees and Markov models, enable healthcare providers to assess the potential outcomes and costs associated with different treatment paths. This allows for a systematic evaluation of the benefits, risks, and costs of various treatment options, helping healthcare professionals identify the optimal course of action for individual patients. Analysis effectiveness analysis, for example, can help determine the most efficient use of healthcare resources by considering both the clinical effectiveness and economic impact of different interventions.

Resource allocation is a critical challenge in healthcare due to limited resources and increasing demands. Decision science provides tools and methodologies to optimize resource allocation, ensuring that healthcare resources are distributed in the most efficient and equitable manner. By employing techniques such as mathematical modeling, simulation, and optimization, healthcare organizations can analyze patient flow, bed capacity, staffing levels, and other factors to improve operational efficiency and patient access to care.

In summary, decision science brings significant benefits to the healthcare and medicine field by supporting clinical decision-making, treatment planning, and resource allocation. By applying rigorous analytical techniques and incorporating evidence-based approaches, decision science enhances the quality of patient care, promotes cost-effective interventions, and improves the overall efficiency of healthcare systems [20, 21].

1.7.4 Financial Decision-Making

Financial decision-making plays a crucial role in the world of finance, and decision science provides valuable tools and techniques to support this process. Decision science combines mathematical models, statistical analysis, and optimization techniques to help investors, financial institutions, and asset managers make informed and rational decisions in complex and uncertain financial markets. One area where decision science is applied in financial decision-making is portfolio optimization. Investors aim to construct portfolios that maximize returns while minimizing risk. Decision science techniques, such as mean–variance analysis and modern portfolio theory, help investors identify the optimal combination of assets to achieve their desired risk-return tradeoff. By considering historical data, expected returns, and risk measures, decision science assists in allocating investments across various assets to build well-diversified portfolios [22].

Risk assessment is another critical aspect of financial decision-making, and decision science provides tools to evaluate and manage risks effectively. Techniques like value-at-risk (VaR) and stress testing help assess the potential downside risks associated with investment portfolios and financial products. By quantifying the potential losses at different confidence levels, decision science aids in determining appropriate risk management strategies and setting risk limits.

Moreover, decision science supports investment analysis by incorporating various models and techniques. Fundamental analysis, technical analysis, and quantitative models are commonly used to evaluate investment opportunities and make informed investment decisions. Decision science provides frameworks for analyzing financial statements, assessing the intrinsic value of securities, and identifying patterns and trends in market data. These quantitative approaches help investors gain insights into the potential risks and returns of investment options, facilitating better decision-making.

In addition to these specific applications, decision science also contributes to financial decision-making through its emphasis on data-driven analysis and evidence-based reasoning. It promotes the use of historical data, statistical analysis, and optimization techniques to support financial decisions. By incorporating quantitative methods and rigorous analysis, decision science helps reduce subjective biases and enhances the objectivity and reliability of financial decision-making processes [23].

Overall, decision science is a valuable discipline in financial decision-making. It enables investors, financial institutions, and asset managers to utilize mathematical models, statistical analysis, and optimization techniques to make informed decisions, optimize portfolios, assess risks, and evaluate investment opportunities. By leveraging the power of decision science, financial professionals can navigate complex and uncertain financial markets with greater confidence and improve their chances of achieving their financial objectives.

1.7.5 Project Management and Scheduling

Project management and scheduling play a crucial role in decision making, particularly in complex and time-sensitive endeavors. Effective project management involves identifying goals, breaking them down into manageable tasks, allocating resources, and coordinating activities to achieve desired outcomes within defined constraints. Scheduling, on the other hand, focuses on creating a timeline and sequencing tasks to optimize efficiency and meet project deadlines. Both project management and scheduling are integral components of decision making in various domains.

Here are some key points highlighting the relationship between project management, scheduling, and decision making:

- Resource Allocation: Project management requires making decisions about the allocation of resources, such as human capital, finances, equipment, and materials. Effective decision making in this aspect involves evaluating the availability and capabilities of resources, determining the optimal allocation strategy, and ensuring that resources are utilized efficiently throughout the project lifecycle. Scheduling comes into play by aligning the availability of resources with project timelines to avoid bottlenecks and optimize productivity [24, 25].
- Risk Assessment and Mitigation: Decision making in project management involves assessing and managing risks. Project managers need to identify potential risks, evaluate their likelihood, and impact, and make decisions on how to mitigate or respond to them. Scheduling can help in this process by allowing for contingency planning, allocating buffer time for potential delays, and identifying critical path activities that require close monitoring and mitigation efforts [16].

- Time Management: Scheduling is a critical aspect of project management, as it involves creating a timeline, sequencing tasks, and establishing deadlines. Effective decision making in time management requires considering task dependencies, resource availability, and project objectives. By making informed decisions about task sequencing, prioritization, and resource allocation, project managers can optimize time utilization, ensure timely completion of tasks, and meet project deadlines [16].
- Stakeholder Communication: Decision making in project management includes effective communication with stakeholders. Project managers must make decisions regarding information sharing, reporting mechanisms, and stakeholder engagement strategies. Clear and timely communication of project schedules, progress updates, and potential deviations from the plan allows stakeholders to make informed decisions and provide necessary input for course corrections, if required.
- Adaptability and Flexibility: Decision making in project management often requires adaptability and flexibility. As projects progress, unforeseen circumstances, changing requirements, or new information may necessitate adjustments to the original plan. Effective project managers make informed decisions on whether to stick to the plan or modify it to accommodate changing circumstances. Scheduling provides a framework for evaluating the impact of changes, identifying potential bottlenecks, and making informed decisions on adjusting timelines, resources, or task priorities [25].

In summary, project management and scheduling are intertwined with decision making in numerous ways. Effective decision making in project management involves resource allocation, risk assessment, time management, stakeholder communication, and adaptability. Scheduling serves as a tool to visualize project timelines, sequence tasks, and optimize resource utilization, enabling project managers to make informed decisions that align with project objectives and constraints. By integrating decision making, project management, and scheduling, organizations can improve project outcomes, enhance efficiency, and deliver successful results.

1.7.6 Environmental Planning and Management

Decision science plays a crucial role in environmental planning and management by providing valuable tools and techniques for evaluating and addressing complex environmental challenges. From assessing the impacts of policies and projects to optimizing resource allocation for conservation efforts, decision science offers a systematic and evidence-based approach to achieving sustainable environmental practices.

One key application of decision science in environmental planning is the evaluation of policy impacts. Environmental policies and regulations often have far-reaching consequences, and decision science provides methods for assessing their potential effects. Techniques such as environmental impact assessment (EIA) and strategic environmental assessment (SEA) help decision-makers identify and evaluate the environmental, social, and economic impacts of proposed projects or policies. By considering various factors and using quantitative analysis, decision science assists in making informed decisions that minimize negative impacts and maximize positive outcomes.

Optimal resource allocation is another critical aspect of environmental management. Decision science techniques, including optimization models and multi-criteria decision analysis, aid in determining the most effective and efficient allocation of resources for conservation efforts. These tools consider factors such as biodiversity value, ecosystem services, budget constraints, and stakeholder preferences to identify the best strategies for protecting and managing natural resources. By optimizing resource allocation, decision science supports sustainable environmental practices and helps achieve conservation goals [26].

Risk analysis is an integral part of decision science in environmental planning and management. Environmental decisions often involve inherent uncertainties and potential risks. Decision science provides methodologies for assessing and managing these risks through techniques such as probabilistic modeling, scenario analysis, and decision tree analysis. By quantifying and evaluating risks associated with different courses of action, decision-makers can make informed choices that minimize potential harm to the environment and human well-being.

Cost–benefit analysis is another important tool in decision science for environmental planning and management. It helps assess the economic feasibility and environmental impact of projects or policies by comparing the costs and benefits associated with different alternatives. By quantifying the monetary and non-monetary values of environmental outcomes, decision-makers can weigh the trade-offs and make decisions that optimize both economic and environmental considerations.

In summary, decision science plays a vital role in environmental planning and management by providing rigorous methodologies for evaluating policy impacts, optimizing resource allocation, managing risks, and conducting cost–benefit analyses. By integrating these tools and approaches, decision-makers can make informed and sustainable environmental decisions that balance economic development with environmental conservation and promote long-term environmental well-being.

1.7.7 Supply Chain and Operations Management

Decision science plays a crucial role in optimizing supply chain and operations management processes, enabling organizations to make informed decisions that enhance efficiency, reduce costs, and meet customer demands effectively.

One key area where decision science is applied is inventory management. Organizations need to strike a balance between carrying sufficient inventory to meet customer demand and minimizing holding costs. Decision models such as Economic Order Quantity (EOQ) and Just-In-Time (JIT) help determine optimal inventory levels based on factors such as demand variability, lead times, and cost considerations. By utilizing these models, organizations can reduce excess inventory, minimize stockouts, and optimize their working capital [18, 27, 28].

Production scheduling is another critical aspect of supply chain and operations management where decision science is employed. Decision models and optimization techniques assist in developing efficient production schedules by considering factors such as production capacity, resource availability, and order fulfillment requirements. By utilizing techniques like linear programming and simulation, organizations can identify the most effective production sequences, allocate resources optimally, and streamline their operations.

Distribution network design is another area where decision science plays a significant role. Organizations need to determine the optimal configuration of their distribution network to minimize transportation costs, reduce delivery lead times, and improve customer service. Decision models and network optimization techniques aid in identifying the most cost-effective locations for warehouses, distribution centers, and transportation routes. By optimizing the distribution network, organizations can enhance their responsiveness to customer demands, improve order fulfillment, and reduce overall logistics costs [29, 30].

Simulation techniques are widely used in supply chain and operations management to analyze various scenarios, evaluate performance, and assess the impact of different decisions. Through simulation, organizations can model complex systems, test different strategies, and identify bottlenecks or areas for improvement. This enables them to make informed decisions and implement changes that optimize their operations and improve overall performance.

By leveraging decision science in supply chain and operations management, organizations can achieve significant benefits such as improved operational efficiency, reduced costs, enhanced customer satisfaction, and increased competitive advantage. The utilization of data-driven decision models and simulation techniques empowers organizations to make informed choices, mitigate risks, and adapt their operations to dynamic market conditions, ultimately leading to better overall supply chain performance and operational success.

1.7.8 Engineering and Technology

In the field of engineering and technology, decision-making holds a fundamental role in ensuring the success of projects and achieving desired outcomes. Professionals in these fields regularly engage in decision-making processes to assess various design options, select appropriate materials, and optimize processes to meet specific objectives.

One crucial aspect of decision-making in engineering and technology is the evaluation of design options. Engineers and technologists often encounter multiple feasible design alternatives when developing new products, structures, or systems. Through careful analysis, they weigh factors such as functionality, cost-effectiveness, efficiency, durability, and environmental impact to make informed decisions that align with project requirements and stakeholder needs. By selecting the most suitable design option, engineers can drive innovation, improve product performance, and enhance user experiences [18].

Additionally, decision-making plays a critical role in project management within engineering and technology domains. Project managers are responsible for making strategic decisions regarding resource allocation, scheduling, budgeting, and risk management. They rely on decision-making frameworks and tools to assess project complexities, evaluate trade-offs, and determine optimal strategies. Effective decision-making in project management ensures efficient utilization of resources, timely project completion, and successful achievement of project goals.

Furthermore, decision-making is vital in risk analysis within engineering and technology fields. Professionals assess potential risks and uncertainties associated with projects or technological advancements. By conducting thorough risk assessments and considering factors such as probability, severity, and mitigation measures, they can make informed decisions to minimize risks and maximize project success. Decisions regarding risk allocation, risk mitigation strategies, and contingency plans are crucial in safeguarding project outcomes and minimizing adverse effects [29].

Another critical area where decision-making is essential in engineering and technology is systems design. Engineers and technologists design complex systems, such as manufacturing processes, transportation networks, or information systems, by considering various interdependencies, performance requirements, and constraints. Decision-making is integral to determining system architectures, component selection, process flows, and system integration strategies. Through effective decision-making in systems design, engineers can optimize system performance, enhance efficiency, and ensure compatibility with intended objectives.

In summary, decision-making is foundational in engineering and technology fields. It enables professionals to assess design options, select appropriate materials, optimize processes, manage projects, analyze risks, and design complex systems. By employing systematic and analytical approaches to decision-making, engineers and technologists can drive innovation, achieve project success, and contribute to advancements in their respective fields.

1.7.9 Decision Making in Maintenance and Reliability

Decision-making in maintenance involves making choices regarding the optimal strategies, actions, and resource allocation to ensure the reliability, availability, and efficiency of assets. Maintenance professionals evaluate factors such as asset condition, failure patterns, maintenance costs, and performance indicators to make informed decisions on maintenance activities. These decisions may include determining the appropriate maintenance approach (preventive, predictive, corrective), scheduling maintenance tasks, prioritizing maintenance activities, and selecting maintenance techniques or technologies.

- Reliability-Centered Maintenance (RCM): Reliability-centered maintenance is a decision-making framework specifically focused on optimizing maintenance strategies to achieve desired levels of reliability, safety, and cost-effectiveness. RCM involves systematically analyzing asset functions, failure modes, and consequences to identify appropriate maintenance actions. Through RCM, maintenance decision-makers can determine the most suitable maintenance tasks, intervals, and techniques for each asset, considering factors such as criticality, operational context, and risk tolerance [22].
- Risk-Based Maintenance (RBM): Risk-based maintenance is a decision-making approach that prioritizes maintenance activities based on risk assessments. Maintenance decisions are guided by the level of risk associated with asset failures, considering the potential consequences and likelihood of failure events. RBM involves evaluating risk factors, such as failure probability, criticality, safety, environmental impact, and financial implications, to optimize maintenance decisions and allocate resources effectively [31].
- Condition-Based Maintenance (CBM): Condition-based maintenance relies on real-time monitoring and analysis of asset condition data to make maintenance decisions. By continuously monitoring parameters such as vibration, temperature, pressure, or performance indicators, maintenance professionals can assess the actual health and performance of assets. Based on these condition assessments, decisions can be made to trigger maintenance actions when specific thresholds or conditions are met, optimizing maintenance interventions, and minimizing unnecessary maintenance activities [32].
- Spare Parts Management: Decision-making in spare parts management involves determining the optimal stocking levels, inventory control strategies, and procurement practices for spare parts. Maintenance decision-makers need to consider factors such as asset criticality, historical usage patterns, lead times, and cost implications to ensure the availability of necessary spare parts while minimizing inventory costs. Effective spare parts management decision-making contributes to reducing downtime, improving maintenance response time, and minimizing costs associated with spare parts [33].
- Asset Replacement and Life-Cycle Management: Decision-making regarding asset replacement and life-cycle management involves evaluating the economic and technical feasibility of replacing aging or obsolete assets. Maintenance professionals assess factors such as asset condition, performance, reliability, maintenance costs, and technological advancements to determine the optimal time for asset replacement. These decisions aim to balance the costs of continued maintenance, potential risks, and benefits associated with upgrading or replacing assets.
- Reliability Improvement Strategies: Decision-making in reliability improvement focuses on identifying and implementing strategies to enhance asset reliability and performance. Maintenance decision-makers assess data and information related

to asset failures, maintenance history, and reliability metrics to identify patterns and root causes of failures. Based on this analysis, decisions can be made to implement proactive measures such as redesigning components, improving maintenance practices, enhancing asset monitoring, or implementing condition-based maintenance programs to improve asset reliability and minimize failures [34].

Decision making and Reliability models refer to mathematical or statistical techniques used to assess the reliability or performance characteristics of systems, products, or processes. These models aim to predict the behavior of components or systems over time and provide insights into their failure probabilities, failure modes, and other reliability-related metrics.

- The Reliability Redundancy Allocation Problem (RAP) is a decision-making problem in reliability engineering that involves determining the optimal allocation of redundant components or subsystems within a system to achieve desired levels of system reliability, availability, or performance. RAP aims to find the best configuration of redundancies that maximizes system reliability while considering constraints such as cost, weight, space, and other factors [31].
  The goal of RAP is to identify the optimal allocation of redundancy resources, such as spare components, backup systems, or parallel subsystems, to improve system reliability or minimize downtime. The problem involves selecting the appropriate number and types of redundancies, their locations within the system, and the level of redundancy required for each component or subsystem [34].
  RAP typically considers both functional and physical redundancies. Functional redundancy involves duplicating critical components or subsystems, while physical redundancy involves having backup systems or spare parts readily available to replace failed components. The allocation decisions in RAP involve balancing the benefits of increased reliability against the costs and constraints associated with redundancy [35].
  Solving RAP involves mathematical modeling and optimization techniques to find the optimal allocation strategy. Various approaches, such as mathematical programming, heuristic methods, and evolutionary algorithms, are employed to address the complexity of the problem and find near-optimal solutions. The objective function in RAP can be formulated to maximize system reliability, availability, or other performance measures while considering constraints such as budget, weight, or space limitations.
  RAP has significant practical applications in industries where system reliability is crucial, such as aerospace, telecommunications, power systems, transportation, and critical infrastructure. By optimizing the allocation of redundancies, organizations can enhance system resilience, reduce downtime, improve maintenance strategies, and minimize the impact of failures or disruptions.
- Choices K out of N and Reliability Redundancy Allocation Problem:
  The concept of "choices K out of N" and the Reliability Redundancy Allocation Problem (RAP) are related in the sense that they both involve selecting subsets or combinations of elements from a larger set. However, they address different aspects of decision-making in the context of reliability engineering.

"Choices K out of N" refers to determining the number of distinct subsets that can be formed by selecting K elements from a set of N elements, without considering the order of selection. It is a combinatorial concept used to calculate the number of combinations or subsets. This concept is more focused on counting the possibilities rather than optimization.

On the other hand, the Reliability Redundancy Allocation Problem (RAP) involves optimizing the allocation of redundant components or subsystems within a system to achieve desired levels of system reliability, availability, or performance. RAP goes beyond counting combinations and aims to find the best configuration of redundancies that maximizes system reliability while considering constraints and objectives.

While both concepts involve making decisions about subsets or combinations, "choices K out of N" is a more general concept used for counting possibilities, whereas RAP is a specific problem in reliability engineering that addresses the optimal allocation of redundancies for improving system reliability and performance.

Briefly, "choices K out of N" focuses on counting combinations, while the Reliability Redundancy Allocation Problem (RAP) focuses on optimizing the allocation of redundancies to improve system reliability [36, 37].

In summary, the relation between decision making and reliability models lies in their shared goal of improving outcomes and reducing uncertainty in different contexts. While decision making focuses on selecting the best course of action among alternatives, reliability models aim to assess and predict the performance and failure characteristics of systems, products, or processes.

1.7.10 Human Resources and Talent Management

Decision science plays a significant role in human resources (HR) and talent management by providing valuable insights and analytical tools for various HR practices. It aids in making informed decisions related to recruitment, selection, performance evaluation, and succession planning, among others. By incorporating decision models and data analytics, organizations can optimize their HR processes and make effective people-related decisions.

In the context of recruitment and selection, decision science techniques help HR professionals identify the most suitable candidates for job positions. By analyzing job requirements, candidate profiles, and historical data, decision models can be developed to assess the likelihood of a candidate's success in a specific role. These models can consider factors such as education, experience, skills, and cultural fit, enabling organizations to make more objective and evidence-based hiring decisions. Performance evaluation is another area where decision science contributes to HR practices. By leveraging data analytics, organizations can assess employee performance more accurately and identify areas of improvement. Decision models can be utilized to analyze performance metrics, track progress, and provide feedback to employees. These models can help identify high-performing employees, address skill gaps, and develop tailored performance improvement plans [21].

Succession planning, which involves identifying and developing future leaders within an organization, can also benefit from decision science. By analyzing workforce demographics, skill sets, and potential candidates, organizations can create data-driven succession plans. Decision models can assist in identifying high-potential employees, evaluating their readiness for leadership roles, and designing development programs to groom them for future responsibilities.

Additionally, decision science supports workforce planning by providing insights into labor supply and demand dynamics. By analyzing historical data, market trends, and organizational goals, organizations can make informed decisions regarding workforce size, composition, and deployment. These decisions contribute to effective resource allocation, optimal staffing levels, and improved operational efficiency.

Overall, decision science enhances HR and talent management practices by leveraging decision models, data analytics, and evidence-based approaches. By applying these techniques, organizations can make more informed decisions related to recruitment, selection, performance evaluation, and succession planning. This data-driven approach enables organizations to optimize their HR processes, identify high-potential employees, address skill gaps, and align their workforce with organizational objectives. Ultimately, decision science empowers HR professionals to make more effective people-related decisions, leading to improved employee performance, engagement, and organizational success.

1.7.11 Crisis Management and Emergency Response

Decision science is highly relevant in the field of crisis management and emergency response, providing valuable tools and techniques to aid decision-makers in assessing risks, developing contingency plans, and allocating resources effectively during critical situations. By utilizing decision models and simulation techniques, decision science assists in evaluating different scenarios, prioritizing actions, and coordinating responses to minimize the impact of crises. One of the key contributions of decision science in crisis management is risk assessment. Decision models and analytical approaches enable decision-makers to identify and evaluate potential risks and vulnerabilities associated with different types of crises. By analyzing historical data, conducting risk assessments, and utilizing predictive modeling, decision science helps in understanding the likelihood and potential impact of various crises. This information is crucial in developing proactive measures and response plans to mitigate risks and enhance preparedness [57].

Contingency planning is another area where decision science plays a vital role. Decision models can be used to assess the effectiveness of different contingency plans and strategies by considering various factors such as available resources, response times, and potential outcomes. These models aid in determining optimal courses of action and identifying the most effective and efficient response options to manage crises. By utilizing decision science techniques, decision-makers can anticipate potential challenges, develop robust contingency plans, and allocate resources in a way that maximizes the effectiveness of the response efforts.

During emergencies, decision science supports decision-making by providing simulation techniques and scenario analysis. Decision models allow decision-makers to simulate different crisis scenarios, evaluate potential consequences, and explore the effectiveness of different response strategies. These simulations provide insights into the dynamics of the crisis, potential bottlenecks, and resource requirements. By considering multiple scenarios and their outcomes, decision-makers can make informed decisions in real-time, adapting their response strategies as needed.

Furthermore, decision science aids in resource allocation during emergencies. By considering available resources, logistical constraints, and priorities, decision models assist in optimizing resource allocation to address critical needs. This includes allocating personnel, equipment, supplies, and financial resources in a way that maximizes their impact and supports effective emergency response.

Overall, decision science serves as a valuable tool in crisis management and emergency response. By utilizing decision models, simulation techniques, and analytical approaches, decision-makers can assess risks, develop contingency plans, and allocate resources effectively. This data-driven approach enhances preparedness, response capabilities, and the ability to mitigate the impact of crises. Decision science enables decision-makers to make informed decisions under high-pressure situations, resulting in more effective crisis management and improved outcomes.

1.7.12 Public Policy and Governance

Decision science is highly relevant in the realm of public policy and governance, providing valuable tools and techniques to support evidence-based policymaking. By utilizing decision science, policymakers can analyze the potential impacts of different policy options, consider stakeholder preferences, and evaluate trade-offs to make informed decisions. One of the key contributions of decision science in public policy is in policy analysis. Decision models and quantitative techniques can be employed to evaluate the potential outcomes and impacts of different policy options. By considering factors such as costs, benefits, risks, and social implications, decision science helps policymakers assess the effectiveness and feasibility of various policy interventions. This analysis enables policymakers to make evidence-based decisions and select policies that are likely to achieve desired outcomes.

Furthermore, decision science supports policy design by considering stakeholder preferences and objectives. Through techniques like multi-criteria decision analysis, decision science facilitates the integration of diverse perspectives and values in the policy design process. By incorporating stakeholders' input and preferences, policymakers can develop policies that are more inclusive and responsive to societal needs. Implementation of policies is another area where decision science plays a crucial role.

Decision models can assist in planning and allocating resources to ensure effective policy implementation. These models help policymakers analyze resource requirements, assess potential challenges, and identify strategies to overcome implementation barriers. By utilizing decision science techniques, policymakers can improve the efficiency and effectiveness of policy implementation, leading to better outcomes.

Evaluation of policies is equally important in the decision-making process, and decision science provides tools to assess the performance and impact of policies. Through techniques like cost–benefit analysis, impact evaluation, and policy simulation, decision science enables policymakers to evaluate the effectiveness and efficiency of policies. By analyzing relevant data, measuring outcomes, and comparing against policy objectives, decision science contributes to evidence-based policy evaluation and supports iterative policy improvement [38].

Decision science is applied in various policy domains, such as education, transportation, energy, and social welfare. In education, decision science techniques can aid in optimizing resource allocation, evaluating education programs, and informing curriculum development. In transportation, decision science can support infrastructure planning, traffic management, and transportation policy analysis. In energy, decision science helps in evaluating renewable energy options, optimizing energy distribution, and designing energy efficiency programs. In social welfare, decision science techniques contribute to policy analysis and program evaluation to address social inequality and improve the well-being of communities.

Overall, decision science enhances public policy and governance by providing evidence-based analysis, stakeholder engagement, and evaluation frameworks. By utilizing decision models, quantitative techniques, and data-driven approaches, policymakers can make informed decisions, design effective policies, implement them efficiently, and evaluate their impact. Decision science contributes to the improvement of public policy outcomes, leading to better governance and societal well-being.

1.7.13 The Application of Decision Making Would Not End to Mentioned Area

As a note, the application of decision-making extends far beyond the examples mentioned above. It permeates various aspects of our personal and professional lives, impacting resource allocation, cost management, and overall efficiency. By highlighting specific areas, we aim to emphasize the significance of employing a scientific approach to decision-making and the potential consequences of neglecting it. In business and organizational settings, decisions related to resource allocation can significantly impact the bottom line. Whether it's financial resources, human capital, or physical assets, without a systematic and evidence-based decision-making process, there is a risk of wasteful spending and inefficiencies. For instance, improper budgeting without considering the return on investment or neglecting to conduct a thorough cost–benefit analysis can lead to resource misallocation and unnecessary financial burdens. Furthermore, operational decisions in areas such as production, logistics, and supply chain management can greatly affect costs and operational effectiveness. Failing to apply a scientific approach to these decisions can result in inefficiencies, delays, and increased expenses. For example, choosing suppliers without considering factors like reliability, quality, and cost-effectiveness may lead to subpar products, production delays, or even supply chain disruptions [17].

In the realm of project management, the consequences of poor decision-making can be particularly pronounced. Inadequate risk assessment, failure to engage stakeholders, or ignoring project management best practices can lead to project delays, budget overruns, and substandard outcomes. By not utilizing a scientific approach to decision-making, valuable resources may be wasted, project objectives may be compromised, and overall project success may be jeopardized.

In our personal lives, decision-making plays a crucial role as well. Choices related to financial investments, career paths, education, and health can have long-lasting implications. Without a systematic and informed approach, we risk making decisions based on intuition, hearsay, or emotions, potentially leading to missed opportunities, financial setbacks, or compromised well-being.

By illustrating these examples, we hope to emphasize the importance of applying a scientific approach to decision-making. Such an approach involves gathering relevant data, analyzing information objectively, considering different alternatives, and weighing potential outcomes. By incorporating evidence-based decision-making techniques, we can enhance resource allocation, optimize costs, and improve overall efficiency across various domains, minimizing wastage and maximizing value.

Ultimately, recognizing the significance of a scientific approach to decision-making serves as a reminder that informed choices lead to better outcomes, while haphazard decision-making can have costly consequences.

1.8 The Reputable and Helpful Models and Techniques of Decision Making

There are several reputable and helpful models and techniques of decision-making that have been developed and utilized in various fields. These models and techniques provide structured frameworks for analyzing problems, evaluating options, and making informed decisions. Here are some widely recognized models and techniques.

1.8.1 *Rational Decision-Making Model*

This is a systematic, step-by-step approach that involves identifying the problem, generating alternative solutions, evaluating options based on criteria and objectives, selecting the best alternative, implementing the decision, and evaluating the outcome. The rational decision-making model emphasizes a logical and analytical approach to decision-making.

1.8.2 *Decision Trees*

Decision trees are a powerful tool used in decision analysis and machine learning to model and visualize decision-making processes. They provide a graphical representation of decisions and their potential outcomes in a tree-like structure, resembling a flowchart.

In a decision tree, each node represents a decision point or event, and branches emanating from the nodes represent the possible choices or outcomes. The structure of the tree helps to organize and display the sequence of decisions and events that can occur. The tree branches typically split based on specific criteria or factors relevant to the decision problem.

One of the key advantages of decision trees is their ability to handle uncertainty and probabilistic outcomes. At each branch, probabilities can be assigned to different outcomes, reflecting the likelihood of each event occurring. This allows decision-makers to consider the range of possible outcomes and their associated probabilities, enabling a more comprehensive evaluation of decision paths.

Decision trees also facilitate the calculation of expected values, which represent the weighted average of the potential outcomes considering their probabilities. Expected values help decision-makers assess the potential payoff or value associated with different choices at each decision point. By evaluating the expected values along different decision paths, decision-makers can compare and select the most favorable course of action.

Additionally, decision trees support sensitivity analysis and "what-if" scenarios. By altering the probabilities or values assigned to different events, decision-makers can assess the robustness of their decisions and understand the impact of different scenarios on the expected outcomes.

Decision trees are widely used in various domains, including finance, business, healthcare, and engineering. They can assist in strategic planning, risk analysis, resource allocation, and problem-solving. The visual nature of decision trees makes them intuitive and easy to interpret, enabling effective communication of complex decision processes to stakeholders and team members.

With advancements in machine learning and artificial intelligence, decision trees have also been incorporated into algorithms for automated decision-making. They serve as a fundamental building block for more complex models such as random forests, boosting algorithms, and ensemble methods [39].

In summary, decision trees provide a structured and visual representation of decision-making processes, allowing decision-makers to analyze and compare different decision paths, consider probabilities, and expected values, and make informed choices. Their versatility and ability to handle uncertainty make them valuable tools for decision analysis in situations involving multiple sequential decisions and probabilistic outcomes.

1.8.3 Cost–Benefit Analysis

Cost–benefit analysis is a systematic approach used to evaluate the financial implications of different alternatives. It involves quantifying and comparing the costs and benefits associated with each option to determine their relative economic feasibility and potential return on investment.

In cost–benefit analysis, decision-makers assign monetary values to both the costs and the benefits of each alternative. Costs typically include expenses such as initial investment, operating costs, maintenance, and any other relevant expenditures. Benefits, on the other hand, encompass the positive outcomes or advantages that can be gained from choosing a particular option.

To conduct a cost–benefit analysis, decision-makers follow several steps:

- Identify and list all relevant costs and benefits: This step involves identifying the various costs and benefits associated with each alternative. It is important to consider both direct and indirect costs and benefits, as well as short-term and long-term implications.
- Assign monetary values: Decision-makers assign monetary values to the identified costs and benefits. This may involve estimating future cash flows, conducting market research, consulting experts, or using historical data to determine appropriate monetary figures.
- Calculate the net value: The net value is calculated by subtracting the total costs from the total benefits. A positive net value indicates that the benefits outweigh the costs, while a negative net value suggests the opposite.
- Evaluate trade-offs: Decision-makers assess the trade-offs between costs and benefits. They consider the magnitude and timing of the costs and benefits, as well as their relative importance in achieving the desired objectives.
- Compare alternatives: Decision-makers compare the net values of different alternatives to determine which option provides the highest overall benefit in relation to its costs. This allows for a quantitative assessment of the financial viability and potential return on investment of each alternative.

Cost–benefit analysis provides decision-makers with a framework for making financially informed decisions by evaluating trade-offs between costs and benefits.

It helps to objectively assess the economic feasibility of different alternatives, identify the most cost-effective option, and allocate resources efficiently. However, it's important to note that cost–benefit analysis has limitations, as it focuses primarily on monetary values and may not capture all qualitative or non-monetary aspects of a decision. Therefore, it is often used in conjunction with other decision-making techniques to ensure a comprehensive evaluation.

1.8.4 SWOT Analysis

SWOT analysis is a widely used framework for assessing the internal and external factors that can impact a decision-maker or organization. It helps to identify strengths, weaknesses, opportunities, and threats, hence the acronym SWOT. By examining these factors, decision-makers gain valuable insights into strategic planning, risk management, and decision-making processes.

The key components of SWOT analysis are:

- Strengths: These are the internal attributes and capabilities that give an individual or organization a competitive advantage. They can include factors such as unique expertise, valuable resources, strong brand reputation, skilled personnel, or efficient processes. Identifying strengths helps decision-makers understand their core competencies and areas where they have an advantage over competitors.
- Weaknesses: Weaknesses refer to internal limitations or areas where an individual or organization may be at a disadvantage. These could be factors like inadequate resources, outdated technology, lack of expertise in certain areas, or inefficient processes. Recognizing weaknesses is crucial for decision-makers to address areas that need improvement or strategic changes.
- Opportunities: Opportunities are external factors or trends in the environment that can be advantageous for an individual or organization. They could include emerging markets, technological advancements, changing consumer preferences, or regulatory changes. Identifying opportunities allows decision-makers to capitalize on favorable conditions and make informed choices to drive growth and success.
- Threats: Threats are external factors that pose risks or challenges to an individual or organization. These can include intense competition, economic downturns, changing market trends, legal or regulatory constraints, or emerging disruptive technologies. By identifying threats, decision-makers can develop strategies to mitigate risks and adapt to the changing environment effectively.

SWOT analysis helps decision-makers gain a holistic understanding of their internal strengths and weaknesses, as well as the external opportunities and threats they face. It provides a structured framework for assessing the current situation and informs decision-making processes. Based on the insights gained from a SWOT analysis, decision-makers can develop strategies that leverage strengths, address weaknesses, capitalize on opportunities, and mitigate threats.

SWOT analysis is often used in strategic planning, marketing, project management, and risk assessment. It enables decision-makers to make more informed and proactive decisions, align their actions with their capabilities and the external environment, and maximize their chances of success. However, it's important to recognize that SWOT analysis should be complemented with additional analysis and evaluation techniques to ensure a comprehensive understanding of the decision-making context [40].

1.8.5 Pareto Analysis

Pareto Analysis, also referred to as the 80/20 rule or the principle of factor sparsity, is a technique used to identify and prioritize the vital few factors that have the most significant impact on a problem or desired outcome. The analysis is based on the observation that a significant majority of effects often come from a small number of causes.

The name "Pareto Analysis" stems from Vilfredo Pareto, an Italian economist, who observed that approximately 80% of the wealth in a society is typically held by 20% of the population. This principle was later generalized and applied to various fields beyond economics. The process of Pareto Analysis involves the following steps:

- Define the problem or outcome: Clearly articulate the problem or desired outcome that requires analysis and improvement.
- Collect data: Gather relevant data related to the problem or outcome. This could include factors, causes, occurrences, or any other relevant information.
- Categorize and quantify the data: Group the data into categories or factors and quantify their occurrences, frequency, or impact.
- Calculate the cumulative impact: Rank the factors in descending order based on their occurrence or impact. Calculate the cumulative impact of each factor by summing up the occurrences or impacts from the highest-ranked factor downwards.
- Plot the data on a Pareto chart: Create a Pareto chart, which is a bar graph with the factors plotted on the x-axis and their cumulative impact on the y-axis. The factors are presented in descending order from left to right, and a line graph shows the cumulative impact.
- Identify the vital few factors: Analyze the Pareto chart to identify the vital few factors that contribute to most of the problem or outcome. Typically, the top 20% of factors will account for approximately 80% of the impact.
- Prioritize actions and allocate resources: Once the vital few factors are identified, focus on addressing or improving these factors as they have the most significant impact. Allocate resources, time, and effort accordingly to maximize the effectiveness of interventions.

Pareto Analysis is a valuable tool for decision-makers as it helps prioritize efforts, resources, and actions by highlighting the factors that have the most significant impact. By focusing on the vital few, decision-makers can efficiently allocate resources, solve problems, and improve outcomes. However, it's important to note that Pareto Analysis is a starting point and should be supplemented with further analysis and evaluation to gain a comprehensive understanding of the problem or outcome and develop effective strategies [41].

1.8.6 Linear Programming (LP), Non-Linear Programming (NLP), and Integer Programming (IP)

LP, NLP, and IP have various applications in decision-making across different domains. Here are some common applications of these optimization techniques:

- Linear Programming (LP) Applications: a. Resource Allocation: LP is used to allocate limited resources such as labor, materials, and capital optimally. It helps determine the best utilization of resources to maximize productivity and minimize costs. b. Production Planning: LP assists in optimizing production levels and resource allocation to meet demand while minimizing costs, considering factors like capacity constraints, production limitations, and inventory management. c. Transportation and Logistics: LP is applied to optimize transportation routes, shipment assignments, and distribution strategies, considering factors like distance, capacity, and cost to ensure efficient logistics operations. d. Portfolio Optimization: LP aids in selecting the optimal mix of investments in a portfolio by considering risk-return trade-offs and constraints, enabling decision-makers to maximize portfolio performance. e. Diet and Nutrition Planning: LP helps in developing optimal meal plans by balancing nutritional requirements with cost constraints, ensuring a well-balanced diet at the lowest cost.
- Non-linear Programming (NLP) Applications: a. Engineering Design: NLP is used to optimize complex engineering designs, considering non-linear relationships between design variables and performance objectives, leading to optimal design solutions. b. Financial Modeling: NLP assists in portfolio optimization, option pricing, risk management, and asset-liability management by considering non-linear relationships and complex financial constraints. c. Marketing and Pricing: NLP is applied to determine optimal pricing strategies, considering demand curves, market competition, and profitability objectives, leading to optimal pricing decisions. d. Process Optimization: NLP helps optimize processes in manufacturing, energy, and chemical industries by considering non-linear relationships and constraints, aiming to maximize efficiency and minimize costs.
- Integer Programming (IP) Applications: a. Project Scheduling: IP is used to optimize project schedules considering discrete variables like activity sequencing, resource allocation, and precedence relationships, leading to efficient project planning. b. Network Optimization: IP helps optimize routing and flow problems in

transportation and communication networks, considering integer constraints on decisions like facility locations, route selection, and capacity planning. c. Production Planning: IP is applied to address discrete decisions in production planning, such as selecting production levels, determining batch sizes, and deciding which products to produce, aiming to minimize costs and meet demand requirements. d. Supply Chain Management: IP assists in optimizing supply chain decisions such as facility locations, inventory management, and distribution network design, considering discrete variables and constraints.

These are just a few examples of how linear programming, non-linear programming, and integer programming are applied in decision-making processes across various industries and domains. The optimization techniques help decision-makers find optimal solutions, make efficient use of resources, improve operational efficiency, and achieve desired objectives while considering constraints and trade-offs [42, 43].

1.8.7 *Queuing Theory*

Queuing theory plays a crucial role in decision-making by providing insights into the behavior and performance of waiting systems. It helps decision-makers optimize resource allocation, improve service levels, and make informed decisions regarding capacity planning and process design. Here are some key aspects of queuing theory in the context of decision-making [44]:

- Performance Evaluation: Queuing theory allows decision-makers to evaluate the performance of a queuing system by analyzing important metrics such as waiting time, queue length, service utilization, and system throughput. By understanding these performance measures, decision-makers can identify bottlenecks, assess system efficiency, and make decisions to improve overall performance.
- Resource Allocation: Queuing theory helps decision-makers determine the optimal allocation of resources such as servers, staff, or equipment to minimize waiting times and maximize system utilization. By analyzing different queuing models and scenarios, decision-makers can assess the impact of resource allocation decisions and make informed choices to balance costs and service levels.
- Capacity Planning: Queuing theory assists decision-makers in determining the appropriate capacity of a system to meet customer demand. It allows them to estimate the required number of servers or service points based on expected arrival rates and service times. By considering factors such as peak demand periods, variability in arrival rates, and desired service levels, decision-makers can make capacity planning decisions that ensure efficient operations and customer satisfaction.
- Service Level Management: Queuing theory provides decision-makers with tools to manage service levels and customer expectations. By analyzing queuing

models, they can determine the appropriate number of servers or service points needed to achieve specific service level targets, such as average waiting time or probability of queuing. This information enables decision-makers to set realistic service level goals and make operational decisions to meet customer expectations.
- Decision Optimization: Queuing theory allows decision-makers to optimize their decisions by considering various factors and constraints. For example, they can evaluate different queuing configurations, staffing levels, or service policies to identify the most cost-effective and efficient approach. By using queuing models and performance measures, decision-makers can assess trade-offs and make decisions that align with their objectives and constraints.
- Scenario Analysis: Queuing theory enables decision-makers to conduct scenario analysis by modeling different queuing configurations and scenarios. They can explore "what-if" scenarios to assess the impact of changes, such as process redesign, system upgrades, or changes in customer demand patterns. This helps decision-makers understand the potential outcomes of different decisions and make informed choices based on a deeper understanding of the system's behavior.

In summary, queuing theory provides decision-makers with valuable insights and analytical tools to optimize resource allocation, improve service levels, and make informed decisions regarding capacity planning and process design. By understanding the behavior of waiting systems and analyzing key performance measures, decision-makers can optimize their decisions, enhance operational efficiency, and ultimately improve customer satisfaction Performance Evaluation: Queuing theory allows decision-makers to evaluate the performance of a queuing system by analyzing important metrics such as waiting time, queue length, service utilization, and system throughput. By understanding these performance measures, decision-makers can identify bottlenecks, assess system efficiency, and make decisions to improve overall performance.

1.8.8 Simulation Approaches

Simulation is a powerful decision-making tool that involves creating a model or representation of a real-world system and conducting experiments or scenarios to gain insights and inform decision-making. It allows decision-makers to explore different alternatives, assess potential outcomes, and make informed decisions based on the results of simulated scenarios. Here are some key aspects of simulation and its application in decision-making:

- Modeling Complex Systems: Simulation enables decision-makers to model complex systems that are difficult to analyze analytically. By capturing the interactions and dynamics of the system, simulation provides a more comprehensive understanding of how different components and variables affect overall performance. This is particularly useful in decision-making when dealing with intricate systems with numerous interdependencies.

- Scenario Analysis: Simulation allows decision-makers to conduct scenario analysis by simulating different scenarios or what-if situations. By manipulating variables and parameters, decision-makers can explore the potential outcomes of different decisions or events. This helps in evaluating risks, identifying optimal strategies, and making decisions that are robust and resilient to uncertainties.
- Decision Optimization: Simulation can be used for decision optimization by simulating different alternatives and determining the best course of action. Decision-makers can test different decision variables, constraints, and objectives in the simulation model to identify the combination that yields the most favorable outcomes. This aids in making decisions that maximize performance, minimize costs, or achieve specific goals.
- Performance Evaluation: Simulation provides a platform for evaluating the performance of systems or processes under different conditions. Decision-makers can analyze key performance indicators, such as throughput, waiting times, resource utilization, or customer satisfaction, and compare the results across different scenarios. This helps in assessing the effectiveness of existing strategies and making decisions to improve system performance.
- Risk Assessment: Simulation allows decision-makers to assess and manage risks associated with decision-making. By introducing random variations or uncertainties into the simulation model, decision-makers can analyze the impact of uncertainties on outcomes and evaluate the likelihood of different risk events. This helps in identifying high-risk areas, developing risk mitigation strategies, and making decisions that account for uncertainties.
- Sensitivity Analysis: Simulation facilitates sensitivity analysis by systematically varying input parameters and observing the corresponding changes in output variables. Decision-makers can assess the sensitivity of their decisions to different factors and understand which variables have the most significant impact on outcomes. This helps in identifying critical factors, refining decision variables, and making decisions that are robust to changes in the system.

In summary, simulation provides decision-makers with a valuable tool for exploring alternatives, assessing risks, optimizing decisions, and evaluating system performance. By creating a virtual representation of real-world systems and conducting experiments, decision-makers can gain insights, make informed decisions, and improve outcomes in complex and uncertain environments.

1.8.9 Data Envelopment Analysis (DEA)

Data Envelopment Analysis is a mathematical technique used to evaluate the relative efficiency and performance of decision-making units (DMUs) based on multiple inputs and outputs. DEA has various applications in decision-making, particularly in assessing and improving the efficiency of organizations or entities. Here are some common applications of DEA in decision-making:

- Performance Evaluation: DEA allows decision-makers to assess and compare the performance of multiple entities or organizations operating in the same industry or sector. By considering multiple inputs and outputs, DEA provides a comprehensive evaluation of efficiency, helping identify the best-performing units and highlighting areas for improvement [45, 46].
- Benchmarking: DEA is used for benchmarking purposes, where organizations compare their performance against the most efficient or best-practice units in the industry. DEA identifies the sources of inefficiency and provides insights into the practices or strategies employed by the benchmarked units, enabling decision-makers to adopt or adapt those practices to enhance their own efficiency [47].
- Resource Allocation: DEA assists in making informed decisions regarding resource allocation. By evaluating the efficiency of different units or departments within an organization, DEA helps identify areas where resources can be reallocated to improve overall efficiency. It aids decision-makers in optimizing resource allocation based on the performance and efficiency of different units [48].
- Mergers and Acquisitions: DEA is used in the evaluation and decision-making process of mergers and acquisitions. It helps assess the efficiency and compatibility of potential partners or target companies. DEA analysis provides insights into the relative efficiency of the entities involved, assisting decision-makers in making informed decisions about potential mergers or acquisitions [49].
- Policy Evaluation: DEA is applied in evaluating the effectiveness of policies or interventions implemented by governments or organizations. By comparing the performance of different units before and after policy implementation, DEA helps decision-makers determine the impact of policies and identify areas where further improvements can be made [50].
- Efficiency Improvement: DEA aids in identifying and targeting areas for efficiency improvement within an organization. By analyzing the efficiency scores and identifying the inefficient units or departments, decision-makers can focus their efforts on improving specific processes, practices, or resource allocation strategies to enhance overall efficiency [51].
- Sustainable Development: DEA can be used to evaluate the efficiency of organizations in terms of sustainable development goals, such as environmental impact, energy consumption, or social responsibility. DEA analysis helps decision-makers identify areas where organizations can improve their sustainability practices and make more environmentally and socially responsible decisions [52].

Overall, Data Envelopment Analysis provides decision-makers with a quantitative approach to assess the efficiency and performance of entities, benchmark against best practices, optimize resource allocation, and identify areas for improvement. Its versatility makes it a valuable tool in various decision-making contexts, contributing to performance improvement and informed decision-making.

Please be noted the models and techniques are not limited to above-mentioned items but, due to

Rather, to focus on the main purpose of the book, we suffice with these models in this number.

To delve into the main topic of this book, which is "Multiple Attribute Decision Making," we will provide a concise introduction to it in Sect. (1.11) of this chapter.

1.9 The Hierarchy of Decisions

The hierarchy of decisions refers to the classification and organization of decisions based on their level of importance, scope, time frame, and impact within an organization or system. It represents the structured arrangement of decisions according to their strategic, tactical, and operational nature.

The concept recognizes that decisions vary in their significance and the level of authority required to make them. It acknowledges that decision-making is not a uniform process but rather occurs at different levels within an organizational hierarchy, with each level having its own focus, time horizon, and responsibilities.

At the top of the hierarchy, strategic decisions are made by senior leaders or decision-makers with the authority to shape the long-term direction and goals of the organization. These decisions typically involve analyzing the external environment, assessing risks, and setting overarching strategies.

In the middle of the hierarchy, tactical decisions are made by middle-level managers or officials who bridge the gap between strategic decisions and their implementation. These decisions involve translating strategic plans into actionable steps and determining the allocation of resources to achieve specific objectives.

At the bottom of the hierarchy, operational decisions are made by frontline managers or decision-makers responsible for the day-to-day activities and processes. These decisions focus on executing plans, managing resources, and ensuring the smooth operation of the organization.

The hierarchy of decisions provides a framework for understanding the interdependence and coordination between different levels of decision-making. It helps establish clear roles, responsibilities, and decision-making authority within an organization, ensuring that decisions are aligned with the overall objectives and strategies.

Overall, the concept of the hierarchy of decisions emphasizes the need for a structured approach to decision-making that considers the varying levels of impact, time frame, and scope of decisions, enabling effective and coordinated decision-making throughout the organization [16].

1. Strategic Decisions: Strategic decisions are long-term in nature, with a planning horizon spanning several years. They are made by individuals or groups with sufficient authority and responsibility and involve significant risk and uncertainty. Strategic decisions require thorough analysis of the external environment and aim to establish long-term goals and plans. These decisions have a broad impact on the system or organization and may influence multiple interest groups.

Examples of strategic decisions include diversifying revenue streams, acquiring new businesses, or implementing large-scale infrastructure projects.

2. Tactical Decisions: Tactical decisions bridge the gap between strategic decisions and their implementation. They are medium-term in nature and involve translating strategic decisions into action plans. Senior and middle-level executives or government officials responsible for policy implementation often make tactical decisions. These decisions focus on acquiring or augmenting technology, capacity, and facilities, as well as determining their deployment. Examples of tactical decisions include allocating resources among competing entrepreneurs or choosing alternative technologies for value addition.
3. Operational Decisions: Operational decisions are short-term in nature and are made by frontline decision-makers responsible for implementing tactical decisions. These decisions pertain to the actual deployment and management of resources to carry out day-to-day operations. They are focused on production planning, job scheduling, inventory management, equipment replacement, transportation logistics, and other similar operational aspects. In academic institutions, operational decisions could involve class scheduling, test administration, placement services, and organizing events.

It is important to note that the hierarchy of decisions reflects the different levels of decision-making within an organization or system. Strategic decisions set the overall direction, tactical decisions translate strategy into actionable plans, and operational decisions execute those plans at a detailed level. Each level of decision-making plays a crucial role in achieving organizational objectives and responding to the challenges and opportunities presented by the environment. Proper coordination and alignment between these levels of decisions are essential for effective decision-making and the overall success of the organization.

1.10 A Historical Review About Decision Making

Decision-making has evolved over time, shaped by various factors and influences. Here is a brief history of decision-making, divided into eras, as described in the Harvard Business Review:

1. Pre-Industrial Era:

During the Pre-Industrial Era, which spanned centuries prior to the Industrial Revolution, decision-making was primarily intuitive and based on personal judgment. Here are some notable theorists who contributed to decision-making during this era, along with the approximate time periods when their ideas gained prominence:

- Niccolò Machiavelli (15–16th century): Machiavelli, an Italian Renaissance political philosopher, explored decision-making in his influential work, "The Prince." Published in 1532, his ideas emphasized the importance of pragmatic decision-making and the need for leaders to be flexible, adaptive, and willing to make tough choices to maintain power and achieve their goals.

- Thomas Hobbes (17th century): Hobbes, an English philosopher, discussed decision-making in his seminal work, "Leviathan." Published in 1651, he argued that decisions made by a sovereign authority were essential for establishing order and stability in society. Hobbes' ideas laid the foundation for social contract theory and the concept of centralized decision-making.
- Francis Bacon (17th century): Bacon, an English philosopher and statesman, advocated for a more systematic and empirical approach to decision-making. In his work "Novum Organum," published in 1620, he stressed the importance of collecting and analyzing data to inform decisions. Bacon's ideas laid the groundwork for the scientific method and the emergence of evidence-based decision-making.
- René Descartes (17th century): Descartes, a French philosopher and mathematician, emphasized rationality and logic in decision-making. In his philosophical writings, particularly "Discourse on the Method" published in 1637, Descartes developed the concept of Cartesian doubt, encouraging individuals to question their assumptions and beliefs before making decisions. His ideas influenced the development of rational decision-making models.
- Adam Smith (18th century): Smith, a Scottish economist and philosopher, explored decision-making in the context of economic systems. In his seminal work, "The Wealth of Nations," published in 1776, he highlighted the importance of individual decision-making in free markets and the role of self-interest and competition in driving economic outcomes.

These theorists and their ideas contributed to shaping early thinking on decision-making during the Pre-Industrial Era. Their works laid the groundwork for subsequent developments in decision theory and provided insights into the nature of decision-making that influenced decision-making practices of their time.

2. Industrial Era:

During the Industrial Era, which spanned from the late eighteenth century to the early twentieth century, decision-making underwent significant changes due to the Industrial Revolution and the rise of large-scale organizations. Here are some notable theorists who contributed to decision-making during this era, along with the years in which their work gained prominence:

- Frederick Winslow Taylor (late 19th to early 20th century): Taylor, an American engineer, is considered the father of scientific management. In the late 19th and early 20th centuries, he developed principles and methods for optimizing efficiency in the workplace. Taylor's work emphasized the scientific analysis of tasks, standardization of processes, and the separation of decision-making authority between managers and workers.
- Henri Fayol (late 19th to early 20th century): Fayol, a French mining engineer and management theorist, introduced the concept of administrative management. His ideas gained prominence in the early 20th century. Fayol emphasized the importance of managerial functions, such as planning, organizing, coordinating,

commanding, and controlling. His work laid the foundation for hierarchical decision-making structures in organizations.

- Max Weber (early 20th century): Weber, a German sociologist, explored decision-making within the context of bureaucracy. His work gained prominence in the early 20th century and emphasized the rationality and efficiency of bureaucratic organizations. Weber advocated for clear rules, hierarchical authority, and impersonal decision-making processes.
- Chester Barnard (1930s): Barnard, an American management theorist, focused on decision-making within the context of organizational behavior. His work gained prominence in the 1930s. Barnard emphasized the importance of effective communication, cooperation, and acceptance of authority in decision-making processes. He highlighted the role of informal organizations and the social dynamics that influence decision-making.
- Elton Mayo (1930s): Mayo, an Australian psychologist, conducted influential studies at the Hawthorne Works of Western Electric in the 1930s. His research highlighted the impact of social factors on productivity and decision-making. Mayo's work emphasized the importance of human relations, motivation, and employee participation in decision-making processes.

These theorists and their ideas significantly influenced decision-making practices during the Industrial Era. Their work laid the foundation for hierarchical structures, standardized procedures, and a focus on efficiency and productivity. The principles and concepts developed during this era continue to shape decision-making processes in modern organizations, although they have evolved to accommodate changing technologies and organizational dynamics.

3. Information Age Era:

The rise of technology and the availability of information significantly impacted decision-making in this era. Decision-makers had access to vast amounts of data, enabling them to make more informed choices. Decision support systems and analytical tools were developed to aid in decision-making processes.

During the Information Age Era, which began in the late 20th century and continues to the present, decision-making has been profoundly impacted by the widespread availability of technology and access to vast amounts of information. Here are some notable theorists who have contributed to decision-making during this era, along with the years in which their work gained prominence:

- Michael Porter (1980s): Porter, an American economist and professor, introduced the concept of competitive strategy. His work focused on analyzing industry dynamics, competitive forces, and value creation. Porter's frameworks, such as the Five Forces Model and Value Chain Analysis, have helped organizations make strategic decisions in an increasingly competitive business landscape.
- Peter Senge (1990s): Senge, an American systems scientist and author, explored the concept of the learning organization. In his influential book "The Fifth Discipline," Senge emphasized the importance of systems thinking, personal mastery, mental models, shared vision, and team learning in decision-making. His

ideas promoted a holistic and collaborative approach to decision-making within organizations.

- James March (1990s): March, an American political scientist and organizational theorist, contributed to the understanding of decision-making processes in organizations. He introduced concepts such as "garbage can model" and "bounded rationality." March's work highlighted that decision-making is often a complex and messy process influenced by organizational politics, ambiguity, and limited rationality.
- John Kotter (1990s): Kotter, an American organizational change expert and author, explored decision-making in the context of organizational leadership. His work emphasized the importance of creating a sense of urgency, building coalitions, and empowering employees to make decisions. Kotter's change management models provided frameworks for effective decision-making during periods of organizational transformation.
- Daniel Kahneman and Amos Tversky (1970s onwards): Kahneman, an American psychologist, and Tversky, an American cognitive psychologist, conducted groundbreaking research in behavioral economics and decision-making. Their work, which gained prominence in the 1970s and continued into the Information Age Era, shed light on cognitive biases, heuristics, and the influence of psychological factors on decision-making processes.

These theorists and their contributions have shaped the understanding of decision-making during the Information Age Era. Their insights have guided organizations in leveraging technology, managing information overload, and making informed decisions in complex and dynamic environments.

4. Globalization Era

Globalization brought about increased interconnectedness and complexity in decision-making. Organizations expanded their operations across borders, and decision-makers had to navigate cultural, political, and economic differences. Collaborative decision-making and cross-functional teams became more prevalent.

The era of globalization, which gained momentum in the late 20th century and continues to the present day, has significantly influenced decision-making processes across various domains. While globalization is a complex phenomenon influenced by numerous factors, including technological advancements and economic integration, several theorists have contributed to understanding decision-making during this era. Here are some notable theorists and the approximate years when their ideas gained prominence:

- Thomas Friedman (1990s):

Friedman, an American journalist, and author, introduced the concept of the "flat world" in his book "The World Is Flat," published in 2005. He argued that advancements in technology, particularly in communication and transportation, have facilitated global interconnectedness, resulting in a more level playing field for business and decision-making.

- Joseph Stiglitz (1990s):

Stiglitz, an American economist, and Nobel laureate explored the implications of globalization on economic policy and decision-making. In his book "Globalization and Its Discontents," published in 2002, Stiglitz critically examined the effects of global economic integration and advocated for policies that address inequality and ensure sustainable development.

- Manuel Castells (1990s):

Castells, a Spanish sociologist, focused on the impact of globalization on society and decision-making. In his influential trilogy, "The Information Age," published between 1996 and 1998, he analyzed the role of communication networks, information flows, and global capitalism in shaping decision-making processes at the individual, organizational, and societal levels.

- Ulrich Beck (1990s):

Beck, a German sociologist, coined the term "cosmopolitanism" and explored the implications of global risks and uncertainties on decision-making. In his book "Risk Society: Towards a New Modernity," published in 1992, he discussed how globalization has created new challenges and decision dilemmas related to environmental, economic, and social risks.

- Peter Drucker (1990s):

Drucker, an influential management consultant and author, examined the impact of globalization on organizations and management practices. In his book "The Global Economy and the Nation-State," published in 1998, he discussed the need for organizations to adapt to global markets, make strategic decisions in a global context, and navigate the complexities of international competition.

These theorists, among others, have contributed to our understanding of decision-making in the context of globalization. Their works have shed light on the challenges and opportunities that arise from increased interconnectedness and have influenced decision-makers' approaches to managing global risks, harnessing technological advancements, and navigating the complexities of a globalized world.

5. Digital Era:

The digital era is characterized by the proliferation of digital technologies and the rise of artificial intelligence and machine learning. Decision-making processes are increasingly automated, with algorithms and data-driven insights playing a significant role. Real-time data analytics and predictive modeling have become integral to decision-making.

The Digital Era, also known as the Information Age or the Computer Age, refers to the period characterized by the widespread adoption and integration of digital technologies into various aspects of society and decision-making processes. Here are some notable theorists who have contributed to understanding decision-making during this era, along with the approximate years when their ideas gained prominence:

- Nicholas Negroponte (1980s): Negroponte, a Greek American architect and computer scientist, founded the MIT Media Lab and explored the implications of digital technologies on society. In his book "Being Digital," published in 1995, he discussed how digitalization impacts decision-making, information access, and the transformation of industries and institutions.
- Don Norman (1990s): Norman, an American researcher and author, focused on the design and usability of digital products and their influence on decision-making. In his book "The Design of Everyday Things," published in 1988 (though his ideas continued to gain prominence in the 1990s), he emphasized the importance of user-centered design and the role of affordances and constraints in guiding decision-making processes.
- Sherry Turkle (1990s onwards): Turkle, an American psychologist and sociologist, examined the impact of digital technologies on human behavior and decision-making. In her book "Alone Together," published in 2011, she discussed how the constant connectivity of digital devices affects social interactions, self-perception, and decision-making in various contexts.
- Ray Kurzweil (2000s): Kurzweil, an American inventor and futurist, explored the potential of emerging digital technologies and artificial intelligence. In his book "The Singularity Is Near," published in 2005, he discussed the impact of accelerating technological progress on decision-making, forecasting the convergence of human and machine intelligence.
- Cass R. Sunstein (2000s): Sunstein, an American legal scholar, examined decision-making in the context of the digital age and the challenges posed by information overload. In his book "Republic.com," published in 2001, he discussed the implications of digital media for democracy and decision-making processes, focusing on the need for diverse information sources and avoiding filter bubbles.

These theorists have contributed to our understanding of decision-making in the Digital Era. Their works have shed light on the transformative effects of digital technologies on information access, human cognition, user experience, and the challenges and opportunities that arise in decision-making processes influenced by digitalization.

6. Behavioral Era:

In recent years, there has been a growing recognition of the influence of cognitive biases and psychological factors on decision-making. Behavioral economics and psychology have contributed to understanding how individuals and organizations make decisions. This era emphasizes the importance of considering human behavior, emotions, and social dynamics in the decision-making process.

The Behavioral Era, also known as the Behavioral Revolution, refers to a period in the field of economics where the focus shifted from purely rational decision-making models to incorporating insights from psychology and behavioral sciences. Here are some notable theorists who have contributed to the Behavioral Era, along with the approximate years when their ideas gained prominence:

- Daniel Kahneman and Amos Tversky (1970s onwards): Kahneman, an American psychologist, and Tversky, an American cognitive psychologist, are considered pioneers of behavioral economics and decision-making. Their collaboration in the 1970s and subsequent works introduced concepts such as cognitive biases, heuristics, and prospect theory. Their research challenged the assumption of rationality in decision-making and highlighted the role of psychological factors.
- Richard Thaler (1980s onwards): Thaler, an American economist, made significant contributions to behavioral economics and decision-making. In the 1980s, he introduced the concept of "mental accounting" and later developed the field of "nudge theory" with Cass Sunstein. Thaler's work emphasized the importance of understanding biases and designing decision environments to improve individual and societal outcomes.
- Herbert A. Simon (1950s onwards): Simon, an American economist and Nobel laureate, played a crucial role in the development of behavioral economics. In the 1950s, he introduced the concept of "bounded rationality," suggesting that decision-makers have cognitive limitations and rely on heuristics to make decisions. Simon's work emphasized the importance of satisficing and the recognition of the role of uncertainty in decision-making.
- Robert Shiller (1980s onwards): Shiller, an American economist and Nobel laureate, focused on the role of human psychology in shaping economic decisions and market behavior. In the 1980s, he conducted research on behavioral finance, exploring how investor sentiment and irrational behavior influence financial markets. Shiller's work contributed to understanding the impact of psychological biases on decision-making in the realm of finance and investments.
- George Akerlof (1970s onwards): Akerlof, an American economist and Nobel laureate, contributed to behavioral economics by examining how individual motivations and perceptions impact economic decisions. In 1970, he published the seminal paper "The Market for Lemons," which discussed how asymmetric information influences market outcomes. Akerlof's work highlighted the role of behavioral factors in market failures and decision-making.

These theorists, among others, have significantly influenced decision-making during the Behavioral Era. Their ideas have challenged the assumptions of traditional economic models and emphasized the role of psychology, cognitive biases, and social factors in decision-making processes. The Behavioral Era has led to a better understanding of the complexities of human decision-making and has influenced fields such as economics, finance, public policy, and marketing.

7. The contemporary Era:

In The contemporary encompasses the present time and continues to evolve rapidly. Numerous theorists and thinkers have contributed to decision-making in this era, reflecting the diverse range of fields and perspectives. Here are some notable theorists who have made significant contributions to decision-making during the contemporary era, along with the approximate years when their ideas gained prominence:

- Nassim Nicholas Taleb (2000s onwards): Taleb, a Lebanese American scholar, explored decision-making under uncertainty and the concept of "black swan" events. In his book "The Black Swan: The Impact of the Highly Improbable," published in 2007, he discussed the importance of recognizing and accounting for rare and unpredictable events in decision-making processes.
- Dan Ariely (2000s onwards): Ariely, an American behavioral economist, has focused on understanding irrational decision-making and the factors that influence human behavior. In his book "Predictably Irrational," published in 2008, he discussed various cognitive biases and social influences that affect decision-making in everyday life.
- Malcolm Gladwell (2000s onwards): Gladwell, a Canadian journalist and author, has explored decision-making in the context of social psychology and popular culture. In his book "Blink: The Power of Thinking Without Thinking," published in 2005, he examined the role of rapid, intuitive decision-making and the factors that influence our snap judgments.
- Richard H. Thaler and Cass R. Sunstein (2000s onwards): Thaler and Sunstein collaborated on the concept of "nudge theory," which emphasizes designing decision environments to encourage better choices. In their book "Nudge: Improving Decisions About Health, Wealth, and Happiness," published in 2008, they explored how small changes in the choice architecture can influence decision-making without restricting individual freedom.
- Angela Duckworth (2010s onwards): Duckworth, an American psychologist, focused on the concept of grit and its influence on decision-making and achievement. In her book "Grit: The Power of Passion and Perseverance," published in 2016, she discussed how resilience, perseverance, and long-term commitment can impact decision-making and success.

1.11 Multi Attribute Decision Making (MADM)

In a complex and interconnected world, decision-making plays a pivotal role in every aspect of our personal and professional lives. From strategic business choices to everyday life decisions, the ability to evaluate multiple attributes and make informed judgments is crucial. This is where Multi-Attribute Decision Making (MADM) comes into play, offering a comprehensive framework to handle decision problems with multiple criteria or attributes.

In this book, titled "Multi-Attribute Decision Making (MADM)," we delve into the fascinating world of decision-making under multiple criteria. We explore the principles, methods, and applications of MADM, providing a comprehensive guide to mastering this powerful approach to decision analysis.

The book begins by laying a solid foundation, introducing the fundamental concepts of decision-making and highlighting the challenges posed by multi-criteria decision problems. We delve into the various types of decision criteria, such as cost, quality, time, and risk, and examine how these factors interact and impact the decision-making process.

Criteria serve as measures that decision-makers consider enhancing their satisfaction and achieve their goals. Criteria can be both qualitative and quantitative in nature. For example, when choosing a house, quantitative criteria such as size, proximity to the workplace, price, and building design, as well as qualitative criteria like having good neighbors, contribute to the decision-maker's satisfaction.

Comparing criteria to each other can be challenging due to their different scales and potential contradictions. Increasing one criterion may lead to a decrease in another. Therefore, in multi-criteria decision-making, the goal is to identify options that provide the greatest advantage across all criteria.

Criteria can be viewed as indicators or objectives in decision-making. As a result, multi-criteria decision-making problems are classified into two categories: multi-criteria decision-making and multi-objective decision-making.

Overall, the growing complexity of decision-making necessitates the use of multi-criteria decision-making techniques to handle multiple criteria, evaluate options, and ultimately support decision-makers in making well-informed choices that align with their objectives and preferences.

1.11.1 Multi-Criteria Decision-Making Problems

The goal is to choose one option among the available options. For example, hiring one person from a pool of applicants based on criteria such as responsibility, individual skills, and relevant work experience. Or choosing a car from a set of cars considering criteria such as price, model, and quality. In multi-criteria decision-making, options are first ranked based on evaluation criteria. The importance and priority of each criterion and option are examined using pairwise comparison matrices, and ultimately, the top option is selected. Since the set of problem solutions corresponds to the number of options and is therefore countable, it is also called discrete multi-criteria decision-making.

Multi-criteria planning problems are classified into two compensatory and non-compensatory methods:

(1) Non-compensatory methods: These methods do not allow for trade-offs between criteria, meaning that a weakness in one criterion cannot be compensated by an advantage in another criterion. Therefore, each criterion is individually considered, and comparisons are made based on the criterion. For example, obtaining a driver's license, where the criterion for success in both the theory and driving exams is considered, and one cannot compensate for the other.
(2) Compensatory methods: These methods allow for trade-offs between criteria, meaning that a small change in one criterion can be compensated by an opposing change in another criterion. For example, compensating for high costs with better product quality.

1.11.2 Multi-Objective Decision-Making Problems

Multi-objective decision-making aims to achieve multiple goals simultaneously. For example, in a production system, multiple objectives include reducing production time, reducing consumption of raw materials, reducing labor, increasing profits, improving product quality, and increasing customer satisfaction. Achieving these objectives optimally is not possible simultaneously because some objectives may be conflicting, meaning that increasing one may lead to a decrease in another. Multi-objective decision-making seeks solutions that bring all objectives as close to their optimum as possible, thus providing a set of solutions called Pareto optimal solutions. The decision-maker selects a solution from the set of Pareto optimal solutions based on their own policies [53].

1.11.3 Design Models in Conditions of Uncertainty

Decision-making depends on the availability of necessary information. The more complete, up-to-date, and current the information is, the greater the possibility of making accurate and appropriate decisions. When managers are aware of the occurrence and details of an activity, decision-making regarding the implementation of activities becomes easier. However, if some aspects of the activities are unclear, decision-making becomes challenging. In conditions of uncertainty and when the available information is insufficient, making decisions about choosing and implementing activities is not as straightforward. Decision-making under risky conditions is also one type of decision-making where the dimensions of the activity are unclear, and there is not enough information available. The existence of data in probabilistic, random, and fuzzy forms is also examples of decision-making under uncertainty. When parameters and constraints of the problem are not precisely and deterministically available, fuzzy logic comes to the rescue. Fuzzy decision-making enables the definition of imprecise and approximate parameters and constraints of the problem. Factors such as lack of precise information and the existence of information in a mental and linguistic form, which are expressed to some extent and to varying degrees in real life, open the way for decision-making using fuzzy logic. With the emergence of fuzzy thinking, decision-makers can engage in decision-making under ambiguous conditions without the need for complete information and definite numbers, by employing the techniques provided in fuzzy logic. In multi-criteria linear programming problems, the data may also be fuzzy. For example, if the indicators for employment are age, level of responsibility, and experience. The measurement of the responsibility indicator linguistically represents the degree of responsibility, such as completely responsible, somewhat responsible, or for experience, inexperienced, slightly experienced, moderately experienced, and fully experienced. Similarly, the age indicator is expressed as young, very young, not too young, and old, all of which are imprecise and uncertain concepts [54].

Considering that most decisions are multi-objective or multi-indicator, and fuzzy logic is based on human natural language, it is simple and flexible. Fuzzy multi-indicator decision-making problems have wide-ranging applications. This book focuses on studying multi-criteria decision-making problems with fuzzy data [55].

1.12 Scale Measurements of Data

Data can be categorized into four different types based on their scale of measurement [56]:

1. Nominal Data:
   - Description: Nominal data consists of categories or labels without any inherent order or ranking. Each category is distinct and unrelated to other categories.
   - Examples: Gender (Male/Female), Eye Color (Blue/Brown/Green), Marital Status (Single/Married/Divorced)
   - Challenges: The challenges with nominal data include handling data entry errors, ensuring consistency in category definitions, and preventing bias in data collection and categorization.
2. Ordinal Data:
   - Description: Ordinal data involves categories that have a specific order or ranking. The categories have a relative position, but the differences between them may not be uniform or quantifiable.
   - Examples: Educational Levels (Elementary/High School/College/Post-graduate), Rating Scales (Strongly Disagree/Disagree/Neutral/Agree/Strongly Agree)
   - Challenges: The challenges with ordinal data include determining an appropriate scale for ranking, interpreting the magnitude of differences between categories, and ensuring consistent allocation of ranks.
3. Interval Data:
   - Description: Interval data represents measurements where the differences between values are meaningful and consistent. However, there is no true zero point, and ratios between values are not applicable.
   - Examples: Temperature (measured in Celsius or Fahrenheit), Calendar Dates (e.g., July 15th, 2023)
   - Challenges: The challenges with interval data include understanding that the zero point is arbitrary, making sure the measurement scale is consistent, and avoiding incorrect ratio interpretations.
4. Ratio Data:
   - Description: Ratio data is like interval data but includes a meaningful zero point. Ratios between values are meaningful and can be calculated.

- Examples: Height, Weight, Time, Income
- Challenges: The challenges with ratio data are relatively minimal, but they can include dealing with outliers, ensuring consistency in measurement units, and handling missing or incomplete data.

It's important to understand the scale of measurement for data as it affects the types of statistical analysis that can be applied and the interpretations that can be made from the data.

In the next chapters, we will address the computational challenges associated with handling imprecise data in the form of Nominal and Ordinal variables. We will explore the application of fuzzy and interval number concepts to provide effective solutions for incorporating these types of data into decision models.

1.13 Qualitative Data and Ordinal Numbers

Measurement or data can be expressed using natural language descriptions instead of numbers. In statistics, qualitative data is often used interchangeably with "categorical" data. For instance, we can describe someone's favorite color as "blue" or their height as "tall." While these data may fall into categories, those categories may or may not have a specific order or structure. When the categories do not have a natural ordering, we refer to them as nominal categories. Examples of nominal categories include gender, race, religion, or sport.

On the other hand, when the categories can be ordered, they are referred to as ordinal variables. Ordinal variables are categorical variables that involve judging size or ranking. For example, using terms like small, medium, and large to describe sizes or using responses such as strongly disagree, disagree, neutral, agree, and strongly agree to measure attitudes. However, it's important to note that in ordinal variables, we may not know the exact distance or magnitude between the categories. We can assign numerical codes such as 1, 2, 3, 4, and so on to these qualitative data for the purpose of analysis, even though the numerical values themselves do not represent measurable distances.

Here are a few examples to illustrate qualitative data and ordinal variables:

1. Restaurant Ratings: Suppose you are reviewing restaurants and rating them based on their service quality. You use the following ordinal scale: "poor," "fair," "good," and "excellent." These ratings are qualitative and can be assigned numerical codes like 1, 2, 3, and 4 for analysis. By analyzing the numerical codes, you can assess the overall satisfaction levels of customers with different restaurants.
2. Education Levels: Imagine you are studying the education levels of individuals in a survey. The categories for education levels can be ordered from "elementary school," "high school," "college," and "post-graduate studies." These categories represent an ordinal variable, and numerical codes can be assigned accordingly. Analyzing the numerical codes can help understand the distribution and trends in educational attainment among the surveyed population.

3. Pain Intensity: In a medical study, participants rate their pain intensity using the following ordinal scale: "mild," "moderate," and "severe." These qualitative ratings can be coded as 1, 2, and 3 to analyze the severity of pain experienced by individuals. By analyzing the numerical codes, researchers can compare pain levels among different groups or assess the effectiveness of pain management strategies.
4. Customer Satisfaction: In a customer satisfaction survey, respondents may rate their satisfaction levels using an ordinal scale such as "very unsatisfied," "unsatisfied," "neutral," "satisfied," and "very satisfied." These qualitative ratings can be assigned numerical codes (e.g., 1–5) to quantify customer satisfaction levels. Analyzing the numerical data allows organizations to identify areas for improvement and track changes in customer satisfaction over time.

In these examples, the qualitative data is represented using ordinal numbers to facilitate analysis while acknowledging that the numerical values do not capture the precise magnitude or distance between categories. By assigning numerical codes to qualitative data, researchers and analysts can organize, compare, and derive insights from the data, considering the order or ranking implied by the categories.

1.14 Quantitative Data and Cardinal Numbers

In this book, we will delve into the concept of quantitative data and its measurement scales, particularly focusing on cardinal numbers. Quantitative data is expressed using numerical values rather than descriptive language. However, not all numbers have the same properties in terms of their mathematical operations and measurability. Cardinal numbers represent a specific type of quantitative data that allows for meaningful mathematical operations within the numbers themselves.

Let's explore the various scales of quantitative data:

Ratio Scale: The ratio scale is the most common and versatile type of quantitative scale. Measurements on a ratio scale possess a meaningful zero point and consistent intervals. This means that the difference between two values can be accurately interpreted, and ratios between values have practical significance. For example, consider the measurement of age. If a person is 10 years old and another person is 5 years old, we can say that the former is twice as old as the latter. Time is another example of a ratio-scale variable, where zero represents a complete absence (e.g., 0 s), and equal intervals (e.g., 10 s) have consistent meanings. Money is also measured on a ratio scale, where zero represents the absence of funds.

Interval Scale: The interval scale is another type of quantitative measurement that exhibits consistent intervals but lacks a meaningful zero point. Temperature measured in Celsius or Fahrenheit is an example of an interval scale. While the difference between 10 and 20° is the same as the difference between 100 and 110°, we cannot interpret these values in terms of ratios. A temperature of 50 °C is not half as hot as 100 °C. However, the interval of 10° indicates the same temperature difference regardless of the position on the scale. On the other hand, the Kelvin temperature scale is a ratio scale because it has an absolute zero point, representing the absence of heat. Therefore, we can assert that 200 K is twice as hot as 100 K.

In conclusion, quantitative data is expressed using numerical values, and the scale of measurement determines the properties and interpretation of the data. Cardinal numbers are employed when mathematical operations within the numbers hold meaning. Ratio scales have a meaningful zero and consistent intervals, while interval scales possess consistent intervals but lack a meaningful zero. Understanding the scale of measurement is essential for accurate data analysis and interpretation.

1.15 Scientometrics in the field of Fuzzy Multi Attribute Decision Making

A comprehensive exploration of fuzzy multi-attribute decision making is essential in understanding the field's current significance and growth. A search in the Scopus citation database yielded a substantial repository of 5,173 scientific documents, comprising articles, conference papers, catalogs, and more. This wealth of literature is a testament to the active and dynamic nature of research and applications in the realm of fuzzy multi-attribute decision making.

Intriguingly, a perusal of the database not only revealed the extensive number of documents but also highlighted an upward trend in the production of scientific content in this field. The graph displaying the frequency of science production in the domain of fuzzy multi-attribute decision making over the years exhibits a notable ascent (Fig. 1.1). This upsurge in research output underscores the increasing importance and relevance of fuzzy multi-attribute decision making in diverse domains and the growing interest of researchers in contributing to this evolving landscape.

As we delve deeper into the subject matter of this book, it is imperative to recognize the significance of the extensive literature available in the Scopus database and the positive trajectory in research output. These insights affirm the timeliness and importance of our exploration into fuzzy multi-criteria decision making, as it continues to gain prominence in the academic and practical spheres

In Fig. 1.2, countries are depicted based on their utilization of articles, books, and other documents related to fuzzy MADM. Notably, China, India, Pakistan, Iran, and Turkey rank as the top five nations in terms of their usage of these documents.

Figure 1.3 illustrates the distribution of document types within the field of fuzzy MADM.

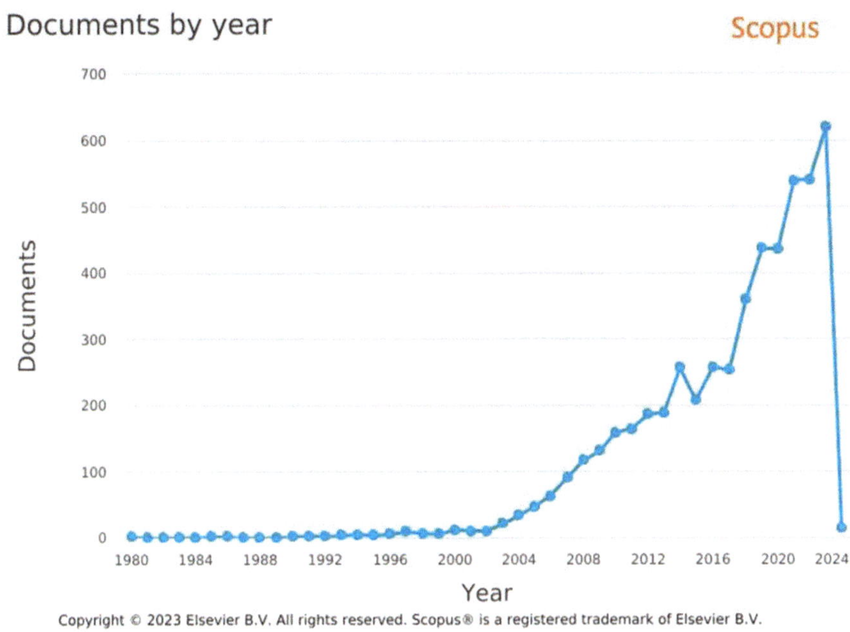

Fig. 1.1 Documents by year

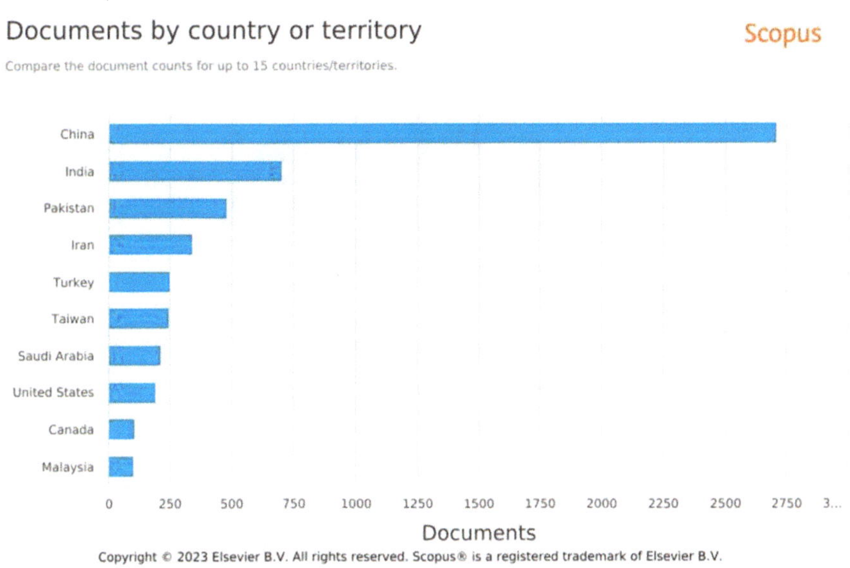

Fig. 1.2 Documents by country

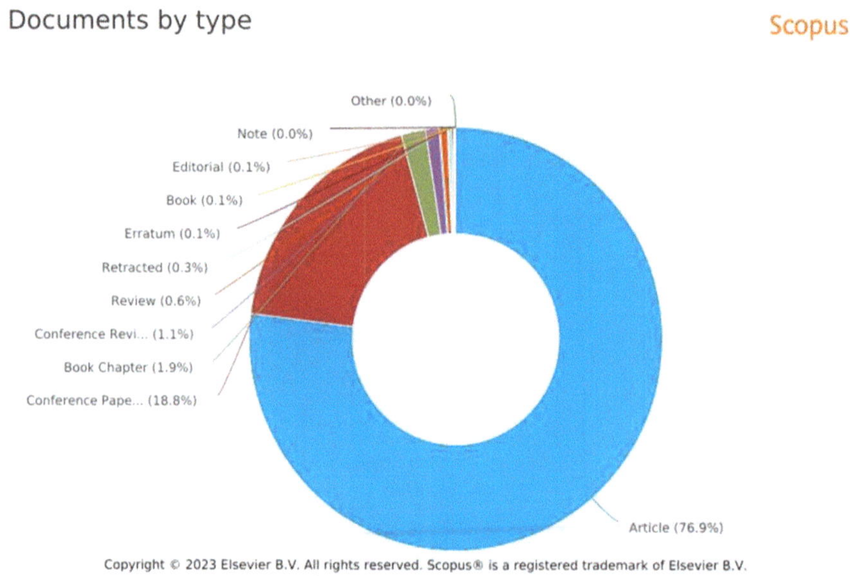

Fig. 1.3 Documents by type

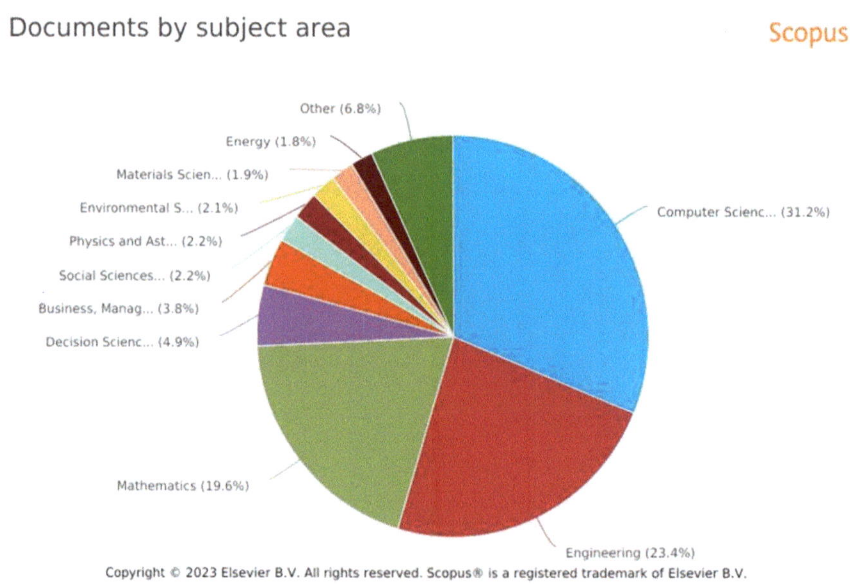

Fig. 1.4 Documents by subject area

Figure 1.4 provides a visual representation of the composition of documents within the Fuzzy MADM field. It reveals that a significant portion of these documents, specifically 31.2%, is closely related to the field of computer science. Additionally, 23.4% of the documents are affiliated with engineering, and 19.6% are primarily focused on mathematical aspects. This data sheds light on the diverse range of disciplines and domains that intersect with or contribute to Fuzzy Multi-Attribute Decision Making, indicating the multidisciplinary nature of this field.

1.16 Conclusion

In conclusion, the first chapter of this book has provided a comprehensive overview of the foundational aspects of Multiple Attribute Decision Making (MADM). We have explored various key topics, including Decision Theory, Decision Science, Decision-Making Theories, reputable domains, and applications of Decision Making, reputable Models and Techniques of Decision Making, as well as a historical review about Decision Making.

Throughout this chapter, we have gained insights into the theoretical frameworks and practical applications of decision-making, highlighting its significance in diverse domains such as business, engineering, finance, public policy, and more. We have discussed the importance of understanding decision theory and the various models and techniques available for effective decision-making. Additionally, we have explored the measurement of data using scale measurements, recognizing its role in the decision-making process.

By examining the historical development of decision-making theories, we have gained a deeper appreciation for the evolution of decision science and its contributions to modern decision-making practices. We have also acknowledged the wide-ranging applications of decision-making in different domains, underscoring its relevance and impact in various fields.

As we conclude this chapter, it is evident that MADM is a multidimensional and dynamic field, encompassing a rich array of theories, models, and applications. The foundation laid in this chapter provides a solid framework for the subsequent chapters, where we will delve deeper into the intricacies of Multiple Attribute Decision Making and explore its practical implementation in real-world scenarios.

By understanding the theories, concepts, and models discussed in this chapter, readers are well-equipped to embark on a journey of exploring and applying MADM techniques to complex decision-making problems. The knowledge gained from this chapter will serve as a strong basis for further exploration and analysis, empowering readers to make informed and effective decisions in their personal and professional lives.

In the upcoming chapters, we will delve into specific MADM methods, case studies, and practical examples, allowing readers to deepen their understanding and develop the necessary skills to navigate the challenges of decision-making in a multifaceted world.

Overall, the first chapter has set the stage for a comprehensive exploration of Multiple Attribute Decision Making, laying the groundwork for the subsequent chapters, and providing readers with a solid foundation to build upon. It is an invitation to embark on a journey of discovery, where the complexities of decision-making will be unraveled, and the tools and techniques to make better decisions will be explored in depth.

Acknowledgement A special thanks to the Iranian DEA society for their unwavering spiritual support and consensus in the writing of this book. Your invaluable support has been truly remarkable, and we are deeply grateful for the opportunity to collaborate with such esteemed professionals.

References

1. Soltanifar, M., Hosseinzadeh Lotfi, F., Sharafi, H., & Lozano, S. (2022). Resource allocation and target setting: A CSW–DEA based approach. *Annals of Operations Research, 318*(1), 557–589.
2. Tversky, A., & Kahneman, D. (1989). Rational choice and the framing of decisions. In *Multiple criteria decision making and risk analysis using microcomputers* (pp. 81–126). Berlin: Springer.
3. Deng, J., Zhan, J., Ding, W., Liu, P., & Pedrycz, W. (2023). A novel prospect-theory-based three-way decision methodology in multi-scale information systems. *Artificial Intelligence Review, 56*(7), 6591–6625.
4. Hammond, K. R., Hamm, R. M., Grassia, J., & Pearson, T. (1987). Direct comparison of the efficacy of intuitive and analytical cognition in expert judgment. *IEEE Transactions on Systems, Man, and Cybernetics, 17*(5), 753–770.
5. Kahneman, D. (2003). A perspective on judgment and choice: Mapping bounded rationality. *American Psychologist, 58*(9), 697.
6. Daniel, K. (2017). *Thinking, fast and slow.*
7. Kahneman, D., & Klein, G. (2009). Conditions for intuitive expertise: A failure to disagree. *American Psychologist, 64*(6), 515–526.
8. Kahneman, D. (2003). Maps of bounded rationality: Psychology for behavioral economics. *American Economic Review, 93*(5), 1449–1475.
9. Dijksterhuis, A., & Nordgren, L. F. (2006). A theory of unconscious thought. *Perspectives on Psychological Science, 1*(2), 95–109.
10. Epstein, S. (1994). Integration of the cognitive and the psychodynamic unconscious. *American Psychologist, 49*(8), 709–724.
11. Gigerenzer, G., & Gaissmaier, W. (2011). Heuristic decision making. *Annual review of psychology, 62*, 451–482.
12. Klein, G. (2003). Intuition at work: Why developing your gut instincts will make you better at what you do. Crown Business.
13. Qin, J., Martínez, L., Pedrycz, W., Ma, X., & Liang, Y. (2023). An overview of granular computing in decision-making: Extensions, applications, and challenges. *Information Fusion*, 101833.
14. Liu, Z. L., Liu, F., Zhang, J. W., & Pedrycz, W. (2023). Optimizing consistency and consensus in group decision making based on relative projection between multiplicative reciprocal matrices. *Expert Systems with Applications, 224*, 119948.
15. Kahneman, D., & Lovallo, D. (1993). Timid choices and bold forecasts: A cognitive perspective on risk taking. *Management Science, 39*(1), 17–31.
16. Mukherjee, S. P. (2021). *Decision-making: Concepts, methods and techniques.* SAGE Publishing India.

17. Wieder, B., & Ossimitz, M. L. (2015). The impact of Business Intelligence on the quality of decision making–a mediation model. *Procedia Computer Science, 64*, 1163–1171.
18. Vahdani, B., Behzadi, S. S., Mousavi, S. M., & Shahriari, M. R. (2016). A dynamic virtual air hub location problem with balancing requirements via robust optimization: Mathematical modeling and solution methods. *Journal of Intelligent & Fuzzy Systems, 31*(3), 1521–1534.
19. Xiao, F., Wen, J., & Pedrycz, W. (2022). Generalized divergence-based decision-making method with an application to pattern classification. In *IEEE transactions on knowledge and data engineering.*
20. Kahneman, D., Lovallo, D., & Sibony, O. (2019). A structured approach to strategic decisions. *MIT Sloan Management Review.*
21. Kahneman, D. (2019). Human engineering of decisions. In *Ethics in an age of pervasive technology* (pp. 190–192). Routledge.
22. Shahriari, M., & Asoodeh, M. H. (2020). Developing a new model based on artificial intelligence techniques for predicting bitcoin fluctuations. *Kepes, 18*(4), 108–116.
23. Shahriari, M., & Asoodeh, M. H. (2021). Predicting long-term deposit openings of bank customers using decision tree and random forest classification. *Kepes, 19*(3), 70–81.
24. Shadab, R., Shahriari, M., Esfeden, G. A., & Lotfi, F. H. (2022). Providing a heuristic model based on data envelopment analysis to improve the solution to the time-cost-quality trade-off problem considering risk and efficiency. *Computer Integrated Manufacturing Systems, 28*(12), 2608–2625.
25. Shafiee, M. A., & Shahriari, M. R. (2021). A mathematical optimization model for integrating the problems of discrete time-cost tradeoff (DTCTP) and multi-mode resource-constrained project scheduling (MRCPSP). *International Journal of Industrial Mathematics, 13*(4), 489–505.
26. Zhang, Q., Yang, Y., Ma, H., & Wu, Y. N. (2019). Interpreting cnns via decision trees. In *Proceedings of the IEEE/CVF Conference on Computer Vision and Pattern Recognition* (pp. 6261–6270).
27. Pilevari, N., Hasanzade, M., & Shahriari, M. (2014). A hybrid fuzzy multiple attribute decision making approach for identification and ranking influencing factors on Bullwhip Effect in supply chain: Real case of Steel industry. *International Journal of Industrial Mathematics, 8*(1), 49–63.
28. Shafiee, M., Lotfi, F. H., & Saleh, H. (2014). Supply chain performance evaluation with data envelopment analysis and balanced scorecard approach. *Applied Mathematical Modelling, 38*(21–22), 5092–5112.
29. Shahriari, M. R. (2017). Soft computing based on a modified MCDM approach under intuitionistic fuzzy sets. *Iranian Journal of Fuzzy Systems, 14*(1), 23–41.
30. Sharafi, H., Lotfi, F. H., Jahanshahloo, G. R., & Razipour-GhalehJough, S. (2020). Fair allocation fixed cost using cross-efficiency based on Pareto concept. *Asia-Pacific Journal of Operational Research, 37*(01), 1950036.
31. Sharifi, M., Saadvandi, M., & Shahriari, M. R. (2020). Presenting a series-parallel redundancy allocation problem with multi-state components using recursive algorithm and meta-heuristic. *Scientia Iranica, 27*(2), 970–982.
32. Shahriari, M. (2022). Using genetic algorithm to optimize a system with repairable components and multi-vacations for repairmen. *International Journal of Nonlinear Analysis and Applications, 13*(2), 3139–3144.
33. Sharifi, M., Shahriyari, M., Khajehpoor, A., & Mirtaheri, S. A. (2022). Reliability optimization of a k-out-of-n series-parallel system with warm standby components. *Scientia Iranica, 29*(6), 3523–3541.
34. Sharifi, M., Cheragh, G., Dashti Maljaii, K., Zaretalab, A., & Shahriari, M. (2021). Reliability and cost optimization of a system with k-out-of-n configuration and choice of decreasing the components failure rates. *Scientia Iranica, 28*(6), 3602–3616.
35. Shahriari, M. (2022). Using a hybrid NSGA-II to solve the redundancy allocation m model of series-parallel systems. *International Journal of Industrial Mathematics, 14*(4), 503–513.
36. Sharifi, M., Shahriari, M. R., & Zaretalab, A. (2019). The effects of technical and organizational activities on redundancy allocation problem with choice of selecting redundancy strategies using the memetic algorithm. *International Journal of Industrial Mathematics, 11*(3), 165–176.

37. Simon, H. A. (1957). Models of man: Social and rational. Wiley & Sons.
38. Yazdi, M., Khan, F., Abbassi, R., & Rusli, R. (2020). Improved DEMATEL methodology for effective safety management decision-making. *Safety Science, 127*, 104705.
39. Zdražil, P., & Applová, P. (2017). Visual evaluation of changes in regional growth and disparities: usage of a pareto chart. Scientific papers of the University of Pardubice. Series D, Faculty of Economics and Administration. 41/2017.
40. Yasodai, P., & Ritha, W. A PARAMETRIC PROGRAMMING APPROACH TO AN INTUITIONISTIC FUZZY QUEUING MODEL. *a a, 2*(3), 4.
41. Živković, Ž, Nikolić, Đ, Đorđević, P., Mihajlović, I., & Savić, M. (2015). Analytical network process in the framework of SWOT analysis for strategic decision making (Case study: Technical faculty in Bor, University of Belgrade, Serbia). *Acta Polytechnica Hungarica, 12*(7), 199–216.
42. Allahviranloo, T., Ghanbari, M., Hosseinzadeh, A. A., Haghi, E., & Nuraei, R. (2011). A note on "Fuzzy linear systems." *Fuzzy sets and systems, 177*(1), 87–92.
43. Allahviranloo, T., Shamsolkotabi, K. H., Kiani, N. A., & Alizadeh, L. (2007). Fuzzy integer linear programming problems. *International Journal of Contemporary Mathematical Sciences, 2*(4), 167–181.
44. Shahriari, M. (2023). Redundancy allocation optimization based on the fuzzy universal generating function approach in the series-parallel systems. *International Journal of Industrial Mathematics, 15*(1), 69–77.
45. Jahanshahloo, G. R., Hosseinzadeh, F., Shoja, N., & Tohidi, G. (2003). A method for solving 0–1 multiple objective linear programming problem using dea. *Journal of the Operations Research Society of Japan, 46*(2), 189–202.
46. Mohagheghi, V., Mousavi, S. M., Vahdani, B., & Shahriari, M. R. (2017). R&D project evaluation and project portfolio selection by a new interval type-2 fuzzy optimization approach. *Neural Computing and Applications, 28*, 3869–3888.
47. Eslami, R., Khodabakhshi, M., Jahanshahloo, G. R., Lotfi, F. H., & Khoveyni, M. (2012). Estimating most productive scale size with imprecise-chance constrained input–output orientation model in data envelopment analysis. *Computers & Industrial Engineering, 63*(1), 254–261.
48. Najafi, E., Aryanegad, M. B., Lotfi, F. H., & Ebnerasould, A. (2009). Efficiency and effectiveness rating of organization with combined DEA and BSC. *Applied Mathematical Sciences, 3*(25–28), 1249–1264.
49. Nozari, H., Najafi, E., Fallah, M., & Hosseinzadeh Lotfi, F. (2019). Quantitative analysis of key performance indicators of green supply chain in FMCG industries using non-linear fuzzy method. *Mathematics, 7*(11), 1020.
50. Razipour-GhalehJough, S., Lotfi, F. H., Rostamy-Malkhalifeh, M., & Sharafi, H. (2021). Benchmarking bank branches: A dynamic DEA approach. *Journal of Information and Optimization Sciences, 42*(6), 1203–1236.
51. Soltanifar, M., Sharafi, H., Hosseinzadeh Lotfi, F., Pedrycz, W., & Allahviranloo, T. (2023). Hybrid multi-attribute decision-making methods based on preferential voting. *Preferential voting and applications: approaches based on data envelopment analysis* (pp. 133–164). Cham: Springer International Publishing.
52. Sharifi, M., Moghaddam, T. A., & Shahriari, M. (2019). Multi-objective redundancy allocation problem with weighted-k-out-of-n subsystems. *Heliyon, 5*(12), e02346.
53. Song, M., Han, L., & Pedrycz, W. (2022). A comprehensive study on effect of multi-subgroup background in group decision-making. *Soft Computing, 26*(24), 13543–13566.
54. Shahriari, M. R., Pilevari, N., & Gholami, Z. (2016). The effect of information systems on the supply chain sustainability using DEMATEL method. *Communications on Advanced Computational Science with Applications, 1*, 47–56.
55. Trieu, V. H. (2017). Getting value from Business Intelligence systems: A review and research agenda. *Decision Support Systems, 93*, 111–124.
56. Krosnick, J. A. (2018). Questionnaire design. *The Palgrave Handbook of Survey Research*, 439–455.
57. XiVedung, E. (2017). *Public policy and program evaluation.* Routledge.

Chapter 2
Fuzzy Introductory Concepts

2.1 Introduction

Fuzzy sets were introduced to the world of science and knowledge by Lotfi Zadeh in 1965 [22]. These sets provide a mathematical model for representing ambiguity and imprecise concepts. In fuzzy thinking, there are no clear boundaries, and the membership of different elements is relative to various concepts and topics. Thus, this type of thinking is more compatible with the nature of humans and our world. It seems that since fuzzy thinking introduces a fresh perspective and extends Aristotelian logic, there should be no opposition to this approach. However, the important point is that according to this mathematical viewpoint, classical mathematics, which is based on Aristotelian logic, also comes into question, and that is where the disagreement begins. Many believe that fuzzy concepts are a generalization of crisp sets, which are precise and definite sets.

Classical mathematics is a suitable tool for expressing various concepts in cases where dealing with a binary world. However, with the growth of human thought and scientific and technological advancements, there is a need for more suitable scientific tools that have a greater ability to adapt to real phenomena to express more complex concepts in life and the human environment. Concepts that are not adequately and feasibly represented using conventional mathematics, which is based on binary and multi-valued criteria.

Scientists in the past used to analyze their environment based on the assumption that everything is either true or false. Although they were sometimes uncertain about the truth or falsehood of certain things. With this mindset, they believed that in a phenomenon, it is either "true" or "false". The mistake of some scientists in the past was that they did not pay attention to the fact that all phenomena cannot be categorized into two categories: "true" or "false"; instead, everything should be assessed relatively and assigned a degree of truth. Most things that appear "true" are "relatively true," and therefore, there are degrees of uncertainty regarding the accuracy and validity of real phenomena. In other words, phenomena are not always black or white, and real

F. Hosseinzadeh Lotfi et al., *Fuzzy Decision Analysis: Multi Attribute Decision Making Approach*, Studies in Computational Intelligence 1121,
https://doi.org/10.1007/978-3-031-44742-6_2

phenomena have varying degrees of ambiguity and imprecision. The applications of fuzzy logic in many of the tools we use demonstrate the high capability and effectiveness of this concept and the proper and precise development of fuzzy sets [1, 2, 20, 22, 27, 28].

2.2 Fuzzy Set Theory: Basic Concepts

In this section, fuzzy concepts, fuzzy sets, and their properties will be discussed.

Definition 2.1 Membership Function (Characteristic)

If P is a well-defined property over the universal set X, then the set A is defined as follows

$$A = \{x|x \in X \& p(x)\}$$

Therefore, set A can be shown as follows:

$$A\{(x, \mu_A(x))|x \in X \,\&\, \mu_A : X \to \{0.1\}\}$$

where,

$$\mu_A(x) = \begin{cases} 1, x \in A \\ 0, x \notin A \end{cases} = \begin{cases} 1, p(x) \\ 0, \sim p(x) \end{cases}$$

Definition 2.2 Fuzzy Set

Let X be the universal set. $\tilde{A}$ is called a fuzzy subset of X if and only if

$$\tilde{A} = \{(x, \mu_{\tilde{A}}(x))|x \in X \,\&\, \mu_{\tilde{A}} : X \to [0, 1]\}$$

where $\mu_{\tilde{A}}$ is the membership function. If $\mu_{\tilde{A}}$ is greater at a point, then the degree of membership of that point in $\tilde{A}$ will be greater. With this definition, it can be observed that crisp sets and membership functions in them are special cases of fuzzy sets and fuzzy membership functions [4, 21].

Definition 2.3 Support

The support of a fuzzy set $\tilde{A}$ is a crisp set and is defined as follows

$$Supp(\tilde{A}) = \{x|x \in X \& \mu_{\tilde{A}}(x) > 0\}$$

Definition 2.4 Null Fuzzy Sets

$\tilde{A}$ is called a null subset of the set X if and only if

$$\forall x\big(x \in X \Rightarrow \mu_{\tilde{A}}(x) = 0\big)$$

Definition 2.5 The Height of a Fuzzy Set

The height of the fuzzy set $\tilde{A}$ of the set X is defined as follows

$$H(\tilde{A}) = Sup\big\{\mu_{\tilde{A}}(x)|x \in X\big\}$$

Definition 2.6 Normal Fuzzy Set

The fuzzy set $\tilde{A}$ of X is called normal if and only if $\exists x\big(x \in X \,\&\, \mu_{\tilde{A}}(x) = 1\big)$.

Example 2.1 Let X = {1, 2, ⋯ , 10}, Consider the fuzzy subset $\tilde{A}$ of X as follows

$$\tilde{A} = \{(1, 0.2), (2, 0.7), (3, 1), (7, 0.2), (5, 0.4)\}$$

Then, $Supp(\tilde{A}) = \{1, 2, 3, 5, 7\}$ and $H(\tilde{A}) = 1$; and considering that $(3, 1) \in \tilde{A}$ therefore $\tilde{A}$ is normal.

Definition 2.7 Complete Fuzzy Set

$\tilde{A}$ of X is called complete if and only if $\forall x(x \in X \Rightarrow \mu_{\tilde{A}}(x) = 1)$.

Obviously, the fuzzy set is called empty if and only if $\forall x(x \in X \Rightarrow \mu_{\tilde{A}}(x) = 0)$.

Note that in Example 2.1, the fuzzy set $\tilde{A}$ is not complete.

Definition 2.8 α-cut of a Fuzzy Set

Let $\tilde{A}$ be a fuzzy set on X. In this case, the α-cut (weak) of $\tilde{A}$ is defined as follows: $\tilde{A}_\alpha = \big\{\mathrm{x}|\mathrm{x} \in \mathrm{X} \,\&\, \mu_{\tilde{A}}(x) \geq \alpha\big\}$ and strong α-cut of $\tilde{A}$ is defined as follows $\tilde{A}_{\overline{\alpha}} = \big\{\mathrm{x}|\mathrm{x} \in \mathrm{X} \,\&\, \mu_{\tilde{A}}(x) > \alpha\big\}$.

Obviously, according to Example 2.1, we can write $\tilde{A}_{0.3} = \{2, 3, 5\}$.

Let $\tilde{A}$ and $\tilde{B}$ be two fuzzy subsets of X. In this case, the operations between the fuzzy sets will be as follows.

Definition 2.9 Two Equal Fuzzy Sets

Two sets $\tilde{A}$ and $\tilde{B}$ are called equal if and only if $\forall x\big(x \in X \Rightarrow \mu_{\tilde{A}}(x) = \mu_{\tilde{B}}(x)\big)$.

Definition 2.10 Union of two fuzzy sets

The union of two fuzzy sets $\tilde{A}$ and $\tilde{B}$ of X is defined as follows:

$\forall x\big(x \in X \Rightarrow \mu_{\tilde{A}\cup\tilde{B}}(x) = Max\{\mu_{\tilde{A}}(x), \mu_{\tilde{B}}(x)\}\big).$

Definition 2.11 Intersection of Two Fuzzy Sets

The membership of two fuzzy sets $\tilde{A}$ and $\tilde{B}$ with the membership function $\mu_{\tilde{A}\cap\tilde{B}}$ is defined as follows

$\forall x\big(x \in X \Rightarrow \mu_{\tilde{A}\cap\tilde{B}}(x) = Min\{\mu_{\tilde{A}}(x), \mu_{\tilde{B}}(x)\}\big).$

Definition 2.12 Complement of a Fuzzy Set

Let $\tilde{A}$ be a fuzzy set on X. In this case, the complement of the fuzzy set with membership function $\mu'_{\tilde{A}}$ is defined as $\forall \mathrm{x}\big(\mathrm{x} \in \mathrm{X} \Rightarrow \mu_{\tilde{A}'}(\mathrm{x}) = 1 - \mu_{\tilde{A}}(\mathrm{x})\big)$.

Definition 2.13 Fuzzy Subsets

Suppose $\tilde{A}$ and $\tilde{B}$ are two fuzzy sets of X then $\tilde{A} \subseteq \tilde{B} \Leftrightarrow$ $\forall x\big(x \in X \Rightarrow \mu_{\tilde{A}}(x) \leq \mu_{\tilde{B}}(x)\big)$.

Definition 2.14 Disjoint Fuzzy Sets

Two fuzzy sets $\tilde{A}$ and $\tilde{B}$ from X are said to be disjoint if and only if $\tilde{A} \cap \tilde{B} = \emptyset$. In other words,

$\tilde{A} \cap \tilde{B} = \emptyset \Leftrightarrow \forall \mathrm{x}\big(\mathrm{x} \in \mathrm{X} \Rightarrow \mu_{\tilde{A}}(\mathrm{x}) = 0 \vee \mu_{\tilde{B}}(\mathrm{x}) = 0\big).$

Example 2.2 Let $\mathrm{X} = \{1, 2, \cdots, 10\}$, considering the fuzzy sets $\tilde{A}$ and $\tilde{B}$ from X as follows:

$\tilde{A} = \{(1, 0.7), (3, 0.4), (5, 0.6), (6, 1), (10, 0.2)\}$

$\tilde{B} = \{(1, 0.4), (2, 1), (5, 0.4), (8, 0.6), (10, 0.7)\}$

Then,

$\tilde{A} \cup \tilde{B} = \{(1, 0.7), (2, 1), (3, 0.4), (5, 0.6), (6, 1), (8, 0.6), (10, 0.7)\}$

$\tilde{A} \cap \tilde{B} = \{(1, 0.4), (5, 0.4), (10, 0.2)\}$

$\tilde{A}' = \{(1, 0.3), (2, 1), (3, 0.6), (4, 1), (5, 0.4), (7, 1), (8, 1), (9, 1), (10, 0.8)\}$

It follows that $\tilde{A}$ and $\tilde{B}$ are not disjoint since $\tilde{A} \cap \tilde{B} \neq \emptyset$, and in this example, $\tilde{A} \not\subset \tilde{B}$ because $\mu_{\tilde{A}}(3) = 0.4 \not\leq \mu_{\tilde{B}}(3) = 0$, and similarly, $\tilde{B} \not\subset \tilde{A}$ because $\mu_{\tilde{B}}(2) = 1 \not\leq \mu_{\tilde{A}}(2) = 0$.

Definition 2.15 Partition of Fuzzy Set

Let X be a universal set, and $\tilde{A}_1, ..., \tilde{A}_n$ be fuzzy subsets of X. We call $\left\{\tilde{A}_1, ..., \tilde{A}_n\right\}$ a fuzzy partition of X if and only if

I. $\forall i\left(i \in \{1, \ldots, n\} \Rightarrow \left(\tilde{A}_i \neq \emptyset \,\&\, \tilde{A}_i \neq X\right)\right)$

II. $\forall x\left(x \in X \Rightarrow \sum_{i=1}^{n} \mu_{\tilde{A}_i}(x) = 1\right)$

In other words $\left\{\tilde{A}_1, ..., \tilde{A}_n\right\}$ is a fuzzy partition of X if every element in X belongs to exactly one of the fuzzy subsets $\tilde{A}_1, ..., \tilde{A}_n$.

Example 2.3 Let X = $\{1, 2, \cdots, 10\}$,

Set $\left\{\tilde{A}_1, \tilde{A}_2, \tilde{A}_3\right\}$ is a fuzzy partition of X that

$\tilde{A}_1 = \{(1, 0.7), (2, 1), (3, 1), (4, 0.2), (6, 0.2), (10, 0.7)\}$

$\tilde{A}_2 = \{(1, 0.3), (4, 0.3), (5, 1), (7, 0.2), (9, 1)\}$

$\tilde{A}_3 = \{(4, 0.5), (6, 0.8), (7, 0.8), (8, 1), (10, 0.3)\}$

Definition 2.16 Cardinal Number of Fuzzy Set

Suppose $\tilde{A}$ is a fuzzy subset of X, in this case the main number $\tilde{A}$ is defined as follows

$$|\tilde{A}| = \sum_{x \in X} \mu_{\tilde{A}}(x)$$

The ratio $\frac{|\tilde{A}|}{|x|}$ is called the relative number $\tilde{A}$.

Suppose $\tilde{A}$ and $\tilde{B}$ are fuzzy sets of X. The following properties hold for the α-cut of the sets:

I. $\forall\alpha\forall\beta\left(0 < \alpha \leq \beta < 1 \Rightarrow \tilde{A}_\beta \leq \tilde{A}_\alpha\right)$

II. $\forall\alpha\left(\alpha \in [0, 1] \Rightarrow \left(\tilde{A} \subseteq \tilde{B} \Leftrightarrow \tilde{A}_\alpha \subseteq \tilde{B}_\alpha\right)\right)$

III. $\forall\alpha\left(\alpha \in [0, 1] \Rightarrow \left(\left(\tilde{A} \cap \tilde{B}\right)_\alpha = \tilde{A}_\alpha \cap \tilde{B}_\alpha\right)\right)$

IV. $\forall\alpha\left(\alpha \in [0, 1] \Rightarrow \left(\left(\tilde{A} \cup \tilde{B}\right)_\alpha = \tilde{A}_\alpha \cup \tilde{B}_\alpha\right)\right)$

V. $\tilde{A} = \cup_\alpha \tilde{A}_\alpha$

Definition 2.17 Bounded Fuzzy Set

A fuzzy set $\tilde{A}$ of X is called bounded if and only if:

$$\forall\alpha\left(\alpha \in [0, 1] \Rightarrow \tilde{A}_\alpha\, is\, bounded\right)$$

Definition 2.18 Convex Fuzzy Set

A fuzzy set $\tilde{A}$ of X is called convex if and only if:

$$\forall x_1 \forall x_2 \forall\lambda\left(x_1, x_2 \in X\, \&\, \lambda \in [0, 1] \Rightarrow \mu_{\tilde{A}}[\lambda x_1 + (1-\lambda)x_2] \geq Min\{\mu_{\tilde{A}}(x_1), \mu_{\tilde{A}}(x_2)\}\right)$$

In other words, $\tilde{A}$ is called a convex set if and only if for every $\alpha \in [0, 1]$, the set $\tilde{A}_\alpha$ is bounded.

Definition 2.19 Extension Principle

Let $f : X_1 \times ... \times X_n \to Y$ and $\tilde{A}_1, ..., \tilde{A}_n$ be fuzzy subsets of the sets $X_1, ..., X_n$, respectively. The fuzzy subset $\tilde{B}$ of Y is obtained using the function f as follows:

$$\tilde{B} = f\left(\tilde{A}_1, \ldots, \tilde{A}_n\right) = \left\{\left(y, \mu_{\tilde{B}}(y)\right) | y = f(x_1, \ldots, x_n)\, \&\, x_i \in X_i, i = 1, \ldots, n\right\}$$

where,

$$\mu_{\tilde{B}}(y) = \begin{cases} \underset{(x_1,\ldots x_n) f(x_1,\ldots x_n)=y}{Sup} \quad Min\left\{\mu_{\tilde{A}_1}(x_1), \ldots \mu_{\tilde{A}_n}(x_n)\right\}, & f^{-1}(y) \neq \emptyset; \\ 0, & f^{-1}(y) \neq \emptyset. \end{cases}$$

Note that the membership function of the Cartesian product of fuzzy sets $\tilde{A}_1, ..., \tilde{A}_n$ from X is defined as follows:

$$\mu_{\tilde{A}_1 \times \ldots \times \tilde{A}_n}(x_1, \ldots, x_n) = Min\left\{\mu_{\tilde{A}_1}(x_1), \ldots, \mu_{\tilde{A}_n}(x_n)\right\}$$

Example 2.4 Let $X_1 = X_2 = Z$, and the fuzzy subsets $\tilde{A}_1$ and $\tilde{A}_2$ of Z be defined as follows:

$$\tilde{A}_1 = \{(-2, 0.3), (-1, 0.7), (0, 0.2), (1, 0.4), (2, 1)\}$$

$$\tilde{A}_2 = \{(-2, 0.8), (-1, 0.2), (0, 0.5), (1, 0.3), (2, 0.2)\}$$

Furthermore, let $f : X_1 \times ... \times X_n \to Y$ be defined by the equation $f(x_1, x_2) = x_1^2 + x_2^2$. In this case, the Cartesian product of the two fuzzy sets is calculated as follows:

$$\tilde{A}_1 \times \tilde{A}_2 = \{((-2,-2), 0.3), ((-2,-1), 0.2), ((-2,0), 0.3), ((-2,2), 0.2), \ldots, (2,2), 0.2)\}$$

According to the extension principle, we can write:

$$\tilde{B} = f\left(\tilde{A}_1, \tilde{A}_2\right) = \{(8, 0.8), (5, 0.7), (4, 0.5), (2, 0.3), (1, 0.5), (0, 0.2)\}$$

Definition 2.20 The fuzzy set $\tilde{A}$ is called a fuzzy number if and only if:

I. $\tilde{A}$ is convex.
II. $\tilde{A}$ is normal (meaning there is only one x in X such that $\mu_{\tilde{A}}(x) = 1$.
III. $\mu_{\tilde{A}}$ is piecewise continuous.

Example 2.5 The fuzzy set $\tilde{A}$ from X, which is defined as follows, is a fuzzy number.

$$\mu_{\tilde{A}}(x) = \begin{cases} 1 - |x|, & x \in [-1, 1] \\ 0, & o.w \end{cases}$$

The graph of the function $\mu_{\tilde{A}}$ is shown in Fig. 2.1.

Definition 2.21 LR Fuzzy Number

A fuzzy number $\tilde{A}$ is called an LR fuzzy number if and only if there exist functions L and R and scalars $\alpha, \beta > 0$ such that:

$$\mu_{\tilde{A}}(x) = \begin{cases} L\left(\dfrac{m-x}{\alpha}\right), & x \le m \\ R\left(\dfrac{x-m}{\beta}\right), & x \ge m \end{cases}$$

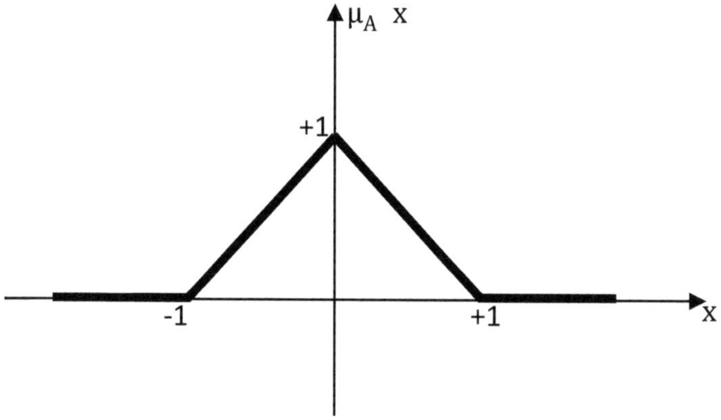

Fig. 2.1 Fuzzy number

m is the median value, and α and β represent the left and right spreads of the LR fuzzy number, respectively.

The functions L and R are functions from R^+ to [0, 1] that satisfy the following conditions:

I. $L(0) = R(0) = 1$.

II. $\forall \mathrm{x}(\mathrm{x} > 0 \Rightarrow \mathrm{L(x)} < 1 \,\&\, \mathrm{R(x)} < 1)$

$\forall \mathrm{x}(\mathrm{x} < 1 \Rightarrow \mathrm{L(x)} > 0 \,\&\, \mathrm{R(x)} > 0)$

III. L and R are non-decreasing functions.

Also, LR fuzzy number can be represented as $\tilde{A} = (m, \alpha, \beta)_{LR}$.

Example 2.6 Let $L(x) = \frac{1}{1+|x|}$ and $R(x) = e^{-x^2}$ then, $\tilde{A} = (2, 2, 1)_{\mathrm{LR}}$, $m = 2$, $\alpha = 2$ and $\beta = 1$.

$$\mu_{\tilde{A}}(x) = \begin{cases} \frac{1}{1+\left|\frac{2-x}{2}\right|}, & x \le 2 \\ e^{-(\frac{x-2}{1})^2}, & x > 2 \end{cases} = \begin{cases} \frac{2}{4-x}, & x \le 2 \\ e^{-(x-2)^2}, & x > 2 \end{cases}$$

Definition 2.22 Interval fuzzy

A fuzzy number $\tilde{A}$ is called an LR Interval fuzzy number if and only if there exist functions L and R and scalars $\alpha, \beta > 0$, $m_1, m_2 \in R$ such that:

$$\mu_{\tilde{A}}(x) = \begin{cases} L\left(\dfrac{m_1 - x}{\alpha}\right), & x \le m_1 \\ 1, & m_1 \le x \le m_2 \\ R\left(\dfrac{x - m_2}{\beta}\right), & x \ge m_2 \end{cases}$$

In this case, the fuzzy interval is shown as $\tilde{A} = (m_1, m_2, \alpha, \beta)_{LR}$.

If $L = R$ and function L is defined as follows:

$$\mathrm{L(x)} = \begin{cases} 1 - \mathrm{x}, & 0 \le \mathrm{x} \le 1 \\ 0, & \mathrm{o.w} \end{cases}$$

Then, $\tilde{A} = (m, \alpha, \beta)_L$ is called triangular fuzzy number.

Definition 2.23 Triangular and Trapezoidal Fuzzy Numbers

The fuzzy number $\tilde{A}$ and $\tilde{B}$ are called triangular and trapezoidal numbers, respectively if and only if

$$\mu_{\tilde{A}}(x) = \begin{cases} \frac{x-l}{m-l}, & l \le x \le m \\ \frac{u-x}{u-m}, & m \le x \le u \\ 0, & o.w \end{cases}, \quad \mu_{\tilde{B}}(x) = \begin{cases} \frac{x-l}{m_1-l}, & l \le x \le m_1 \\ 1, & m_1 \le x \le m_2 \\ \frac{u-x}{u-m_2}, & m_2 \le x \le u \\ 0, & o.w \end{cases}$$

Then, $\tilde{A}$ and $\tilde{B}$ denoted $\tilde{A} = (l, m, u)$ and $\tilde{B} = (l, m_1, m_2, u)$.

Let $\tilde{A} = (l, m, u)$ and $\tilde{B} = (l', m', u')$ are two triangular fuzzy numbers then, the operations between the fuzzy numbers will be as follows.

$$\tilde{A} + \tilde{B} = (l + l', m + m', u + u')$$

$$r\tilde{A} = \begin{cases} (rl, rm, ru) \ r \ge 0 \\ (ru, rm, rl) \ r < 0 \end{cases}$$

$$\tilde{A} - \tilde{B} = (l - u', m - m', u - l')$$

If $\tilde{A}$ and $\tilde{B}$ are two non-negative triangular fuzzy numbers, the best approximation for multiplication and division would be as follows:

$$\tilde{A}.\tilde{B} \simeq (l.l', m.m', u.u')$$

$$\frac{\tilde{A}}{\tilde{B}} \simeq (\frac{l}{u'}, \frac{m}{m'}, \frac{u}{l'})$$

Similarly, if $\tilde{A} = (l, m_1, m_2, u)$ and $\tilde{B} = (l', m_1', m_2', u')$ are two trapezoidal fuzzy numbers, the operations between fuzzy numbers would be as follows:

$$\tilde{A} + \tilde{B} = (l + l', m_1 + m_1', m_2 + m_2', u + u')$$

$$r\tilde{A} = \begin{cases} (rl, rm_1, rm_2, ru) \ , r \ge 0 \\ (ru, rm_2, rm_1, rl) \ , r < 0 \end{cases}$$

$$\tilde{A} - \tilde{B} = (l - u', m_1 - m_2', m_2 - m_1', u - l')$$

If $\tilde{A}$ and $\tilde{B}$ are two non-negative trapezoidal fuzzy numbers, then the best trapezoidal approximation for multiplication and division would be as follows:

$$\tilde{A}.\tilde{B} \simeq (l.l', m_1.m_1', m_2.m_2', u.u')$$

$$\frac{\tilde{A}}{\tilde{B}} \simeq \left(\frac{l}{u'}, \frac{m_1}{m_2'}, \frac{m_2}{m_1'}, \frac{u}{l'}\right)$$

2.3 Ranking of Fuzzy Numbers

In the fuzzy number set, due to the absence of the strong Trichotomy law, there is a need for methods to rank fuzzy numbers. Different methods have been developed for ranking specific numbers, and we will mention some of these methods below.

2.3.1 *Fuzzy Number Ranking Based on α-Cuts*

Let $\tilde{A}$ and $\tilde{B}$ as two fuzzy numbers and let $\tilde{A}_\alpha = [a_\alpha^-, a_\alpha^+]$ and $\tilde{B}_\alpha = [b_\alpha^-, b_\alpha^+]$ be the corresponding α-cut sets of fuzzy numbers $\tilde{A}$ and $\tilde{B}$, respectively. In this case, the ranking of fuzzy numbers is defined according to the following equations:

$$\tilde{A} \le \tilde{B} \Leftrightarrow \forall\alpha\left(\alpha \in [0, 1] \Rightarrow \mathrm{a}_\alpha^- \le \mathrm{b}_\alpha^- \,\&\, \mathrm{a}_\alpha^+ \le \mathrm{b}_\alpha^+\right)$$

2.3.2 *Fuzzy Number Ranking Based on Hamming Distance*

Assume $\tilde{A}$ and $\tilde{B}$ are two fuzzy numbers with membership functions $\mu_{\tilde{A}}(x)$ and $\mu_{\tilde{B}}(x)$. In this case, the ranking of fuzzy numbers is defined as follows.

$$\tilde{A} \le \tilde{B} \Leftrightarrow d\left[Max\left[\tilde{A}, \tilde{B}\right], \tilde{A}\right] \le d\left[Max\left[\tilde{A}, \tilde{B}\right], \tilde{B}\right]$$

where,

$$d\left[\tilde{M}, \tilde{N}\right] = \int_R \left|\mu_{\tilde{M}}(x) - \mu_{\tilde{N}}(x)\right| dx$$

2.4 Type-2 Fuzzy Numbers

The concept of Type-2 fuzzy sets is a generalization of Type-1 fuzzy sets. In fact, Type-2 fuzzy sets are fuzzy sets where the membership degree of each element from the universal set is a Type-1 fuzzy set itself, and they are called Type-2 fuzzy sets [5, 7, 8, 10, 15, 18, 19, 25].

Definition 2.24 Definition of Type-2 Fuzzy Sets, denoted by $\tilde{A}$:

$$\tilde{A} = \left\{\left((x, u), \mu_{\tilde{A}}(x, u)\right)|x \in X\&u \in J_x \subseteq [0, 1]\right\}$$

where $\mu_{\tilde{A}} : X \times U \to V$ and X represents the universal set, $U = [0, 1]$ is the membership domain, and $V = [0, 1]$ is the secondary membership of x. Each value of x from X corresponds to the value of u, which is a member of $U = [0, 1]$, which is called the primary membership degree, and each primary membership degree assigns a secondary membership degree $\mu_{\tilde{A}}(x, u)$, J_x represents the set of primary memberships corresponding to x in X.

Definition 2.25 Suppose $\tilde{A}$ is a type-2 fuzzy set. If the relationship $\mu_{\tilde{A}}(x, u) = 1$ holds for all secondary membership degrees, then $\tilde{A}$ is called an interval type-2 fuzzy set, represented as follows:

$$\tilde{A} = \left\{\left(x, [\underline{u}_x, \overline{u}_x]\right)\middle|x \in X\&[\underline{u}_x, \overline{u}_x] \subseteq U\right\}$$

The following representation is obtained by taking the union over all elements x with their corresponding intervals:

$$\tilde{A} = \bigcup\left(x, \tilde{A}(x)\right) = \bigcup\left(x, [\underline{u}_x, \overline{u}_x]\right)$$

Some logical operators on interval type-2 fuzzy sets were introduced by Zadeh in 1975, which will be defined later [9, 29, 30, 31].

Definition 2.26 Let $\tilde{A}$ and $\tilde{B}$ be two interval type-2 fuzzy sets defined on the universal set X. The union, intersection, and complement are expressed as follows, where $\vee$ and $\wedge$ denote the symbols for maximum and minimum, respectively:

Union: $(\tilde{A} \cup \tilde{B})(x) = \left[\underline{u}_x^A \vee \underline{u}_x^B, \overline{u}_x^A \vee \overline{u}_x^B\right]$.

Intersection: $(\tilde{A} \cap \tilde{B})(x) = \left[\underline{u}_x^A \wedge \underline{u}_x^B, \overline{u}_x^A \wedge \overline{u}_x^B\right]$.

Complement: $\tilde{A}(x) = \left[1 - \overline{u}_x^A, 1 - \underline{u}_x^A\right]$.

All elements of an interval type-2 fuzzy set have interval membership degrees. Therefore, each interval type-2 fuzzy set can be represented using two type-1 fuzzy sets, as shown by Mendel et al. [14]. They demonstrated that the concept of interval type-2 fuzzy sets is equivalent to the notion of footprint of uncertainty, thus each interval type-2 fuzzy set can be expressed using upper and lower membership functions introduced for the footprint of uncertainty, as follows [3, 16, 17, 24, 26, 31].

Definition 2.27 Let $\tilde{A}$ be an interval type-2 fuzzy set, and let type-1 fuzzy sets $\tilde{A}^U$ and $\tilde{A}^L$ ($\overline{\tilde{A}}, \underline{\tilde{A}}$) be the upper and lower membership functions of $\tilde{A}$, respectively. In this case, for every $x \in X$, we have:

$$\tilde{A}(x) = \left[\tilde{A}^L(x), \tilde{A}^U(x)\right]$$

Based on the above definition, the concept of α-cut for a type-2 interval fuzzy set is expressed as follows.

Definition 2.28 Suppose $\tilde{A}$ is a type-2 interval fuzzy set. Then, for every $x \in X$ and $\alpha \in [0, 1]$, the α-cut set $\tilde{A}$ is defined as follows:

$$\tilde{A}_\alpha(x) = \left[\tilde{A}^L_\alpha(x), \tilde{A}^U_\alpha(x)\right]$$

Next, another definition for α-cut set, presented by Kaufmann and Gupta [12], is provided.

Definition 2.29 Let $\tilde{A}$ be a type-2 interval fuzzy set, and let $\tilde{A}^L_\alpha$ and $\tilde{A}^U_\alpha$ be the upper and lower α-cut sets, respectively, defined as $[L_{\overline{x}_\alpha}, R_{\overline{x}_\alpha}]$ and $[L_{\underline{x}_\alpha}, R_{\underline{x}_\alpha}]$, Additionally, let $h(\tilde{A}^U)$ and $h(\tilde{A}^L)$ be the heights of the upper and lower membership functions $\tilde{A}$, respectively. In this case, the Kaufmann-Gupta α-cut set [12] for $\tilde{A}$ is given by:

$$\tilde{A}^{KG}_\alpha = \begin{cases} [L_{\bar{x}\alpha}, R_{\bar{x}\alpha}], \left[L_{\underline{x}\alpha}, R_{\underline{x}\alpha}\right], & \alpha < h\left(\tilde{A}^L\right) \\ [L_{\bar{x}\alpha}, R_{\bar{x}\alpha}], & \alpha \geq h\left(\tilde{A}^U\right) \end{cases}$$

2.5 Type-2 Trapezoidal and Triangular Fuzzy Numbers

Due to the widespread application of type-2 interval fuzzy sets with trapezoidal or triangular upper and lower membership functions, a special type of these sets called type-2 trapezoidal and triangular fuzzy numbers has drawn the attention of researchers [7, 9]. The relevant definitions are provided below.

Definition 2.30 A type-2 interval fuzzy set $\tilde{A}$ defined over the closed real interval $[a^u_1, a^u_4]$ is referred to as a type-2 trapezoidal interval fuzzy number, and it is expressed as:

$$\tilde{A} = \left(\tilde{A}^l, \tilde{A}^u\right) = ((a^l_1, a^l_2, a^l_3, a^l_4; h^l_1, h^l_2), (a^u_1, a^u_2, a^u_3, a^u_4; h^u_1, h^u_2))$$

Its membership functions are also in the form of the following relationships:

$$\mu^l_{\tilde{A}}(x) = \begin{cases} h^l_1\left(\dfrac{x - a^l_1}{a^l_2 - a^l_1}\right), & a^l_1 \le x \le a^l_2 \\ \left(h^l_2 - h^l_1\right)\left(\dfrac{x - a^l_1}{a^l_3 - a^l_2}\right) + h^l_1, & a^l_2 \le x \le a^l_3 \\ h^l_2\left(\dfrac{a^l_4 - x}{a^l_4 - a^l_3}\right), & a^l_3 \le x \le a^l_4 \\ 0, & x \le a^l_1, x \ge a^l_4 \end{cases}$$

and,

$$\mu^u_{\tilde{A}}(x) = \begin{cases} h^u_1\left(\dfrac{x - a^u_1}{a^u_2 - a^u_1}\right), & a^u_1 \le x \le a^u_2 \\ \left(h^u_2 - h^u_1\right)\left(\dfrac{x - a^u_1}{a^u_3 - a^u_2}\right) + h^u_1, & a^u_2 \le x \le a^u_3 \\ h^u_2\left(\dfrac{a^u_4 - x}{a^u_4 - a^u_3}\right), & a^u_3 \le x \le a^u_4 \\ 0, & x \le a^u_1, x \ge a^u_4 \end{cases}$$

If $h^l_1 = h^l_2 = h^l, h^u_1 = h^u_2 = h^u$ then it is called type-2 trapezoidal interval fuzzy number, and if $h^l_1 = h^l_2 = h^u_1 = h^u_2 = 1$ then it is called complete type-2 trapezoidal interval fuzzy number. Figure 2.2 shows the interval type-2 trapezoidal fuzzy number.

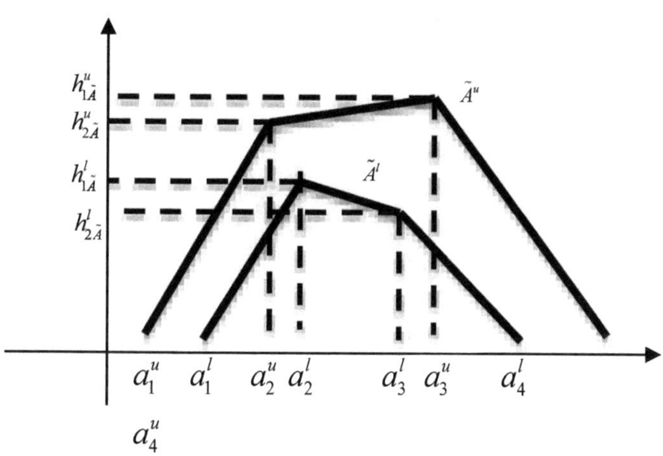

Fig. 2.2 Interval type-2 trapezoidal fuzzy number

Definition 2.31 A type-2 fuzzy set over the closed real interval $[a_1^u, a_3^u]$ is called a type-2 triangular fuzzy set and is represented as follows

$$\tilde{A} = \left(\tilde{A}^l, \tilde{A}^u\right) = \left(\left(a_1^l, a_2^l, a_3^l; h^l\right), \left(a_1^u, a_2^u, a_3^u; h^u\right)\right)$$

The membership functions of these sets are defined by the following two relationships:

$$\mu_{\tilde{A}}^l(x) = \begin{cases} h_1^l\left(\dfrac{x - a_1^l}{a_2^l - a_1^l}\right), & a_1^l \le x \le a_2^l \\ h^l\left(\dfrac{a_3^l - x}{a_3^l - a_2^l}\right), & a_2^l \le x \le a_3^l \\ 0, & x \le a_1^l, x \ge a_3^l \end{cases}$$

and,

$$\mu_{\tilde{A}}^u(x) = \begin{cases} h_1^u\left(\dfrac{x - a_1^u}{a_2^u - a_1^u}\right), & a_1^u \le x \le a_2^u \\ h^u\left(\dfrac{a_3^u - x}{a_3^u - a_2^u}\right), & a_2^u \le x \le a_3^u \\ 0, & x \le a_1^u, x \ge a_3^u \end{cases}$$

If $a_2^l = a_2^u$ and $h^l = h^u = 1$, the type-2 triangular fuzzy number is referred to as a complete type-2 triangular fuzzy number. If $h^l < 1$ and $h^u = 1$, it is called a type-2 triangular fuzzy number. Figure 2.3 illustrates a type-2 triangular fuzzy number.

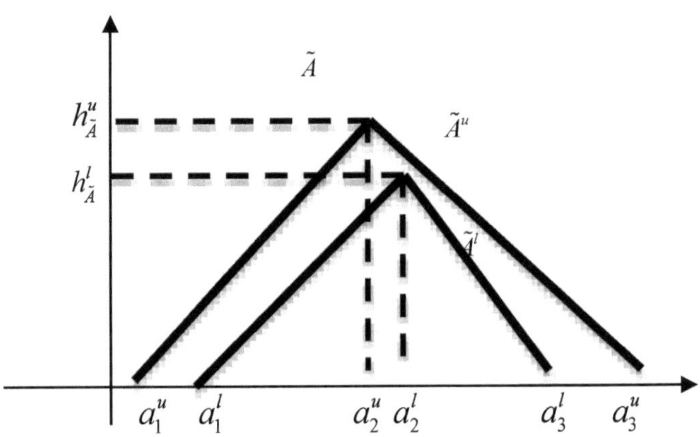

Fig. 2.3 Interval type-2 triangular fuzzy number

2.5.1 Arithmetic Operations on Type-2 Trapezoidal Fuzzy Numbers

Definition 2.32 Let $\tilde{A}$ and $\tilde{B}$ be two type-2 trapezoidal fuzzy numbers, and k be a constant real number. The arithmetic operations such as addition, subtraction, multiplication, and scalar multiplication are defined as follows:

Addition:

$$\tilde{A} \oplus \tilde{B} = \left(\tilde{A}^l \oplus \tilde{B}^l, \tilde{A}^u \oplus \tilde{B}^u\right)$$
$$= \begin{pmatrix} \left(a_1^l + b_1^l, a_2^l + b_2^l, a_3^l + b_3^l, a_4^l + b_4^l; \min\left\{h^l_{1\tilde{A}}, h^l_{1\tilde{B}}\right\}, \min\left\{h^l_{2\tilde{A}}, h^l_{2\tilde{B}}\right\}\right), \\ \left(a_1^u + b_1^u, a_2^u + b_2^u, a_3^u + b_3^u, a_4^u + b_4^u; \min\left\{h^u_{1\tilde{A}}, h^u_{1\tilde{B}}\right\}, \min\left\{h^u_{2\tilde{A}}, h^u_{2\tilde{B}}\right\}\right) \end{pmatrix}$$

Subtraction:

$$\tilde{A} - \tilde{B} = \left(\tilde{A}^l - \tilde{B}^l, \tilde{A}^u - \tilde{B}^u\right)$$
$$= \begin{pmatrix} \left(a_1^l - b_1^l, a_2^l - b_2^l, a_3^l - b_3^l, a_4^l - b_4^l; \min\left\{h^l_{1\tilde{A}}, h^l_{2\tilde{B}}\right\}, \min\left\{h^l_{2\tilde{A}}, h^l_{1\tilde{B}}\right\}\right), \\ \left(a_1^u - b_1^u, a_2^u - b_2^u, a_3^u - b_3^u, a_4^u - b_4^u; \min\left\{h^u_{1\tilde{A}}, h^u_{2\tilde{B}}\right\}, \min\left\{h^u_{2\tilde{A}}, h^u_{1\tilde{B}}\right\}\right) \end{pmatrix}$$

Multiplication:

$$\tilde{A} \otimes \tilde{B} = \left(\left(c_1^l, c_2^l, c_3^l, c_4^l; h^l_{1c}, h^l_{2c}\right), \left(c_1^u, c_2^u, c_3^u, c_4^u; h^u_{1c}, h^u_{2c}\right)\right)$$

$$\begin{aligned} c_1^l &= \min\left\{a_1^l \times b_1^l, a_1^l \times b_4^l, a_4^l \times b_1^l, a_4^l \times b_4^l\right\} \\ c_2^l &= \min\left\{a_2^l \times b_2^l, a_2^l \times b_3^l, a_3^l \times b_2^l, a_3^l \times b_3^l\right\} \\ c_3^l &= \max\left\{a_2^l \times b_2^l, a_2^l \times b_3^l, a_3^l \times b_2^l, a_3^l \times b_3^l\right\} \\ c_4^l &= \min\left\{a_1^l \times b_1^l, a_1^l \times b_4^l, a_4^l \times b_1^l, a_4^l \times b_4^l\right\} \end{aligned}$$

$$\begin{aligned} h^l_{1C} &= \min\left\{h^l_{1\tilde{A}}, h^l_{1\tilde{B}}\right\} \\ h^l_{2C} &= \min\left\{h^l_{2\tilde{A}}, h^l_{2\tilde{B}}\right\} \end{aligned}$$

$$\begin{aligned} c_1^u &= \min\left\{a_1^u \times b_1^u, a_1^u \times b_4^u, a_4^u \times b_1^u, a_4^u \times b_4^u\right\} \\ c_2^u &= \min\left\{a_2^u \times b_2^u, a_2^u \times b_3^u, a_3^u \times b_2^u, a_3^u \times b_3^u\right\} \\ c_3^u &= \max\left\{a_2^u \times b_2^u, a_2^u \times b_3^u, a_3^u \times b_2^u, a_3^u \times b_3^u\right\} \\ c_4^u &= \min\left\{a_1^u \times b_1^u, a_1^u \times b_4^u, a_4^u \times b_1^u, a_4^u \times b_4^u\right\} \end{aligned}$$

$$h^u_{1C} = \min\left\{h^u_{1\tilde{A}}, h^u_{1\tilde{B}}\right\}$$

$$h^u_{2C} = \min\left\{h^u_{2\tilde{A}}, h^u_{2\tilde{B}}\right\}$$

Scalar multiplication:

If $k \geq 0$

$$k\tilde{A} = \left(k\tilde{A}^l, k\tilde{A}^u\right) = \left(\left(ka^l_1, ka^l_2, ka^l_3, ka^l_4; h^l_{1\tilde{A}}, h^l_{2\tilde{A}}\right), \left(ka^u_1, ka^u_2, ka^u_3, ka^u_4; h^u_{1\tilde{A}}, h^u_{2\tilde{A}}\right)\right)$$

If $k \leq 0$

$$k\tilde{A} = \left(k\tilde{A}^l, k\tilde{A}^u\right) = \left(\left(ka^l_4, ka^l_3, ka^l_2, ka^l_1; h^l_{1\tilde{A}}, h^l_{2\tilde{A}}\right), \left(ka^u_4, ka^u_3, ka^u_2, ka^u_1; h^u_{1\tilde{A}}, h^u_{2\tilde{A}}\right)\right)$$

2.5.2 Arithmetic Operations on Type-2 Triangular Fuzzy Numbers

Definition 2.33 Let

$$\tilde{A} = \left(\tilde{A}^l, \tilde{A}^u\right) = \left(\left(a^l_1, a^l_2, a^l_3; h^l_{\tilde{A}}\right), \left(a^u_1, a^u_2, a^u_3; h^u_{\tilde{A}}\right)\right)$$

and,

$$\tilde{B} = \left(\tilde{B}^l, \tilde{B}^u\right) = \left(\left(b^l_1, b^l_2, b^l_3; h^l_{\tilde{B}}\right), \left(b^u_1, b^u_2, b^u_3; h^u_{\tilde{B}}\right)\right)$$

If two triangular fuzzy numbers of interval type-2 and a real number are fixed, arithmetic operations on these numbers are defined as follows [11]:

Addition:

$$\tilde{A} \oplus \tilde{B} = \left(\tilde{A}^l \oplus \tilde{B}^l, \tilde{A}^u \oplus \tilde{B}^u\right) = \begin{pmatrix} \left(a^l_1 + b^l_1, a^l_2 + b^l_2, a^l_3 + b^l_3; \min\left\{h^l_{\tilde{A}}, h^l_{\tilde{B}}\right\}\right), \\ \left(a^u_1 + b^u_1, a^u_2 + b^u_2, a^u_3 + b^u_3; \min\left\{h^u_{\tilde{A}}, h^u_{\tilde{B}}\right\}\right) \end{pmatrix}$$

Subtraction:

$$\tilde{A} - \tilde{B} = \left(\tilde{A}^l - \tilde{B}^l, \tilde{A}^u - \tilde{B}^u\right) = \left(\begin{array}{l} \left(a_1^l - b_1^l, a_2^l - b_2^l, a_3^l - b_3^l; \min\left\{h_{\tilde{A}}^l, h_{\tilde{B}}^l\right\}\right), \\ \left(a_1^u - b_1^u, a_2^u - b_2^u, a_3^u - b_3^u; \min\left\{h_{\tilde{A}}^u, h_{\tilde{B}}^u\right\}\right) \end{array}\right)$$

Multiplication:

$$\begin{aligned}
&\tilde{A} \otimes \tilde{B} = \left(\left(c_1^l, c_2^l, c_3^l; h_C^l\right), \left(c_1^u, c_2^u, c_3^u; h_C^u\right)\right) \\
&c_1^l = \min\left\{a_1^l \times b_1^l, a_1^l \times b_3^l, a_3^l \times b_1^l, a_3^l \times b_3^l\right\} \\
&c_2^l = \left\{a_2^l \times b_2^l\right\} \\
&c_3^l = \min\left\{a_1^l \times b_1^l, a_1^l \times b_3^l, a_3^l \times b_1^l, a_3^l \times b_3^l\right\} \\
&h_C^l = \min\left\{h_{\tilde{A}}^l, h_{\tilde{B}}^l\right\} \\
&c_1^u = \min\left\{a_1^u \times b_1^u, a_1^u \times b_3^u, a_3^u \times b_1^u, a_3^u \times b_3^u\right\} \\
&c_2^u = \left\{a_2^u \times b_2^u\right\} \\
&c_3^u = \max\left\{a_1^u \times b_1^u, a_1^u \times b_3^u, a_3^u \times b_1^u, a_3^u \times b_3^u\right\} \\
&h_C^u = \min\left\{h_{\tilde{A}}^u, h_{\tilde{B}}^u\right\}
\end{aligned}$$

Scalar multiplication:

If $k \geq 0$

$$k\tilde{A} = \left(k\tilde{A}^l, k\tilde{A}^u\right) = \left(\left(ka_1^l, ka_2^l, ka_3^l; h_{\tilde{A}}^l\right), \left(ka_1^u, ka_2^u, ka_3^u; h_{\tilde{A}}^u\right)\right)$$

If $k \leq 0$

$$k\tilde{A} = \left(k\tilde{A}^l, k\tilde{A}^u\right) = \left(\left(ka_3^l, ka_2^l, ka_1^l; h_{\tilde{A}}^l\right), \left(ka_3^u, ka_2^u, ka_1^u; h_{\tilde{A}}^u\right)\right)$$

2.6 Ranking of Type-2 Interval Fuzzy Numbers

Many real-world problems are characterized by complexity and significant uncertainties. Nowadays, ranking methods play a crucial role in evaluating fuzzy numbers. The proposed methods for ranking type-2 interval fuzzy numbers have attracted much attention from researchers compared to type-1 interval fuzzy numbers [7].

Let $F(R)$ be the set of all type-2 interval fuzzy numbers. One approach for ordering the elements of $F(R)$ involves the use of ranking functions such as

$$R : F(R) \to R$$

Which assigns a real number to each type-2 interval fuzzy number. In fact, for any ranking function such as R, it operates on two type-2 interval fuzzy numbers, $\tilde{A}$ and $\tilde{B}$, as follows [13]:

$\tilde{A} \geq_R \tilde{B}$ if and only if $R\left(\tilde{A}\right) \geq R\left(\tilde{B}\right)$.

$\tilde{A} \leq_R \tilde{B}$ if and only if $R\left(\tilde{A}\right) \leq R\left(\tilde{B}\right)$.

$\tilde{A} =_R \tilde{B}$ if and only if $R\left(\tilde{A}\right) = R\left(\tilde{B}\right)$.

Fuzzy ranking functions typically extract a single characteristic from each fuzzy number and then proceed to order the fuzzy numbers based on that characteristic. This approach has complicated the subject of fuzzy number ranking, and depending on the method used, different results may be obtained. In the following, we will introduce several examples of ranking functions for type-2 fuzzy numbers [7, 25].

2.6.1 *Some Ranking Methods for Type-2 Interval Fuzzy Numbers*

The first method is an extension of the Yager method [21].

Definition 2.34 Let $\tilde{A}=(\tilde{A}^l, \tilde{A}^u)=((a_1^l, a_2^l, a_3^l, a_4^l; h_1^l, h_2^l), (a_1^u, a_2^u, a_3^u, a_4^u; h_1^u, h_2^u))$ as a trapezoidal fuzzy number of interval type-2 is the corresponding ranking function expressed as follow:

$$R_1\left(\tilde{A}\right) = \frac{1}{2}\left[R_1\left(\tilde{A}^l\right) + R_1\left(\tilde{A}^u\right)\right]$$

where,

$$R_1\left(\tilde{A}^l\right) = \frac{1}{2}\left(\int_0^{h_1^l} \mu_{\tilde{A}_1^l}^{-1}(y)dy + \int_{h_1^l}^{h_2^l} \mu_{\tilde{A}_2^l}^{-1}(y)dy + \int_0^{h_2^l} \mu_{\tilde{A}_3^l}{}^{-1}(y)dy\right)$$

$$R_1\left(\tilde{A}^u\right) = \frac{1}{2}\left(\int_0^{h_1^u} \mu_{\tilde{A}_1^u}^{-1}(y)dy + \int_{h_1^u}^{h_2^u} \mu_{\tilde{A}_2^u}^{-1}(y)dy + \int_0^{h_2^u} \mu_{\tilde{A}_3^u}^{-1}(y)dy\right)$$

In general, for a trapezoidal fuzzy number, we will have an interval type-2.

$$R_1\left(\tilde{A}^l\right) = \frac{a_1^l h_1^l + a_2^l h_2^l + a_3^l\left(2h_2^l - h_1^l\right) + a_4^l h_2^l}{4}$$

and,

$$R_1\left(\tilde{A}^u\right) = \frac{a_1^u h_1^u + a_2^u h_2^u + a_3^u\left(2h_2^u - h_1^u\right) + a_4^u h_2^u}{4}$$

Hence,

$$R_1\left(\tilde{A}\right) = \frac{a_1^l h_1^l + a_2^l h_2^l + a_3^l\left(2h_2^l - h_1^l\right) + a_4^l h_2^l + a_1^u h_1^u + a_2^u h_2^u + a_3^u\left(2h_2^u - h_1^u\right) + a_4^u h_2^u}{8}$$

If $h_1^l = h_2^l = h^l$ *and* $h_1^u = h_2^u = h^u$ then, $\tilde{A} = \left(\tilde{A}^l, \tilde{A}^u\right) = \left(\left(a_1^l, a_2^l, a_3^l, a_4^l; h^l\right), \left(a_1^u, a_2^u, a_3^u, a_4^u; h^u\right)\right)$

That is, if $\tilde{A}$ is an interval type-2 flat trapezoidal fuzzy number, then

$$R_1\left(\tilde{A}\right) = \frac{\left(\left(a_1^l + a_2^l + a_3^l + a_4^l\right)h^l + \left(a_1^u + a_2^u + a_3^u + a_4^u\right)h^u\right)}{8}$$

Also, if $h^l = h^u$, then $\tilde{A}$ is a type-2 trapezoidal fuzzy number of complete interval and its ranking is obtained as follows:

$$R_1\left(\tilde{A}\right) = \frac{a_1^l + a_2^l + a_3^l + a_4^l + a_1^u + a_2^u + a_3^u + a_4^u}{8}$$

Similarly, the above method for interval type 2 triangular fuzzy number $\tilde{A} = \left(\tilde{A}^l, \tilde{A}^u\right) = \left(\left(a_1^l, a_2^l, a_3^l; h^l\right), \left(a_1^u, a_2^u, a_3^u; h^u\right)\right)$ is equal to

$$R_1\left(\tilde{A}\right) = \frac{\left(\left(a_1^l + 2a_2^l + a_3^l\right)h^l + \left(a_1^u + 2a_2^u + a_3^u\right)h^u\right)}{8}$$

Also, if $h^l = h^u$ and $a_2^l = a_2^u = a_2$ then $\tilde{A}$, a type-2 triangular fuzzy number, will be a complete interval and its ranking is:

$$R_1\left(\tilde{A}\right) = \frac{a_1^l + a_3^l + a_1^u + a_2^u + 4a_2}{8}$$

In fact, the introduced ranking method will vary depending on the shape of the upper and lower membership functions associated with the type-2 interval fuzzy number. Furthermore, this method can be used for special cases when the target number is a type-1 trapezoidal or triangular fuzzy number, especially when the target number is a type-1 triangular fuzzy number. This method is expressed by Yager [21].

The second method is an extension of the Chen and Hsieh method [6] and it is expressed as follows:

Definition 2.35 Let $\tilde{A}=\left(\tilde{A}^l,\tilde{A}^u\right)=\left(\left(a_1^l,a_2^l,a_3^l,a_4^l;h_1^l,h_2^l\right),\left(a_1^u,a_2^u,a_3^u,a_4^u;h_1^u,h_2^u\right)\right)$ as a trapezoidal fuzzy number of interval type-2 is the corresponding ranking function expressed as follow [6]:

$$R_2\left(\tilde{A}\right)=\frac{1}{2}\left[R_2\left(\tilde{A}^l\right)+R_2\left(\tilde{A}^u\right)\right]$$

where,

$$R_2\left(\tilde{A}^l\right)=\frac{2}{h_1^l+h_2^l}\left(\int_0^{h_1^l} y\mu_{\tilde{A}_1^l}^{-1}(y)dy+\int_{h_1^l}^{h_2^l} y\mu_{\tilde{A}_2^l}^{-1}(y)dy+\int_0^{h_2^l} y\mu_{\tilde{A}_3^l}^{-1}(y)dy\right)$$

$$R_2\left(\tilde{A}^u\right)=\frac{1}{2}\left(\int_0^{h_1^u} y\mu_{\tilde{A}_1^u}^{-1}(y)dy+\int_{h_1^u}^{h_2^u} y\mu_{\tilde{A}_2^u}^{-1}(y)dy+\int_0^{h_2^u} y\mu_{\tilde{A}_3^u}^{-1}(y)dy\right)$$

In general, for a trapezoidal fuzzy number, we will have an interval type-2.

$$R_2\left(\tilde{A}^l\right)=\frac{a_1^l+a_2^l\left(\left(h_2^l\right)^2+h_1^lh_2^l\right)+a_3^l\left(4\left(h_2^l\right)^2-h_1^lh_2^l-\left(h_1^l\right)^2\right)+a_4^l}{3\left(h_1^l+h_2^l\right)}$$

and,

$$R_2\left(\tilde{A}^u\right)=\frac{a_1^u+a_2^u\left(\left(h_2^u\right)^2+h_1^uh_2^u\right)+a_3^u\left(4\left(h_2^u\right)^2-h_1^uh_2^u-\left(h_1^u\right)^2\right)+a_4^u}{3\left(h_1^u+h_2^u\right)}$$

Hence,

$$R_2\left(\tilde{A}\right)=\frac{a_1^l+a_2^l\left(\left(h_2^l\right)^2+h_1^lh_2^l\right)+a_3^l\left(4\left(h_2^l\right)^2-h_1^lh_2^l-\left(h_1^l\right)^2\right)+a_4^l}{6\left(h_1^u+h_2^u\right)}+\frac{a_1^u+a_2^u\left(\left(h_2^u\right)^2+h_1^uh_2^u\right)+a_3^u\left(4\left(h_2^u\right)^2-h_1^uh_2^u-\left(h_1^u\right)^2\right)+a_4^u}{6\left(h_1^u+h_2^u\right)}$$

If $h_1^l = h_2^l = h^l$ *and* $h_1^u = h_2^u = h^u$ then, $\tilde{A} = \left(\tilde{A}^l, \tilde{A}^u\right) = \left(\left(a_1^l, a_2^l, a_3^l, a_4^l; h^l\right), \left(a_1^u, a_2^u, a_3^u, a_4^u; h^u\right)\right)$

That is, if $\tilde{A}$ is an interval type-2 flat trapezoidal fuzzy number, then

$$R_2\left(\tilde{A}^l\right) = \frac{a_1^l + 2a_2^l\left(h^l\right)^2 + 2a_3^l\left(h^l\right)^2 + a_4^l 12h^l}{12h^l} + \frac{a_1^u + 2a_2^u(h^u)^2 + 2a_3^u(h^u)^2 + a_4^u}{12h^u}$$

Also, if $h^l = h^u$, then $\tilde{A}$ is a type-2 trapezoidal fuzzy number of complete interval and its ranking is obtained as follows:

$$R_2\left(\tilde{A}\right) = \frac{a_1^l + 2a_2^l + 2a_3^l + a_4^l + a_1^u + 2a_2^u + 2a_3^u + a_4^u}{12}$$

Similarly, the above method for interval type 2 triangular fuzzy number $\tilde{A} = \left(\tilde{A}^l, \tilde{A}^u\right) = \left(\left(a_1^l, a_2^l, a_3^l; h^l\right), \left(a_1^u, a_2^u, a_3^u; h^u\right)\right)$ is equal to

$$R_2\left(\tilde{A}\right) = \frac{a_1^l + 4a_2^l\left(h^l\right) + a_3^l}{12h^l} + \frac{a_1^u + 2a_2^u + a_3^u}{12h^u}$$

Also, if $h^l = h^u$ and $a_2^l = a_2^u = a_2$ then $\tilde{A}$, a type-2 triangular fuzzy number, will be a complete interval and its ranking is:

$$R_2\left(\tilde{A}\right) = \frac{a_1^l + a_3^l + a_1^u + a_3^u + 8a_2}{12}$$

In fact, the introduced ranking method will vary depending on the shape of the upper and lower membership functions associated with the type-2 interval fuzzy number. Furthermore, this method can be used for special cases when the target number is a type-1 trapezoidal or triangular fuzzy number, especially when the target number is a type-1 triangular fuzzy number. The second method is an extension of the Chen and Hsieh method [6]. The third method is expressed as follows:

Definition 2.36 Let $\tilde{A} = \left(\tilde{A}^l, \tilde{A}^u\right) = \left(\left(a_1^l, a_2^l, a_3^l, a_4^l; h_1^l, h_2^l\right), \left(a_1^u, a_2^u, a_3^u, a_4^u; h_1^u, h_2^u\right)\right)$ as a trapezoidal fuzzy number of interval type-2 is the corresponding ranking function expressed as follow [6]:

$$R_3\left(\tilde{A}\right) = \frac{1}{2}\left[R_3\left(\tilde{A}^l\right) + R_3\left(\tilde{A}^u\right)\right]$$

where,

$$R_3\left(\tilde{A}^l\right) = \frac{1}{3}\left(\int_0^{h_1^l} 2\mu_{\tilde{A}_1^l}^{-1}(y)dy + \int_{h_1^l}^{h_2^l} \mu_{\tilde{A}_2^l}^{-1}(y)dy + \int_0^{h_2^l} 2\mu_{\tilde{A}_3^l}^{-1}(y)dy\right)$$

$$R_3\left(\tilde{A}^u\right) = \frac{1}{3}\left(\int_0^{h_1^u} 2\mu_{\tilde{A}_1^u}^{-1}(y)dy + \int_{h_1^u}^{h_2^u} y\mu_{\tilde{A}_2^u}^{-1}(y)dy + \int_0^{h_2^u} 2\mu_{\tilde{A}_3^u}^{-1}(y)dy\right)$$

In general, for a trapezoidal fuzzy number, we will have an interval type-2.

$$R_3\left(\tilde{A}^l\right) = \frac{2a_1^l h_1^l + a_2^l\left(\frac{h_1^l + h_2^l}{2}\right) + a_3^l\left(\frac{3h_2^l - h_1^l}{2}\right) + 2a_4^l h_2^l}{6}$$

and,

$$R_3\left(\tilde{A}^u\right) = \frac{2a_1^u h_1^u + a_2^u\left(\frac{h_1^u + h_2^u}{2}\right) + a_3^u\left(\frac{3h_2^u - h_1^u}{2}\right) + 2a_4^u h_2^u}{6}$$

Hence,

$$R_3\left(\tilde{A}\right) = \frac{2a_1^l h_1^l + a_2^l\left(\frac{h_1^l + h_2^l}{2}\right) + a_3^l\left(\frac{3h_2^l - h_1^l}{2}\right) + 2a_4^l h_2^l}{12} + \frac{2a_1^u h_1^u + a_2^u\left(\frac{h_1^u + h_2^u}{2}\right) + a_3^u\left(\frac{3h_2^u - h_1^u}{2}\right) + 2a_4^u h_2^u}{12}$$

If $h_1^l = h_2^l = h^l$ *and* $h_1^u = h_2^u = h^u$ then, $\tilde{A} = \left(\tilde{A}^l, \tilde{A}^u\right) = \left(\left(a_1^l, a_2^l, a_3^l, a_4^l; h^l\right), \left(a_1^u, a_2^u, a_3^u, a_4^u; h^u\right)\right)$

That is, if $\tilde{A}$ is an interval type-2 flat trapezoidal fuzzy number, then

$$R_3\left(\tilde{A}\right) = \frac{\left(2a_1^l + a_2^l + a_3^l + 2a_4^l\right)h^l + \left(2a_1^u + a_2^u + a_3^u + 2a_4^u\right)h^u}{12}$$

Also, if $h^l = h^u$, then $\tilde{A}$ is a type-2 trapezoidal fuzzy number of complete interval and its ranking is obtained as follows:

$$R_3\left(\tilde{A}\right) = \frac{2a_1^l + a_2^l + a_3^l + 2a_4^l + 2a_1^u + a_2^u + a_3^u + 2a_4^u}{12}$$

Similarly, the above method for interval type 2 triangular fuzzy number $\tilde{A} = \left(\tilde{A}^l, \tilde{A}^u\right) = \left(\left(a_1^l, a_2^l, a_3^l; h^l\right), \left(a_1^u, a_2^u, a_3^u; h^u\right)\right)$ is equal to

$$R_3\left(\tilde{A}\right) = \frac{\left(a_1^l + a_2^l + a_3^l\right)h^l + \left(a_1^u + a_2^u + a_3^u\right)h^u}{6}$$

Also, if $h^l = h^u$ and $a_2^l = a_2^u = a_2$ then $\tilde{A}$, a type-2 triangular fuzzy number, will be a complete interval and its ranking is:

$$R_3\left(\tilde{A}\right) = \frac{a_1^l + a_3^l + a_1^u + a_3^u + 2a_2}{6}$$

Note: A ranking method will have value and credibility when it satisfies certain well-defined properties.

Let $\tilde{A}, \tilde{B}, \tilde{C} \in F(R)$ then relation $\geq_{R_i}$; $i = 1, 2, 3$ is a partial order relation on the set $F(R)$. In fact, the ranking function must possess the properties of reflexivity, anti-symmetry, and transitivity.

Reflexivity: Suppose $\tilde{A} \in F(R)$ then if $\tilde{A} \geq_{R_i} \tilde{A}$, the reflexivity property holds.

Antisymmetry: If $\tilde{A}, \tilde{B} \in F(R)$ is desired and assuming $\tilde{A} \geq_{R_i} \tilde{B}$ and $\tilde{B} \geq_{R_i} \tilde{A}$, then if $\tilde{B} =_{R_i} \tilde{A}$, the property of antisymmetry holds.

Transitivity: let $\tilde{A}, \tilde{B}, \tilde{C} \in F(R)$, if $\tilde{A} \geq_{R_i} \tilde{B}$ and $\tilde{B} \geq_{R_i} \tilde{C}$, then if $\tilde{A} \geq_{R_i} \tilde{C}$, the property of transitivity holds.

Well-defined properties holds for all the above ranking methods.

Definition 2.37 Let $\tilde{A}$ and $\tilde{B}$ are two fuzzy numbers and $k \in \{R - 0\}$, the linear ranking function is as $R\left(\tilde{A} + k\tilde{B}\right) = R\left(\tilde{A}\right) + kR\left(\tilde{B}\right)$.

Note: All three ranking functions for fuzzy trapezoidal and triangular complete interval type-2 numbers are linear functions.

Note: Let $\tilde{A} = \left(\tilde{A}^l, \tilde{A}^u\right)$ and $\tilde{B} = \left(\tilde{B}^l, \tilde{B}^u\right)$ are fuzzy trapezoidal complete interval type-2 numbers:

(1) If $\inf \ \sup(\tilde{A}^l) \geq \inf \ \sup(\tilde{B}^l)$ and $\inf \ \sup(\tilde{A}^u) \geq \inf \ \sup(\tilde{B}^u)$ then $\tilde{A} \geq_{R_i} \tilde{B}$,
(2) Let $k \in R$ then $R_i\left(\tilde{A} + k\tilde{B}\right) = R_i\left(\tilde{A}\right) + kR_i\left(\tilde{B}\right)$

Definition 2.38 Let $\tilde{A}$ and $\tilde{B}$ are two interval type 2 fuzzy numbers. The steps of ranking and comparing of interval type-2 fuzzy numbers are expressed as follows.

Step 1: Express $\tilde{A}$ and $\tilde{B}$ as two interval type-2 fuzzy numbers, according to their upper and lower membership functions as $\tilde{A} = \left(\tilde{A}^l, \tilde{A}^u\right)$ and $\tilde{B} = \left(\tilde{B}^l, \tilde{B}^u\right)$.

Step 2: Specify the type of shape function representing the lower membership function of both interval type-2 fuzzy numbers $\tilde{A}$ and $\tilde{B}$.

Step 3: Determine the ranking function R corresponding to the shape functions representing the lower membership function of both interval type-2 fuzzy numbers $\tilde{A}$ and $\tilde{B}$.

Step 4: Specify the type of shape function representing the upper membership function of both interval type-2 fuzzy numbers $\tilde{A}$ and $\tilde{B}$.

Step 5: Determine the ranking function R corresponding to the shape functions representing the upper membership function of both interval type-2 fuzzy numbers $\tilde{A}$ and $\tilde{B}$.

Step 6: Calculate $R\left(\tilde{A}^u\right)$, $R\left(\tilde{A}^l\right)$, $R\left(\tilde{B}^l\right)$ and $R\left(\tilde{B}^u\right)$.

Step 7: Compute $R\left(\tilde{A}\right) = \frac{R(\tilde{A}^l)+R(\tilde{A}^u)}{2}$ and $R\left(\tilde{B}\right) = \frac{R(\tilde{B}^l)+R(\tilde{B}^u)}{2}$.

Step 8: According to the results of step 7, rank as follows:

$R\left(\tilde{A}\right) \geq R\left(\tilde{B}\right)$ if and only if $\tilde{A} \geq_R \tilde{B}$.

$R\left(\tilde{A}\right) \leq R\left(\tilde{B}\right)$ if and only if $\tilde{A} \leq_R \tilde{B}$.

$R\left(\tilde{A}\right) = R\left(\tilde{B}\right)$ if and only if $\tilde{A} =_R \tilde{B}$.

2.7 Conclusions

The theory of fuzzy sets offers valuable tools for dealing with imprecision and uncertainty in real-world problems. By utilizing fuzzy sets, particularly Type-1, Type-2, and Type-2 interval fuzzy sets, we can effectively capture and model the inherent vagueness and ambiguity present in many scenarios. Type-2 interval fuzzy sets, in particular, provide a significant advantage over Type-1 fuzzy sets by offering a broader coverage of uncertainty and additional information. They strike a balance between expressive power and computational complexity, making them well-suited for modeling optimization problems in imprecise environments. Compared to Type-2 interval fuzzy sets, Type-1 fuzzy sets are computationally simpler but may not capture the full extent of uncertainty.

To solve fuzzy problems, a defuzzification process is essential. Ranking functions serve as one effective method for defuzzifying fuzzy problems. These functions assign crisp numbers to fuzzy sets or fuzzy numbers, transforming fuzzy problems into regular linear problems that can be easily solved. Throughout this chapter, we have provided a comprehensive overview of the concepts and definitions associated with Type-1, Type-2, and Type-2 interval fuzzy sets. We have also explored trapezoidal and triangular Type-2 fuzzy numbers, examining their properties and practical applications.

Furthermore, we have discussed various existing ranking methods for comparing trapezoidal and triangular Type-2 fuzzy numbers, emphasizing their mathematical properties. These ranking methods enable us to make comparative assessments and decisions based on fuzzy information, ultimately aiding in problem-solving and decision-making processes. By understanding and leveraging the power of fuzzy sets, along with defuzzification techniques and ranking methods, we can effectively navigate and address the challenges posed by imprecision and uncertainty in a wide range of real-world applications. The knowledge and insights presented in this chapter lay a strong foundation for readers to apply fuzzy set theory and related tools in their own fields of study and professional endeavors.

Acknowledgement A special thanks to the Iranian DEA society for their unwavering spiritual support and consensus in the writing of this book. Your invaluable support has been truly remarkable, and we are deeply grateful for the opportunity to collaborate with such esteemed professionals.

References

1. Allahviranloo, T. (2004). Numerical methods for fuzzy system of linear equations. *Applied Mathematics and Computation, 155*(2), 493–502.
2. Allahviranloo, T., Lotfi, F. H., Kiasary, M. K., Kiani, N. A., & Alizadeh, L. (2008). Solving fully fuzzy linear programming problem by the ranking function. *Applied Mathematical Sciences, 2*(1), 19–32.
3. Aliev, R. A., Pedrycz, W., Guirimov, B. G., Aliev, R. R., Ilhan, U., Babagil, M., & Mammadli, S. (2011). Type-2 fuzzy neural networks with fuzzy clustering and differential evolution optimization. *Information Sciences, 181*(9), 1591-1608.
4. Armand, A., Allahviranloo, T., & Gouyandeh, Z. (2018). Some fundamental results on fuzzy calculus. *Iranian Journal of Fuzzy Systems, 15*(3), 27–46.
5. Castillo, O., Melin, P., Kacprzyk, J., & Pedrycz, W. (2007, November). Type-2 fuzzy logic: Theory and applications. In *2007 IEEE International Conference on Granular Computing* (*GRC 2007*) (pp. 145–145). IEEE.
6. Chen, S. H., & Hsieh, C. H. (1999). Graded mean integration representation of generalized fuzzy number. *Journal of Chinese Fuzzy Systems, 5*(2), 1–7.
7. Figueroa, J. C. (2009). Solving fuzzy linear programming problems with interval type-2 RHS. In *Proceedings of the 2009 IEEE International Conference on Systems, Man, and Cybernetics San Antonio, TX, USA, October*.
8. Figueroa, J. C. (2011). Interval type-2 fuzzy linear programming: Uncertain constraints. In: *2011 IEEE Symposium Series on Computational Intelligence.*
9. Hisdal, E. (1981). The IF THEN ELSE statement and interval-valued fuzzy sets of higher type. *International Journal of Man-Machine Studies, 15*, 385–455.
10. Javanmard, M., & Mishmast Nehi, H. (2019). Rankings and operations for interval type-2 fuzzy numbers: A review and some new methods. *Journal of Applied Mathematics and Computing, 59*, 597–630.
11. Javanmard, M., & Nehi, H. M. (2017, March). Interval type-2 triangular fuzzy numbers; new ranking method and evaluation of some reasonable properties on it. In *2017 5th Iranian Joint Congress on Fuzzy and Intelligent Systems* (*CFIS*) (pp. 4–6). IEEE.
12. Kaufmann, A., & Gupta, M. (1985). *Introduction to fuzzy arithmetic theory and applications.* Van Nostran Reinhold Co., Inc.

13. Kim, E. H., Oh, S. K., & Pedrycz, W. (2018). Reinforced hybrid interval fuzzy neural networks architecture: Design and analysis. *Neurocomputing, 303*, 20-36.
14. Mendel, J. M., John, R. I., & Liu, F. L. (2006). Interval type-2 fuzzy logical systems made simple. *IEEE Transactions on Fuzzy Systems, 14*, 808–821.
15. Mohagheghi, V., Mousavi, S. M., Vahdani, B., & Shahriari, M. R. (2017). R&D project evaluation and project portfolio selection by a new interval type-2 fuzzy optimization approach. *Neural Computing and Applications, 28*, 3869–3888.
16. Mohamadghasemi, A., Hadi-Vencheh, A., Lotfi, F. H., & Khalilzadeh, M. (2020). An integrated group FWA-ELECTRE III approach based on interval type-2 fuzzy sets for solving the MCDM problems using limit distance mean. *Complex & Intelligent Systems, 6*, 355-389.
17. Mohamadghasemi, A., Hadi-Vencheh, A., & Hosseinzadeh Lotfi, F. (2023). An integrated group entropy-weighted interval type-2 fuzzy weighted aggregated sum product assessment (WASPAS) method in maritime transportation. *Scientia Iranica.*
18. Mokhtari, M., Allahviranloo, T., Behzadi, M. H., & Lotfi, F. H. (2022). Introducing a trapezoidal interval type-2 fuzzy regression model. *Journal of Intelligent & Fuzzy Systems, 42*(3), 1381–1403.
19. Mokhtari, M., Allahviranloo, T., Behzadi, M. H., & Lotfi, F. H. (2021). Interval type-2 fuzzy least-squares estimation to formulate a regression model based on a new outlier detection method using a new distance. *Computational and Applied Mathematics, 40*(6), 207.
20. Pedrycz, W. (1993). *Fuzzy control and fuzzy systems.* Research Studies Press Ltd.
21. Pedrycz, W. (2020). Fuzzy relational calculus. In *Handbook of fuzzy computation* (pp. B3–3). CRC Press.
22. Pedrycz, W., & Gomide, F. (1998). *An introduction to fuzzy sets: Analysis and design.* MIT press.
23. Qin, J., Liu, X., & Pedrycz, W. (2015). An extended VIKOR method based on prospect theory for multiple attribute decision making under interval type-2 fuzzy environment. *Knowledge-Based Systems, 86*, 116–130.
24. Qin, J., Liu, X., & Pedrycz, W. (2017). An extended TODIM multi-criteria group decision making method for green supplier selection in interval type-2 fuzzy environment. *European Journal of Operational Research, 258*(2), 626-638.
25. Shahriari, M. R. (2017). Soft computing based on a modified MCDM approach under intuitionistic fuzzy sets. *Iranian Journal of Fuzzy Systems, 14*(1), 23–41.
26. Shen, Y., Pedrycz, W., & Wang, X. (2019). Approximation of fuzzy sets by interval type-2 trapezoidal fuzzy sets. *IEEE Transactions on Cybernetics, 50*(11), 4722-4734.
27. Yager, R. R. (1978) Ranking fuzzy subsets over the unit interval. In *Proceedings of the 17th IEEE International Conference on Decision and Control, San Diago, California* (pp. 1435–1437).
28. Zadeh, L. A. (1965). Fuzzy sets. *Information and Control, 8*, 338–353.
29. Zadeh, L. A. (1975). The concept of linguistic variable and its application to approximate reasoning-1. *Information Sciences, 8*, 199–249.
30. Zadeh, L. A. (1975). The concept of linguistic variable and its application to approximate reasoning-1. *Information Sciences, 8*, 301–357.
31. Zadeh, L. A. (1975). The concept of linguistic variable and its application to approximate reasoning-1. *Information Sciences, 9*, 43–80.

Chapter 3
Weight Determination Methods in Fuzzy Environment

3.1 Introduction

Weighting the criteria in multi-criteria decision-making is one of the important and fundamental steps. In criteria weighting, it is determined how important each criterion is from the perspective of experts, and comparing them with each other determines the weight value of each criterion. Determining the weights of the criteria is examined in two cases. The first case is when the decision matrix (alternative-criterion) is available, and the value of each criterion for each alternative is known. In this case, the importance of criteria is calculated based on the dispersion (variance) of each criterion in different alternatives, which can be referred to as the Shannon entropy method. The second case involves using the pairwise comparison matrix of the criteria, which is completed by experts, and is calculated using approximate methods or least squares methods. In this chapter, the aim is to present these methods in two different cases assuming the values in the decision matrix or pairwise comparisons are fuzzy. Accordingly, it is assumed that the fuzzy information is triangular. Obviously, all the contents can be expanded to trapezoidal fuzzy numbers. For each method, an attempt has been made to explain the algorithm by providing examples.

3.2 Fuzzy Approximation Methods

In this section, some approximate methods for calculating weights using fuzzy data are discussed.

In all of these methods, two types of fuzzy data (triangular and trapezoidal) are used. A numerical example is provided for each method. The aim of the presented methods is to calculate the weight of each criterion as a fuzzy number similar to the data.

F. Hosseinzadeh Lotfi et al., *Fuzzy Decision Analysis: Multi Attribute Decision Making Approach*, Studies in Computational Intelligence 1121,
https://doi.org/10.1007/978-3-031-44742-6_3

In all methods of this section, assume a pairwise comparison matrix is available as follows:

$$\begin{array}{c} C_1 \\ C_2 \\ \vdots \\ C_n \end{array}\begin{bmatrix} (1,1,1) & (a_{12}^l, a_{12}^m, a_{12}^u) & \dots & (a_{1n}^l, a_{1n}^m, a_{1n}^u) \\ (a_{21}^l, a_{21}^m, a_{21}^u) & (1,1,1) & \dots & (a_{2n}^l, a_{2n}^m, a_{2n}^u) \\ \vdots & \vdots & \ddots & \vdots \\ (a_{n1}^l, a_{n1}^m, a_{n1}^u) & (a_{n2}^l, a_{n2}^m, a_{n2}^u) & \dots & (1,1,1) \end{bmatrix} \quad (3.1)$$

with columns labelled C_1, C_2, …, C_n.

The element $\left(a_{ij}^l, a_{ij}^m, a_{ij}^u\right)$ *where* $a_{ij}^l > 0$ represent the relative importance of criterion "i" with respect to criterion "j". Due to the inverse relationship in the pairwise comparison matrix, the relationships (3.2) hold.

$$a_{ji}^l = \frac{1}{a_{ij}^u},\ a_{ji}^m = \frac{1}{a_{ij}^m},\ a_{ji}^u = \frac{1}{a_{ij}^l} \quad (3.2)$$

3.2.1 *Fuzzy Row Sum Method*

In this method, first add up the elements of each row. Based on formula (3.3), and perform the following steps.

$$\left(b_i^l, b_i^m, b_i^u\right) = \left(\sum_{j=1}^n a_{ij}^l, \sum_{j=1}^n a_{ij}^m, \sum_{j=1}^n a_{ij}^u\right),\ i = 1, \dots, n \quad (3.3)$$

To calculate the final weights, the fuzzy numbers obtained in (3.3) need to be normalized using a normalization process. For this purpose, consider:

$$c = \sum_{i=1}^n \sum_{j=1}^n a_{ij}^u$$

In this case, the fuzzy weights of each criterion are obtained from Eq. (3.4).

$$\left(w_i^l, w_i^m, w_i^u\right) = \left(\frac{\sum_{j=1}^n a_{ij}^l}{c}, \frac{\sum_{j=1}^n a_{ij}^m}{c}, \frac{\sum_{j=1}^n a_{ij}^u}{c}\right),\ i = 1, \dots, n \quad (3.4)$$

Example 3.1 Consider the pairwise comparison matrix for the following criteria.

$$\begin{array}{c} C_1 \\ C_2 \\ C_3 \end{array}\overset{\begin{array}{ccc} C_1 & C_2 & C_3 \end{array}}{\begin{bmatrix} (1,1,1) & (2,5,7) & (1,2,3) \\ \left(\frac{1}{7},\frac{1}{5},\frac{1}{2}\right) & (1,1,1) & \left(\frac{1}{5},\frac{1}{3},1\right) \\ \left(\frac{1}{3},\frac{1}{2},1\right) & (1,3,5) & (1,1,1) \end{bmatrix}}$$

The sum of each row in this matrix will be as follows:

$$\left(b_1^l, b_1^m, b_1^u\right) = (4, 8, 11)$$
$$\left(b_2^l, b_2^m, b_2^u\right) = \left(\frac{47}{35}, \frac{23}{15}, \frac{5}{2}\right)$$
$$\left(b_3^l, b_3^m, b_3^u\right) = \left(\frac{7}{3}, \frac{9}{2}, 7\right)$$

Since, $c = 11 + \frac{5}{2} + 7 = \frac{41}{2}$ then, the weight of each criterion is obtained from the following equations:

$$\left(w_1^l, w_1^m, w_1^u\right) = \left(\frac{4}{\frac{41}{2}}, \frac{8}{\frac{41}{2}}, \frac{11}{\frac{41}{2}}\right) = (0.20, 0.39, 0.54)$$
$$\left(w_2^l, w_2^m, w_2^u\right) = \left(\frac{\frac{47}{35}}{\frac{41}{2}}, \frac{\frac{23}{15}}{\frac{41}{2}}, \frac{\frac{5}{2}}{\frac{41}{2}}\right) = (0.07, 0.07, 0.12)$$
$$\left(w_3^l, w_3^m, w_3^u\right) = \left(\frac{\frac{7}{3}}{\frac{41}{2}}, \frac{\frac{9}{2}}{\frac{41}{2}}, \frac{7}{\frac{41}{2}}\right) = (0.11, 0.22, 0.34)$$

3.2.2 *Fuzzy Column Sum Method*

Once again, consider the pairwise comparison matrix (3.1). In this method, first, sum up the elements of each column. According to the formula (3.5):

$$\left(b_j^l, b_j^m, b_j^u\right) = \left(\sum_{i=1}^{n} a_{ij}^l, \sum_{i=1}^{n} a_{ij}^m, \sum_{i=1}^{n} a_{ij}^u\right), \quad j = 1, \ldots, n \tag{3.5}$$

Assuming

$$c = \sum_{i=1}^{n}\sum_{j=1}^{n} a_{ij}^{u}$$

In this case, the fuzzy weights of each criterion are obtained from Eq. (3.6).

$$\left(w_j^l, w_j^m, w_j^u\right) = \left(\frac{\sum_{i=1}^{n} a_{ij}^l}{c}, \frac{\sum_{i=1}^{n} a_{ij}^m}{c}, \frac{\sum_{i=1}^{n} a_{ij}^u}{c}\right), \; j = 1, ..., n \tag{3.6}$$

Consider Example 3.1 once again. The sum of the columns in this matrix will be as follows:

$$\left(b_1^l, b_1^m, b_1^u\right) = \left(\frac{31}{21}, \frac{17}{10}, \frac{5}{2}\right)$$
$$\left(b_2^l, b_2^m, b_2^u\right) = (4, 9, 13)$$
$$\left(b_3^l, b_3^m, b_3^u\right) = \left(\frac{11}{5}, \frac{10}{3}, 5\right)$$

In this case calculating $c = \frac{5}{2} + 13 + 5 = \frac{41}{2}$, the weights of each criterion are calculated using the following equations:

$$\left(w_1^l, w_1^m, w_1^u\right) = \left(\frac{\frac{31}{21}}{\frac{41}{2}}, \frac{\frac{17}{10}}{\frac{41}{2}}, \frac{\frac{5}{2}}{\frac{41}{2}}\right) = (0.07, 0.08, 0.12)$$

$$\left(w_2^l, w_2^m, w_2^u\right) = \left(\frac{4}{\frac{41}{2}}, \frac{9}{\frac{41}{2}}, \frac{13}{\frac{41}{2}}\right) = (0.20, 0.44, 0.63)$$

$$\left(w_3^l, w_3^m, w_3^u\right) = \left(\frac{\frac{11}{5}}{\frac{41}{2}}, \frac{\frac{10}{3}}{\frac{41}{2}}, \frac{5}{\frac{41}{2}}\right) = (0.11, 0.16, 0.24)$$

3.2.3 *Fuzzy Geometric Mean Method*

Again, consider the pairwise comparison matrix (3.1). In this method, first, calculate the geometric mean of the elements in each row, and then normalize the obtained fuzzy numbers.

$$\left(b_i^l, b_i^m, b_i^u\right) = \left(\sqrt[n]{\prod_{j=1}^{n} a_{ij}^l}, \sqrt[n]{\prod_{j=1}^{n} a_{ij}^m}, \sqrt[n]{\prod_{j=1}^{n} a_{ij}^u}\right), \; i = 1, ..., n \tag{3.7}$$

Assuming $c = \sum_{i=1}^{n} \sqrt[n]{\prod_{j=1}^{n} a_{ij}^{u}}$, the weight of each criterion is obtained from Eq. (3.8).

$$\left(w_i^l, w_i^m, w_i^u\right) = \left(\frac{\sqrt[n]{\prod_{j=1}^{n} a_{ij}^{l}}}{c}, \frac{\sqrt[n]{\prod_{j=1}^{n} a_{ij}^{m}}}{c}, \frac{\sqrt[n]{\prod_{j=1}^{n} a_{ij}^{u}}}{c}\right), \quad i = 1, \dots, n \tag{3.8}$$

For further clarification, consider Example 3.1. The necessary calculations for calculating the weights of each criterion are presented below for this example using fuzzy geometric mean method.

$$\left(b_1^l, b_1^m, b_1^u\right) = \left(\sqrt[3]{2}, \sqrt[3]{10}, \sqrt[3]{21}\right) = (1.26, 2.15, 2.76)$$

$$\left(b_2^l, b_2^m, b_2^u\right) = \left(\sqrt[3]{\frac{1}{35}}, \sqrt[3]{\frac{1}{15}}, \sqrt[3]{\frac{1}{2}}\right) = (0.31, 0.41, 0.79)$$

$$\left(b_3^l, b_3^m, b_3^u\right) = \left(\sqrt[3]{\frac{1}{3}}, \sqrt[3]{\frac{3}{2}}, \sqrt[3]{5}\right) == (0.69, 1.14, 1.71)$$

$$c = 2.76 + 0.79 + 1.71$$

$$\left(w_1^l, w_1^m, w_1^u\right) = \left(\frac{1.26}{5.26}, \frac{2.15}{5.26}, \frac{2.76}{5.26}\right) = (0.24, 0.41, 0.52)$$

$$\left(w_2^l, w_2^m, w_2^u\right) = \left(\frac{0.31}{5.26}, \frac{0.41}{5.26}, \frac{0.79}{5.26}\right) = (0.06, 0.08, 0.15)$$

$$\left(w_3^l, w_3^m, w_3^u\right) = \left(\frac{0.69}{5.26}, \frac{1.14}{5.26}, \frac{1.71}{5.26}\right) = (0.13, 0.22, 0.32)$$

3.3 Fuzzy Shannon Entropy Method

Hosseinzadeh Lotfi et al. [3] developed the Shannon entropy method for imprecise data. They proposed a method based on interval and fuzzy data to determine the weights. Assuming that the decision matrix in general has "m" alternatives and "n" criteria in the form of Table 3.1, this method is discussed.

$\tilde{a}_{ij}$ represents the value of jth criterion for ith alternative, which can be any possible fuzzy number (triangular, trapezoidal, etc.).

Table 3.1 Decision matrix with fuzzy data

Alternative	C_1	C_2	...	C_n
A_1	$\tilde{a}_{11}$	$\tilde{a}_{12}$	...	$\tilde{a}_{1n}$
A_2	$\tilde{a}_{21}$	$\tilde{a}_{22}$	...	$\tilde{a}_{2n}$
$\vdots$	$\vdots$	$\vdots$	$\ddots$	$\vdots$
A_m	$\tilde{a}_{m1}$	$\tilde{a}_{m2}$	...	$\tilde{a}_{mn}$

Step 1: Converting fuzzy numbers to intervals using α-cuts.

Since it is assumed that each $\tilde{a}_{ij}$ is a fuzzy number, the fuzzy set is convex. Therefore, for each $\alpha \in [0, 1]$, consider that the α-cut will be an interval. Thus, this decision matrix will be transformed as follows (Table 3.2):

where,

$$\left[\left(a_{ij}\right)_{\alpha}^{L}, \left(a_{ij}\right)_{\alpha}^{U}\right] = \left[Min\left\{a_{ij} \in R \middle| \mu_{\tilde{a}_{ij}}(\tilde{a}_{ij}) \geq \alpha\right\}, Max\left\{a_{ij} \in R \middle| \mu_{\tilde{a}_{ij}}(\tilde{a}_{ij}) \geq \alpha\right\}\right].$$

Step 2: Normalization

$$\begin{aligned} \left(P_{ij}\right)_{\alpha}^{L} &= \frac{\left(a_{ij}\right)_{\alpha}^{L}}{\sum_{i=1}^{m}\left(a_{ij}\right)_{\alpha}^{U}} \quad ,i = 1, \ldots, m \ \ ,j = 1, \ldots, n \\ \left(P_{ij}\right)_{\alpha}^{U} &= \frac{\left(a_{ij}\right)_{\alpha}^{U}}{\sum_{i=1}^{m}\left(a_{ij}\right)_{\alpha}^{U}} \quad ,i = 1, \ldots, m, \ \ j = 1, \ldots, n \end{aligned} \tag{3.9}$$

Step 3: Calculation of entropy interval bounds

$$\begin{aligned} h_{j}^{L} &= Min\left\{\frac{-1}{\ln m}\sum_{i=1}^{m} P_{ij}^{L} Ln\ P_{ij}^{L}, \frac{-1}{\ln m}\sum_{i=1}^{m} P_{ij}^{U} Ln\ P_{ij}^{U}\right\} \ ,j = 1, \ldots, n \\ h_{j}^{U} &= Max\left\{\frac{-1}{\ln m}\sum_{i=1}^{m} P_{ij}^{L} Ln\ P_{ij}^{L}, \frac{-1}{\ln m}\sum_{i=1}^{m} P_{ij}^{U} Ln\ P_{ij}^{U}\right\}, \ \ j = 1, \ldots, n \end{aligned} \tag{3.10}$$

Table 3.2 Interval decision matrix using α-cuts

Alternative	C_1	C_2	...	C_n
A_1	$\left[(a_{11})_{\alpha}^{L}, (a_{11})_{\alpha}^{U}\right]$	$\left[(a_{12})_{\alpha}^{L}, (a_{12})_{\alpha}^{U}\right]$	...	$\left[(a_{1n})_{\alpha}^{L}, (a_{1n})_{\alpha}^{U}\right]$
A_2	$\left[(a_{21})_{\alpha}^{L}, (a_{21})_{\alpha}^{U}\right]$	$\left[(a_{22})_{\alpha}^{L}, (a_{22})_{\alpha}^{U}\right]$	...	$\left[(a_{2n})_{\alpha}^{L}, (a_{2n})_{\alpha}^{U}\right]$
$\vdots$	$\vdots$	$\vdots$	$\ddots$	$\vdots$
A_m	$\left[(a_{m1})_{\alpha}^{L}, (a_{m1})_{\alpha}^{U}\right]$	$\left[(a_{m2})_{\alpha}^{L}, (a_{m2})_{\alpha}^{U}\right]$	...	$\left[(a_{mn})_{\alpha}^{L}, (a_{mn})_{\alpha}^{U}\right]$

Note that, if $P_{ij}^L = 0$ then put $P_{ij}^L \, Ln P_{ij}^L = 0$ and if $P_{ij}^U = 0$ then put $P_{ij}^U \, Ln P_{ij}^U = 0$.

Step 4: Calculation of the interval of diversification d_j.

In this step, using Eqs. (3.11) the bounds of d_j obtains as follows:

$$\begin{aligned} d_j^L &= 1 - h_j^U, \ j = 1, ..., n \\ d_j^U &= 1 - h_j^L, \ j = 1, ..., n \end{aligned} \tag{3.11}$$

Step 5: Calculation of bounds of weight vector.

Obtain the weight interval for each criterion using Eq. (3.12).

$$\begin{aligned} (w_j)_\alpha^L &= \frac{d_j^L}{\sum_{j=1}^n d_j^U}, \ j = 1, ..., n \\ (w_j)_\alpha^U &= \frac{d_j^U}{\sum_{j=1}^n d_j^U}, \ j = 1, ..., n \end{aligned} \tag{3.12}$$

Example 3.2 Consider the following decision matrix.

$$\begin{array}{c} \\ A_1 \\ A_2 \\ A_3 \end{array} \begin{array}{c} \begin{array}{cccc} C_1 & C_2 & C_3 & C_4 \end{array} \\ \begin{bmatrix} (2,3,5) & (1,3,4) & (2,4,6) & (2,3,6) \\ (1,6,7) & (1,3,6) & (2,5,6) & (3,5,7) \\ (4,6,7) & (5,6,7) & (1,5,7) & (2,6,8) \end{bmatrix} \end{array}$$

Considering $\tilde{a} = (l, m, u), \tilde{a}_\alpha = (l + \alpha(m - l), u - \alpha(u - m))$.

Step 1: Assuming $\alpha = 0.3$, the fuzzy decision matrix will be transformed into the following interval decision matrix.

$$\begin{array}{c} \\ A_1 \\ A_2 \\ A_3 \end{array} \begin{array}{c} \begin{array}{cccc} C_1 & C_2 & C_3 & C_4 \end{array} \\ \begin{bmatrix} [2.3, 4.4] & [1.6, 3.7] & [2.6, 5.4] & [2.3, 5.1] \\ [2.5, 6.7] & [1.6, 5.1] & [2.9, 5.7] & [3.6, 6.4] \\ [4.6, 6.7] & [5.3, 6.7] & [2.2, 6.4] & [3.2, 7.4] \end{bmatrix} \end{array}$$

Step 2: The normalized decision matrix will be as follows.

$$\begin{array}{c} \\ A_1 \\ A_2 \\ A_3 \end{array} \begin{array}{c} \begin{array}{cccc} C_1 & C_2 & C_3 & C_4 \end{array} \\ \begin{bmatrix} [0.13, 0.25] & [0.1, 0.24] & [0.15, 0.31] & [0.12, 0.27] \\ [0.14, 0.38] & [0.1, 0.33] & [0.17, 0.33] & [0.19, 0.34] \\ [0.26, 0.38] & [0.34, 0.43] & [0.13, 0.37] & [0.17, 0.39] \end{bmatrix} \end{array}$$

Step 3: Calculation of entropy interval bounds.

$$\begin{array}{ccccc} & C_1 & C_2 & C_3 & C_4 \\ \left[h_j^L, h_j^U\right] = \big[& [0.81, 0.98], & [0.76, 0.97], & [0.77, 1], & [0.79, 0.99] \big] \end{array}$$

Step 4: Calculation of diversification d_j bounds.

$$\begin{array}{ccccc} & C_1 & C_2 & C_3 & C_4 \\ \left[d_j^L, d_j^U\right] = \big[& [0.02, 0.19], & [0.03, 0.24], & [0, 0.23], & [0.01, 0.21] \big] \end{array}$$

Step 5: Calculation of bounds of weight vector

$$\begin{array}{ccccc} & C_1 & C_2 & C_3 & C_4 \\ \left[(w_j)_{0.3}^L, (w_j)_{0.3}^U\right] = \big[& [0.02, 0.22], & [0.03, 0.28], & [0, 0.27], & [0.01, 0.24] \big] \end{array}$$

3.3.1 Shannon Entropy Method Using Triangular Fuzzy Number

Below are the steps of the Shannon entropy method described in a triangular fuzzy form [8]:

Step 1: Calculate the values of $\tilde{p}_{ij} = (p_{ij}^l, p_{ij}^m, p_{ij}^u)$. Normalize the decision matrix.

$$p_{ij}^l = \frac{a_{ij}^l}{\sum_{i=1}^m a_{ij}^u},\ p_{ij}^m = \frac{a_{ij}^m}{\sum_{i=1}^m a_{ij}^u},\ p_{ij}^u = \frac{a_{ij}^u}{\sum_{i=1}^m a_{ij}^l},\ j = 1, \ldots, n; i = 1, \ldots, m.$$

Step 2: Calculate h_i(fuzzy entropy of each criteria)

$$h_i^l = -(\ln m)^{-1} \sum_{i=1}^m p_{ij}^l \ln p_{ij}^l, \quad i = 1, \ldots, n,$$

$$h_i^m = -(\ln m)^{-1} \sum_{i=1}^m p_{ij}^m \ln p_{ij}^m, \quad i = 1, \ldots, n,$$

$$h_i^u = -(\ln m)^{-1} \sum_{i=1}^m p_{ij}^u \ln p_{ij}^u, \quad i = 1, \ldots, n.$$

That if $p^l_{ij} = 0$ we put $p^l_{ij}\ln p^l_{ij} = 0$, if $p^m_{ij} = 0$ we put $p^m_{ij}\ln p^m_{ij} = 0$ and also if $p^u_{ij} = 0$ we put $p^u_{ij}\ln p^u_{ij} = 0$.

Step 3: interval degree of diversification

$$d^l_i = 1 - h^u_i, d^m_i = 1 - h^m_i, d^u_i = 1 - h^l_i \quad i = 1,...,n.$$

Step 4: Determining the fuzzy weight $\tilde{w}_i = (w^l_i, w^m_i, w^u_i)$

$$w^l_i = \frac{d^l_i}{\sum_{t=1}^{n} d^u_t}, \; w^u_i = \frac{d^m_i}{\sum_{t=1}^{n} d^m_t}, \; w^u_i = \frac{d^u_i}{\sum_{t=1}^{n} d^l_t}, \quad i = 1, ..., n.$$

However, in cases where some values of the decision matrix are fuzzy while others are crisp, it is possible to assign the value of $(x^l_{ij}, x^m_{ij}, x^u_{ij}) = (x_{ij}, x_{ij}, x_{ij})$. This will determine the fuzzy weights. Another approach in this regard is to use alpha cuts, where the values of the decision matrix are transformed into interval values. By using the method mentioned in the previous section, the weights can be extracted. It is important to note that when some indicators have precise values, some have interval values, and others have triangular fuzzy values, utilizing the alpha cut method to convert fuzzy values into interval values and applying the method mentioned in the previous section can be fruitful.

3.4 Fuzzy Least Squares Method

This method is used to calculate the weights of the pairwise comparison matrix. Let's assume that the pairwise comparison matrix is represented by the following table for comparing criteria $C_1, C_2, ..., C_n$.

$$\begin{array}{c} \\ C_1 \\ C_2 \\ \vdots \\ C_n \end{array} \begin{array}{c} \begin{array}{cccc} C_1 & C_2 & \cdots & C_n \end{array} \\ \begin{bmatrix} (1,1,1) & (a^l_{12}, a^m_{12}, a^u_{12}) & \cdots & (a^l_{1n}, a^m_{1n}, a^u_{1n}) \\ (a^l_{21}, a^m_{21}, a^u_{21}) & (1,1,1) & \cdots & (a^l_{2n}, a^m_{2n}, a^u_{2n}) \\ \vdots & \vdots & \ddots & \vdots \\ (a^l_{n1}, a^m_{n1}, a^u_{n1}) & (a^l_{n2}, a^m_{n2}, a^u_{n2}) & \cdots & (1,1,1) \end{bmatrix} \end{array}$$

Element $\left(a^l_{ij}, a^m_{ij}, a^u_{ij}\right)$ represents the relative importance of ith criterion compared to jth criterion. Hence:

$$\begin{aligned} &\frac{\tilde{w}_i}{\tilde{w}_j} = \left(a^l_{ij}, a^m_{ij}, a^u_{ij}\right) \\ &a^l_{ij} = \frac{1}{a^u_{ji}}, \quad a^m_{ij} = \frac{1}{a^m_{ji}}, \quad a^u_{ij} = \frac{1}{a^l_{ji}} \end{aligned} \tag{3.13}$$

If the pairwise comparison matrix is consistent, then:

$$\forall i \ \forall j \quad (\tilde{w}_i = \tilde{a}_{ij}.\tilde{w}_j)$$

Since most pairwise comparison matrices are inconsistent, the goal is to find weights $\tilde{w}_i$ that minimize the difference from $\tilde{a}_{ij}.\tilde{w}_j$. Thus, by using the L2 norm of the difference of two vectors, the optimization problem can be formulated as the well-known Fuzzy Least Squares Method:

$$\begin{aligned} Min \quad & \sum_{i=1}^{n}\sum_{j=1}^{n}\left(\tilde{w}_i - \tilde{a}_{ij}.\tilde{w}_j\right)^2 \\ s.t. \quad & \sum_{i=1}^{n}\tilde{w}_i = 1 \\ & \tilde{w} \geq 0 \end{aligned} \tag{3.14}$$

Assuming $\tilde{w}_i = \left(w_i^l, w_i^m, w_i^u\right)$ as triangular fuzzy numbers, problem (3.14) can be written as (3.15).

$$\begin{aligned} Min \quad & \sum_{i=1}^{n}\sum_{j=1}^{n}\left(\left(w_i^l - a_{ij}^u.w_j^u\right), \left(w_i^m - a_{ij}^m.w_j^m\right), \left(w_i^u - a_{ij}^l.w_j^l\right)\right)^2 \\ s.t. \quad & \sum_{i=1}^{n} w_i^l = 1 \ \ \& \sum_{i=1}^{n} w_i^m = 1 \ \ \& \sum_{i=1}^{n} w_i^u = 1 \\ & 0 \leq w_i^l \leq w_i^m \leq w_i^u, \ i = 1, \ldots, n \end{aligned} \tag{3.15}$$

Considering that the fuzzy term in the objective function is not necessarily non-negative, its square term can be approximated as follows: [7]

$$\begin{aligned} Min \quad & \sum_{i=1}^{n}\sum_{j=1}^{n}\left(\frac{\left(w_i^l - a_{ij}^u.w_j^u\right) + 4\left(w_i^m - a_{ij}^m.w_j^m\right) + \left(w_i^u - a_{ij}^l.w_j^l\right)}{6}\right)^2 \\ s.t. \quad & \sum_{i=1}^{n} w_i^l = 1 \ \ \& \sum_{i=1}^{n} w_i^m = 1 \ \ \& \sum_{i=1}^{n} w_i^u = 1 \\ & 0 \leq w_i^l \leq w_i^m \leq w_i^u, \ i = 1, \ldots, n \end{aligned} \tag{3.16}$$

Since by solving the above model, the obtained weights are equal in all three components, Wang et al. [7] suggest adding the following weight constraints to the problem.

$$
\begin{aligned}
& Min \sum_{i=1}^{n}\sum_{j=1}^{n}\left(\frac{\left(w_i^l - a_{ij}^u w_j^u\right) + 4\left(w_i^m - a_{ij}^m w_j^m\right) + \left(w_i^u - a_{ij}^l w_j^l\right)}{6}\right)^2 \\
& s.t.\ \ w_i^l + \sum_{\substack{j=1 \\ j\neq i}}^{n} w_j^u \geq 1, \quad i = 1, \ldots, n \\
& \sum_{i=1}^{n} w_i^m = 1, \\
& w_i^u + \sum_{\substack{j=1 \\ j\neq i}}^{n} w_j^l \leq 1, \quad i = 1, \ldots, n \\
& \sum_{i=1}^{n} \left(w_i^l + w_i^u\right) = 2 \\
& 0 \leq w_i^l \leq w_i^m \leq w_i^u, \quad i = 1, \ldots, n
\end{aligned} \tag{3.17}
$$

Example 3.3 Suppose the pairwise comparison matrix for four criteria is as follows.

$$
\begin{array}{c} \\ C_1 \\ C_2 \\ C_3 \\ C_4 \end{array}
\begin{array}{c}
\begin{array}{cccc} C_1 & C_2 & C_3 & C_4 \end{array} \\
\begin{bmatrix}
(1,1,1) & (2,3,5) & (3,7,8) & (2,4,5) \\
\left(\frac{1}{5},\frac{1}{3},\frac{1}{2}\right) & (1,1,1) & (1,3,4) & (2,3,4) \\
\left(\frac{1}{8},\frac{1}{7},\frac{1}{3}\right) & \left(\frac{1}{4},\frac{1}{3},1\right) & (1,1,1) & (5,6,9) \\
\left(\frac{1}{5},\frac{1}{4},\frac{1}{2}\right) & \left(\frac{1}{4},\frac{1}{3},\frac{1}{2}\right) & \left(\frac{1}{9},\frac{1}{6},\frac{1}{5}\right) & (1,1,1)
\end{bmatrix}
\end{array}
$$

By solving problem (3.16) for the data in Example (3.3), the fuzzy weights of the criteria are obtained as follows (Table 3.3).

Table 3.3 The result of fuzzy weights

	W_L	W_M	W_U
C1	0.2459	0.3870	0.9700
C2	0.0100	0.2621	0.4032
C3	0.0100	0.2084	0.2084
C4	0.0100	0.1425	0.1425

3.5 BWM Method

In the area of decision-making, particularly when faced with multiple criteria and alternatives, the Best–Worst Method (BWM) offers a novel and effective approach. This method aims to provide a systematic framework for tackling multi-criteria decision-making (MCDM) problems by identifying the best and worst criteria, conducting pairwise comparisons, and deriving reliable and consistent results. MCDM problems arise when decision-makers must evaluate various alternatives based on multiple criteria to determine the most favorable alternatives. Traditionally, methods like the Analytic Hierarchy Process (AHP) have been employed for such tasks. However, the BWM method introduces a fresh perspective, offering distinct advantages over existing approaches [1, 9, 10, 11].

At its core, the BWM method emphasizes the identification of the best and worst criteria according to the decision-maker's judgment. By explicitly highlighting these extreme points, the method helps focus the evaluation process and provides a clear reference for subsequent comparisons. Pairwise comparisons are then conducted between these two criteria and all other criteria to establish their relative importance or desirability. To determine the weights of the different criteria, a maximin problem is formulated and solved. This process ensures that the resulting weights accurately reflect the decision-maker's preferences and priorities. Similarly, the weights of the alternatives with respect to the criteria are derived using the same methodology, ensuring consistency and comparability across evaluations. The final step involves aggregating the weights obtained from different sets of criteria and alternatives. By combining these weights, the method generates a comprehensive evaluation that serves as the basis for selecting the best alternative(s) among the available alternatives. Importantly, the BWM method incorporates a consistency ratio to assess the reliability of the comparisons, further enhancing the robustness of its results. The proposed BWM method has been extensively tested using both numerical examples and real-world decision-making problems, such as mobile phone selection. Comparative analyses with the widely used AHP method have consistently demonstrated the superiority of BWM in terms of the consistency ratio and various evaluation criteria including minimum violation, total deviation, and conformity. Noteworthy advantages of the BWM method over existing MCDM methods include its ability to yield reliable results while requiring fewer comparison data. Moreover, the BWM method promotes consistency throughout the decision-making process, thereby enhancing the validity and credibility of the final outcomes.

In the subsequent sections of this book, we delve deeper into the intricacies of the BWM method, exploring its theoretical foundations, practical applications, and its potential to revolutionize multi-criteria decision-making. Through comprehensive analysis and case studies, we aim to demonstrate the practical value and effectiveness of this innovative approach in real-world decision-making scenarios. The steps of the BWM method are as follows: ([4, 6]).

Step 1: Determining the set of criteria.

In the first step, the research problem needs to be specified, and the factors influencing the problem's objective are extracted. This step should ultimately be confirmed by research experts. Methods such as the Delphi method or fuzzy Delphi can be used in this step to validate and screen the research indicators.

Step 2: Comparing the best criterion with other criteria (BO) and other criteria with the worst criterion (OW).

In this step, the most important and least important criteria, referred to as "best" and "worst," respectively, need to be identified from all the indicators. Then, pairwise comparisons are made between the best criterion and other criteria and between other criteria and the worst criterion. These comparisons are answered using a scale ranging from 1 to 9.

Step 3: Creating a non-linear programming model.

In this step, the optimal non-linear optimization model of the BWM method is established using the following problem:

$$\begin{aligned} &Min \ \ \xi \\ &s.t. \left|\frac{w_B}{w_j} - a_{Bj}\right| \le \xi, \quad j = 1, ..., n, \\ &\left|\frac{w_j}{w_w} - a_{jw}\right| \le \xi, \quad j = 1, ..., n, \\ &\sum_{j=1}^{n} w_j = 1, \\ &w_j \ge 0, \quad j = 1, ..., n. \end{aligned} \tag{3.18}$$

where $a_{Bj}(j = 1, ..., n)$ represents the result of pairwise comparison between the best criterion and other criteria, and $a_{jw}(j = 1, ..., n)$ represents the result of comparison between other criteria and the worst criterion.

The above problem is a non-linear programming problem. However, it should be noted that the objective of solving the problem is essentially to minimize the difference between $\frac{w_B}{w_j}$ and a_{Bj}, which is accomplished using the Chebyshev norm. The objective function has been represented as:

$$Min \ \ Max\left\{\left|\frac{w_B}{w_j} - a_{Bj}\right|, \left|\frac{w_j}{w_w} - a_{jw}\right|\right\}$$

.

Therefore, these two can be considered as $\{|w_B - a_{Bj}w_j|, |w_j - a_{jw}w_w|\}$ and the structure of the problem can be written as below.

Table 3.4 Consistency index (CI) table

a_{BW}	1	2	3	4	5	6	7	8	9
Consistency index	0	0.44	1	1.63	2.3	3	3.73	4.47	5.23

$$
\begin{aligned}
& Min \ \ \xi \\
& s.t. \ \left|w_B - a_{Bj}w_j\right| \ \le \xi, \quad j = 1, \ldots, n, \\
& \qquad \left|w_j - a_{jw}w_w\right| \ \le \xi, \quad j = 1, \ldots, n, \\
& \qquad \sum_{j=1}^{n} w_j = 1, \\
& \qquad w_j \ge 0, \quad j = 1, \ldots, n.
\end{aligned} \tag{3.19}
$$

The next topic to discuss in this method is the examination of the consistency ratio of the pairwise comparison matrix in the weight determination process. Given that ξ^* obtained is actually the difference between $\frac{w_B}{w_j}$ and a_{Bj}, it can be used to determine the inconsistency rate. This involves normalizing this value using the table below and obtaining the relationship (3.20) (Table 3.4).

Where a_{BW} is result of comparing the best criterion with the worst criterion.

$$
Consistency \ Ratio = \frac{\xi^*}{Consistency \ Index} \tag{3.20}
$$

3.5.1 *Fuzzy BWM*

As mentioned in the previous section, the BWM method requires much fewer comparisons due to its designed structure. This simplifies the interactive process between the decision maker and the experts. The structure for determining the weights of criteria in a fuzzy environment is described below, with the following steps: (3.1).

Step 1: Determining the necessary criteria for comparison in the decision structure.

Step 2: Identifying the best and worst criteria (the most important and least important criteria): This step can be determined through expert opinions or other methods.

Step 3: Pairwise comparison of the best criterion with other criteria and other criteria with the worst criterion: In this step, pairwise comparisons can be made using any fuzzy spectrum, but the most common spectrum for the Best Worst Fuzzy Method (FBWM) is the 5-point fuzzy spectrum below. This spectrum is based on linguistic expressions of equal importance (EI), weak importance (WI), somewhat important (FI), very important (VI), and absolutely important (AI) (Table 3.5).

Table 3.5 Converting linguistic variables to fuzzy numbers

Linguistic terms	Membership function
Equality important (EI)	(1,1,1)
Weakly important (WI)	(2/3,1,3/2)
Fairly important (FI)	(3/2,2,5/2)
Very important (VI)	(5/2,3,7/2)
Absolutely important (AI)	(7/2,4,9/2)

Step 4: Solve the fuzzy BWM model and determine the fuzzy weights

$$\begin{aligned}
&Min\ \ Max\left\{\left|\frac{\tilde{w}_B}{\tilde{w}_j}-\tilde{a}_{Bj}\right|,\left|\frac{\tilde{w}_j}{\tilde{w}_w}-\tilde{a}_{jw}\right|\right\}\\
&s.t.\ \sum_{j=1}^{n}R(\tilde{w}_j)=1,\\
&0\le w_i^l\le w_i^m\le w_i^u\\
&w_j\ge 0, \qquad\qquad j=1,\ldots,n.
\end{aligned} \tag{3.21}$$

In model (3.21), we encounter a non-linear structure. The first reason for non-linearity can be resolved by applying a variable change in the objective function. The second reason for non-linearity can be resolved similarly to model (3.19). Therefore, by performing the above steps, we arrive at the following model.([2, 5])

$$\begin{aligned}
&Min\ \ \tilde{\tau}\\
&s.t.\ \left|\tilde{w}_B-\tilde{a}_{Bj}\tilde{w}_j\right|\le\tilde{\tau}, && j=1,\ldots,n,\\
&\left|\tilde{w}_j-\tilde{a}_{jw}\tilde{w}_w\right|\le\tilde{\tau}, && j=1,\ldots,n,\\
&\sum_{j=1}^{n}R(\tilde{w}_j)=1,\\
&0\le w_j^l\le w_j^m\le w_j^u, && j=1,\ldots,n.
\end{aligned} \tag{3.22}$$

$R(\tilde{w}_j)$ refers to the use of a ranking function for fuzzy numbers. It should be noted that the above model is a fuzzy linear programming model, and its extended form is as follows:

$$
\begin{aligned}
& Min \quad \frac{1}{6}(\tau_l + 4\tau_m + \tau_u) \\
& s.t. \quad \frac{1}{6}((w_B^l - a_{Bj}^u w_j^u) + 4(w_B^m - a_{Bj}^m w_j^m) + (w_B^u - a_{Bj}^l w_j^l)) \le \frac{1}{6}(\tau_l + 4\tau_m + \tau_u),\ j = 1, ..., n, \\
& \frac{1}{6}((w_B^l - a_{Bj}^u w_j^u) + 4(w_B^m - a_{Bj}^m w_j^m) + (w_B^u - a_{Bj}^l w_j^l)) \ge -\frac{1}{6}(\tau_u + 4\tau_m + \tau_l),\ j = 1, ..., n, \\
& \frac{1}{6}((w_j^l - a_{jw}^u w_w^u) + 4(w_j^m - a_{jw}^m w_w^m) + (w_j^u - a_{jw}^l w_w^l)) \le \frac{1}{6}(\tau_l + 4\tau_m + 4\tau_u),\ j = 1, ..., n, \\
& \frac{1}{6}((w_j^l - a_{jw}^u w_w^u) + 4(w_j^m - a_{jw}^m w_w^m) + (w_j^u - a_{jw}^l w_w^l)) \ge -\frac{1}{6}(\tau_u + 4\tau_m + \tau_l),\ j = 1, ..., n, \\
& \sum_{j=1}^{n} \frac{1}{6}(w_j^l + 4w_j^m + w_j^u) = 1, \\
& 0 \le w_j^l \le w_j^m \le w_j^u, \quad j = 1, ..., n. \\
& \tau_l \le \tau_m \le \tau_u
\end{aligned}
\tag{3.23}
$$

Given that the objective function of the above model is a coefficient of the Consistency Index, it seems that the decision-maker can use it as an indicator for decision-making about redoing pairwise comparisons.

Example 3.4 Consider pairwise comparison matrix example 3.3, noting that C1 and C4 are the best and worst indices, respectively. $\tilde{a}_{Bj}$, $\tilde{a}_{jw}$ are as follows (Table 3.6):

The fuzzy weights are determined as follows by solving model (3.23) (Table 3.7).

Table 3.6 Pairwise Comparison of the Best and Worst Index with the Rest of the Indices

Criteria	C1	C2	C3	C4
$\tilde{a}_{Bj}$	(1, 1, 1)	(2, 3, 5)	(3, 7, 8)	(2, 4, 5)
$\tilde{a}_{jw}$	(2, 4, 5)	(2, 3, 4)	(5, 6, 9)	(1, 1, 1)

Table 3.7 The fuzzy weights based on fuzzy-BWM

	W_L	W_M	W_U
C1	0.4966	0.5077	0.8482
C2	0.2562	0.2562	0.2562
C3	0.1098	0.1098	0.1601
C4	0.0320	0.0618	0.0993

3.6 Conclusion

In this chapter, we have delved into the topic of fuzzy weight determination methods in multi-criteria decision making. We explored two prominent approaches: the fuzzy least square error method and the fuzzy BWM method. By applying these methods to practical examples, we observed how they can effectively address the uncertainty and imprecision inherent in decision-making processes. The fuzzy least square error method allows for a systematic evaluation of criteria weights based on minimizing the overall error, while the fuzzy BWM method enables the identification of the best and worst criteria for weight assignment. Both methods offer valuable tools for decision makers grappling with complex choices involving multiple criteria. They provide a structured framework for assigning fuzzy weights, taking into account the relative importance and performance of different criteria. By employing these methods, decision makers can enhance the transparency, consistency, and robustness of their decision-making processes. It is worth noting that the selection of a specific fuzzy weight determination method depends on the context, nature of the decision problem, and decision maker's preferences. Different methods may yield varying results, and it is crucial to carefully analyze and interpret the outcomes. Overall, the study of fuzzy weight determination methods in multi-criteria decision making contributes to the advancement of decision science by offering tools to handle uncertainty and subjectivity. It equips decision makers with valuable insights and techniques for effectively navigating complex decision-making scenarios and achieving more informed and reliable outcomes.

Acknowledgement A special thanks to the Iranian DEA society for their unwavering spiritual support and consensus in the writing of this book. Your invaluable support has been truly remarkable, and we are deeply grateful for the opportunity to collaborate with such esteemed professionals.

References

1. Guo, S., & Zhao, H. (2017). Fuzzy best-worst multi-criteria decision-making method and its applications. *Knowledge-Based Systems, 121*, 23–31.
2. Lotfi, F. H., Allahviranloo, T., Jondabeh, M. A., & Alizadeh, L. (2009). Solving a full fuzzy linear programming using lexicography method and fuzzy approximate solution. *Applied Mathematical Modelling, 33*(7), 3151–3156.
3. Lotfi, F. H., & Fallahnejad, R. (2010). Imprecise Shannon's entropy and multi attribute decision making. *Entropy, 12*(1), 53–62.
4. Rezaei, J. (2015). Best-worst multi-criteria decision-making method. *Omega, 53*, 49–57.
5. Sharafi, H., Soltanifar, M., & Lotfi, F. H. (2022). Selecting a green supplier utilizing the new fuzzy voting model and the fuzzy combinative distance-based assessment method. *EURO Journal on Decision Processes, 10*, 100010.
6. Tavana, M., Soltanifar, M., Santos-Arteaga, F. J., & Sharafi, H. (2023). Analytic hierarchy process and data envelopment analysis: A match made in heaven. *Expert Systems with Applications, 223*, 119902.

7. Wang, Y. M., Elhag, T. M., & Hua, Z. (2006). A modified fuzzy logarithmic least squares method for fuzzy analytic hierarchy process. *Fuzzy sets and systems, 157*(23), 3055–3071.
8. Wang, W., & Cui, M. (2007). Hybrid multiple attribute decision making model based on entropy. *Journal of Systems Engineering and Electronics, 18*(1), 72–75.
9. Pedrycz, W., & Song, M. (2011). Analytic hierarchy process (AHP) in group decision making and its optimization with an allocation of information granularity. *IEEE Transactions on Fuzzy Systems, 19*(3), 527-539.
10. Pedrycz, W., & Song, M. (2014). A granulation of linguistic information in AHP decision-making problems. *Information Fusion, 17*, 93-101.
11. Shirouyehzad, H., Lotfi, F. H., Arabzad, S. M., & Dabestani, R. (2013). An AHP/DEA ranking method based on service quality approach: a case study in hotel industry. *International Journal of Productivity and Quality Management, 11*(4), 434-445.

Chapter 4
Non-Compensatory Methods in Uncertainty Environment

4.1 Introduction: Non-Compensatory Fuzzy Methods

Multi-criteria decision-making is an approach for evaluating and selecting the best alternative among available choices based on diverse, different, and conflicting criteria. In many decision-making problems, one or more prioritized alternatives need to be chosen from the available choices. Various criteria intervene in this selection, making the decision-making task challenging. This chapter discusses non-compensatory fuzzy decision-making methods. In these methods, the decision-maker is not willing to trade-off between criteria. The weakness of one criterion cannot be compensated by the strength of another criterion or criteria. Each criterion is evaluated independently of other criteria for assessing competing options. In these methods, criteria are examined independently in the decision-making process. In most real-world problems, there are criteria that have qualitative or fuzzy values. In such cases, the development of multi-criteria decision-making methods with a fuzzy decision matrix is necessary [5, 6].

Non-compensatory models can generally be categorized into three groups. Firstly, when no information about the importance of criteria is available. Secondly, when the criteria have an ordinal importance order. And thirdly, when information regarding the thresholds, standards, and acceptability of each criterion is known. This chapter explores several non-compensatory fuzzy methods based on these categories. Specifically, it delves into the fuzzy dominance methods and fuzzy Max–Min methods from the first category, the fuzzy lexicographic method from the second category, and the fuzzy Conjunctive Satisfying method and fuzzy Disjunctive Satisfying method from the third category [12].

By examining these non-compensatory methods, we aim to provide insights into their applications, advantages, and limitations in MADM. The chapter will present algorithmic procedures and illustrative examples to demonstrate the practicality and effectiveness of these approaches. Understanding and utilizing non-compensatory

F. Hosseinzadeh Lotfi et al., *Fuzzy Decision Analysis: Multi Attribute Decision Making Approach*, Studies in Computational Intelligence 1121,
https://doi.org/10.1007/978-3-031-44742-6_4

Table 4.1 Decision matrix with fuzzy data

Alternative	C_1	C_2	...	C_n
A_1	$(a_{11}^l, a_{11}^m, a_{11}^u)$	$(a_{12}^l, a_{12}^m, a_{12}^u)$	...	$(a_{1n}^l, a_{1n}^m, a_{1n}^u)$
A_2	$(a_{21}^l, a_{21}^m, a_{21}^u)$	$(a_{22}^l, a_{22}^m, a_{22}^u)$	...	$(a_{2n}^l, a_{2n}^m, a_{2n}^u)$
$\vdots$	$\vdots$	$\vdots$	$\ddots$	$\vdots$
A_m	$(a_{m1}^l, a_{m1}^m, a_{m1}^u)$	$(a_{m2}^l, a_{m2}^m, a_{m2}^u)$	...	$(a_{mn}^l, a_{mn}^m, a_{mn}^u)$

methods can enhance decision-making processes, especially in situations where preserving the independence and significance of individual criteria is crucial.

In all methods of this chapter, suppose the fuzzy decision matrix is as follows [7–9]

In Table 4.1, A_i; $i = 1, ..., m$ denotes the alternatives and C_j ; $j = 1, ..., n$ are criteria (indicators), and $\left(a_{ij}^l, a_{ij}^m, a_{ij}^u\right)$ shows the value of ith criterion for j^{th} alternative. Note that some criteria are of the benefit type, while others are of the cost type. If there are qualitative criteria, it is assumed that you have transformed the qualitative alternatives into fuzzy ones. All the methods discussed in this chapter are performed with the assumption of triangular fuzzy numbers for the decision matrix (4.1). It is evident that they can be extended to other types of fuzzy numbers as well [9].

4.2 Fuzzy Lexicographic Method

The Fuzzy Lexicographic method relies on a lexical order or preference hierarchy among the criteria, where certain criteria are considered more important than others. This method allows decision-makers to prioritize specific criteria and focus on their relative importance in the decision-making process. By employing linguistic terms and preference orderings, the Fuzzy Lexicographic method provides a simple and intuitive approach to MADM problems, enabling decision-makers to effectively evaluate and rank alternatives based on their performance on multiple criteria.

Consider the decision matrix as Table 4.1. Suppose there is an absolute priority among the criterion as follows: $(j_1, j_2, ..., j_n)$. In other words, all the alternatives must first be compared based on the C_{j_1} criterion and the best alternative among them should be selected. If the best alternative is unique, then it is chosen as the final alternative. Otherwise, the best alternatives from the first stage are compared relative to the C_{j_2} criterion. These stages are performed in order of absolute priority among the criteria. The termination condition occurs when, in a stage, a unique best alternative is found relative to a criterion, or when we reach the last priority criterion. In this case, all alternatives from the last stage with equal criterion vectors are selected as the final alternative.

Note that in this method, there is no need for scaling or normalization of the decision matrix. The Lexicographic method does not necessarily utilize all the information in the decision matrix for the final decision. It does not require numerical values for the criteria, only the ranking of criteria in the alternatives is needed.

The selection of the best alternative with respect to a criterion depends on whether it represents a benefit or a cost. If the criterion is of the benefit type, then the maximum of that criterion should be calculated among the alternatives. If the indicator is of the cost type, then the minimum of that indicator should be calculated among the alternatives. In any case, we need a ranking function for fuzzy numbers in each column of the Table 4.1. Let's assume the ranking function for fuzzy numbers is denoted as $R : F \rightarrow \Re$, where F represents the set of all fuzzy numbers. Therefore, if $\tilde{a}$ *and* $\tilde{b}$ are two fuzzy numbers, then [2, 3]:

$$\begin{aligned} \tilde{a} \leq \tilde{b} &\Leftrightarrow R(\tilde{a}) \leq R\left(\tilde{b}\right) \\ \tilde{a} < \tilde{b} &\Leftrightarrow R(\tilde{a}) < R\left(\tilde{b}\right) \end{aligned} \tag{4.1}$$

Based on the above information, the fuzzy Lexicographic algorithm is described as follows:

Step 1: Construct the fuzzy decision matrix.

Step 2: Determine the absolute priority among the criteria. Let's assume the absolute priority is determined as follows: $(j_1, j_2, \ldots, j_n)$.

Step3: Put $k = 1$ and $D_{j_1} = \{A_1, \ldots, A_m\}$.

Step 4: Compare all the alternatives in set $D_{j_{k-1}}$ based on criterion C_{j_k}. If only one alternative is better than the others, proceed to step 6. Otherwise, put $D_{j_k} = \left\{A_j \in D_{j_{k-1}} \middle| A_j \text{ is the best in criterion } \mathrm{J_k} \text{ in comparision with members of } D_{j_k}\right\}$.

Step 5: Put $k : k + 1$, if $k \leq n$ go to step4.Step 6: Introduce all the remaining alternatives as the final alternative with the top rank.

It is evident that the selection of the final alternative depends not only on the criteria and their values but also on the ranking function for fuzzy numbers.

Example 4.1 Suppose selecting a contractor for building a bridge in a target area is the problem. Various factors can influence the selection of the contractor. In this example, the factors include reputation (C1), financial capability (C2), execution ability (C3), and proposed execution cost (C4), all of which are fuzzy criteria of triangular type.

The decision matrix for selecting a contractor from 6 contractors is as follows.

Please note that the three indicators of reputation, financial capability, and execution capability are of the same nature as the profit criteria, and the proposed execution cost criterion represents the cost.

Assuming the ranking function rule as $R_1(a, b, c) = \frac{a+4b+c}{6}$ and the absolute priority as (2, 1, 4, 3), applying the Lexicographic Algorithm yields the following results.

Applying function R_1 for elements of Table 4.2 result in Table 4.3.

Now, considering that the first priority is the second criterion, which is a profit-based indicator,

$$D_2 = \{A_2, A_4, A_5\}$$

Note that $Max\{4, 5, 3, 5, 5, 3\} = 5$.

Now, since there is a tie based on the first priority and D_2 has more than one member, we move on to the next step, which is the second priority of criteria. The second priority is the criterion (C1), and the following calculation should be performed.

$$Max\{7.83, 4.67, 5\} = 7.83$$

Therefore, $D_1 = \{A_2\}$.

Note that since only one alternative is selected at this stage, the algorithm concludes, and the selected alternative A_2 is the final choice.

In this example, considering the prioritization between indicators and the values of the indicators for the alternatives, the third and fourth criteria have not been utilized in the final selection. This is despite the possibility that the unselected alternatives may

Table 4.2 Financial capability decision matrix

Alternative	C_1	C_2	C_3	C_4
A_1	(4, 6, 7)	(2, 4, 6)	(7, 8, 9)	(5, 8, 9)
A_2	(5, 8, 10)	(3, 5, 7)	(3, 5, 7)	(3, 6, 10)
A_3	(6, 8, 9)	(1, 3, 5)	(2, 4, 5)	(7, 8, 9)
A_4	(2, 5, 6)	(3, 5, 7)	(3, 5, 6)	(2, 5, 8)
A_5	(3, 5, 7)	(4, 5, 6)	(6, 8, 10)	(3, 5, 7)
A_6	(1, 3, 5)	(2, 3, 4)	(4, 5, 6)	(2, 4, 5)

Table 4.3 Real values of Table 4.2 by applying the function R_1

Alternative	C_1	C_2	C_3	C_4
A_1	5.83	4	8	7.67
A_2	7.83	5	5	6.17
A_3	7.83	3	3.83	8
A_4	4.67	5	4.83	5
A_5	5	5	8	5
A_6	3	3	5	3.83

have excellent values for the third and fourth criteria. This is because Lexicographic method is a non-compensatory approach.

It is evident that changing the ranking function of fuzzy numbers may result in different outcomes.

Now, let's consider the ranking function R_2 applied to triangular fuzzy numbers. Assuming $\tilde{a} = (l, m, u)$ and $\tilde{b} = \left(l', m', u'\right)$, the following would be the case: [1]

$$R(\tilde{a}) \leq R\left(\tilde{b}\right) \Leftrightarrow \|\tilde{a} - M\| \leq \left\|\tilde{b} - M\right\| \tag{4.2}$$

where,

$$\left\|\tilde{a} - \tilde{b}\right\| = \sqrt{(l - l')^2 + (m - m')^2 + (u - u')^2}$$
$$M = (10, 10, 10)$$

Assuming the ranking function rule as R_2 and the absolute priority as $(2, 1, 4, 3)$, applying the Lexicographic Algorithm yields the following results for Table 4.2.

Considering that the first priority is the second criterion (C2), which is a profit-based criteria,

$$\|A_1 - M\| = \sqrt{(2 - 10)^2 + (4 - 10)^2 + (6 - 10)^2} = \sqrt{116} = 10.77$$

$$\|A_2 - M\| = \sqrt{(3 - 10)^2 + (5 - 10)^2 + (7 - 10)^2} = \sqrt{83} = 9.11$$

$$\|A_3 - M\| = \sqrt{(1 - 10)^2 + (3 - 10)^2 + (5 - 10)^2} = \sqrt{155} = 12.45$$

$$\|A_4 - M\| = \sqrt{(3 - 10)^2 + (5 - 10)^2 + (7 - 10)^2} = \sqrt{83} = 9.11$$

$$\|A_5 - M\| = \sqrt{(4 - 10)^2 + (5 - 10)^2 + (6 - 10)^2} = \sqrt{77} = 8.77$$

$$\|A_6 - M\| = \sqrt{(2 - 10)^2 + (3 - 10)^2 + (4 - 10)^2} = \sqrt{149} = 12.21$$

Hence, $D_2 = \{A_3\}$.

Since only one alternative is selected at this stage, the algorithm concludes, and the third alternative A_3 is the final choice.

4.3 Fuzzy Dominance Method

Fuzzy dominance as a non-compensatory method used in Multi-Attribute Decision Making (MADM) to determine the superiority of alternatives based on multiple criteria. Unlike compensatory methods that allow trade-offs between criteria, fuzzy dominance adopts a strict comparison approach without compromising on any criterion. Fuzzy dominance provides decision-makers with a straightforward and intuitive method to identify the most preferred alternative without considering any trade-offs, making it suitable for decision scenarios where strict adherence to criteria is desired. An alternative is considered dominant if it outranks or dominates other alternatives in all criteria.

Let's consider the ranking function accepted in this section to be as follows:

$$\tilde{a} \leq \tilde{b} \Leftrightarrow R(\tilde{a}) \leq R\left(\tilde{b}\right) \Leftrightarrow \frac{l + 4m + u}{6} \leq \frac{l' + 4m' + u'}{6}$$

Definition 4.1 Alternative A_i dominates alternative A_t if and only if

$$A_i \succ A_t \Leftrightarrow \begin{cases} \left(a_{ij}^l, a_{ij}^m, a_{ij}^u\right) \geq \left(a_{tj}^l, a_{tj}^m, a_{tj}^u\right) & \text{if } C_j \text{ is benefitial} \\ \left(a_{ij}^l, a_{ij}^m, a_{ij}^u\right) \leq \left(a_{tj}^l, a_{tj}^m, a_{tj}^u\right) & \text{if } C_j \text{ is cost} \end{cases}$$

and one of the above inequalities hold strictly.

According to the definition, it is possible to reduce the number of alternatives to non-dominated alternatives in a decision-making process. The decision-maker can then choose among those alternatives. This method essentially acts as a filtering process, and no scaling is required.

Example 4.2 Revisit decision matrix (4.2). Based on the ranking function, we can use Table 4.3.

Alternative A_2 dominates alternative A_3 because when comparing second and third rows, we have the following: the first three criteria are profit-based, and the fourth one is a cost-based criterion.

$$\begin{cases} Criterion\ C_1: & 7.83 \geq 7.83 \\ Criterion\ C_2: & 5 \geq 3 \\ Criterion\ C_3: & 5 > 3.83 \\ Criterion\ C_4: & 6.17 < 8 \end{cases}$$

Therefore, third alternative is eliminated.

Alternative A_5 dominates alternative A_4 because comparing fifth and fourth row of Table 4.3:

$$\begin{cases} Criterion\ C_1: & 5 \geq 4.67 \\ Criterion\ C_2: & 5 \geq 5 \\ Criterion\ C_3: & 8 > 4.83 \\ Criterion\ C_4: & 5 \leq 5 \end{cases}$$

Therefore, fourth alternative is eliminated.

By comparing the remaining alternatives, we can observe that there is no dominating alternative other than A_3 and A_4. Therefore, the final choice will be made among alternatives A_1, A_2, A_5, A_6.

It is evident that changing the ranking function may alter the set of independent dominating alternatives. As an exercise, you can apply the ranking function (4.2) and find the set of independent dominating alternatives.

4.4 Fuzzy Max–Min Method

This method, aims to identify the alternative that maximizes the minimum level of satisfaction across multiple criteria. The decision-maker assigns fuzzy membership values to each alternative based on their performance on each criterion. These membership values represent the degree of satisfaction or preference for each alternative. The Fuzzy Max–Min method calculates the minimum membership value for each alternative, indicating the lowest level of satisfaction across the criteria. The alternative with the maximum minimum membership value is considered the most favorable choice. This method allows decision-makers to emphasize the importance of meeting the minimum requirements for each criterion, ensuring that no alternative falls below an acceptable threshold. By considering both the best and worst-case scenarios, the Fuzzy Max–Min method provides a robust and conservative approach to decision-making in uncertain or ambiguous environments [4, 10].

In this method, the decision matrix is first scaled using the infinity norm as follows.

$$\tilde{b}_{ij} = \begin{cases} \dfrac{\tilde{a}_{ij}}{Max\{\tilde{a}_{ij}|i=1,\ldots,m\}} & \text{if } C_j \text{ is benefitial} \\ \dfrac{Min\{\tilde{a}_{ij}|i=1,\ldots,m\}}{\tilde{a}_{ij}} & \text{if } C_j \text{ is cost} \end{cases}$$

With the above scaling, the decision matrix $\left[\tilde{b}_{ij}\right]_{m\times n}$ is generated, where all indicators are considered as profit-based, meaning higher values of criteria are preferred. The fuzzy Max–Min algorithm is as follows:

Step 1: Scale the fuzzy decision matrix using the infinity norm.

Step 2: Calculate the minimum of each row. Let's assume that $\tilde{d}_i$ is the minimum of the ith row relative to all the criteria: $\tilde{d}_i^{min} = Min\left\{\tilde{b}_{ij}|j = 1, \ldots, n\right\}$

Table 4.4 Scaled matrix of Table 4.3 by infinity norm

Alternative	C_1	C_2	C_3	C_4
A_1	0.74	0.8	1	0.5
A_2	1	1	0.62	0.62
A_3	1	0.6	0.48	0.48
A_4	0.6	1	0.6	0.77
A_5	0.64	1	1	0.77
A_6	0.38	0.6	0.62	1

.Step 3: Obtain the alternative A_t from the following relationship.$\tilde{d}_t^{max} = Max\left\{\tilde{d}_i | i = 1, ..., m\right\}$

Note that it is possible to select more than one alternative in Step 3. This method maximizes the minimum profit and is used when the decision-maker is cautious or risk-averse.

Example 4.3 Consider the decision matrix (4.2) transformed into Table 4.3 using the ranking function R_1. First, we scale Table 4.3 using the infinity norm, resulting in the following matrix.

In this step, the minimum of each row of the Table 4.4 obtains as follows:

$$
\begin{aligned}
d_1^{min} &= Min\{0.74, 0.8, 1, 0.5\} = 0.5 \\
d_2^{min} &= Min\{1, 1, 0.62, 0.62\} = 0.62 \\
d_3^{min} &= Min\{1, 0.6, 0.48, 0.48\} = 0.48 \\
d_4^{min} &= Min\{0.6, 1, 0.6, 0.77\} = 0.6 \\
d_5^{min} &= Min\{0.64, 1, 1, 0.77\} = 0.64 \\
d_6^{min} &= Min\{0.38, 0.6, 0.62, 1\} = 0.38
\end{aligned}
$$

The maximum amount of $\tilde{d}_i^{max}$ is $Max\{0.5, 0.62, 0.48, 0.6, 0.64, 0.38\} = 0.64 = d_5^{max}$.

Therefore, alternative A_5 is the final selected alternative.

As an exercise, use the ranking function (4.2) to obtain the final alternative using the Max–Min fuzzy method.

4.5 Fuzzy Conjunctive Satisfying Method

In this method, standard levels are determined by the decision-maker for each criterion, and each alternative is evaluated against the acceptance standard. We either accept or reject an alternative based on its comparison with the specified standard.

This method is applicable when it is necessary for an alternative to have values of criteria at a specific standard level, such as assigning a job to an individual.

Now, let's assume that the standard level for j^{th} criterion is given as $\tilde{b}_j$. If the following condition holds for i^{th} alternative, then alternative A_i will be accepted. Otherwise, it will be rejected. If we assume that all criteria are profit-based, then the acceptance condition for alternative A_i is as follows.

$$\tilde{a}_{i1} \geq \tilde{b}_1 \ \& \ \tilde{a}_{i2} \geq \tilde{b}_2 \ \& \ldots \& \ \tilde{a}_{in} \geq \tilde{b}_n \tag{4.3}$$

Note that if criterion "j" represents cost, then the relation should be $\tilde{a}_{ij} \leq \tilde{b}_j$ instead of $\tilde{a}_{ij} \geq \tilde{b}_j$.

Also, note that if the conditions (4.3) are not satisfied for at least one relation, then alternative "i" will be rejected.

Using this method, the remaining alternatives that meet the acceptance can fall into one of the following scenarios:

(a) The number of accepted alternatives based on the standard is equal to one. In this case, the final alternative is determined.
(b) The number of accepted alternatives based on the standard is greater than one. In this case, the decision-maker can choose one from the final alternatives, or in order to have only one accepted alternative, it is necessary to improve the standard of some criteria. This can be done to the extent that only one alternative is accepted.
(c) The number of accepted alternatives based on the standard is equal to zero. In this case, if the decision-maker necessarily wants to choose one of the available alternatives, some standards must be worsened. This should be done to the extent that only one alternative is accepted.

Example 4.4 Consider the decision matrix (4.2). Suppose the standard levels for selecting alternatives are given by the decision-maker as follows.

$$\tilde{b}_1 = (3, 5, 6), \ \tilde{b}_2 = (2, 5, 6), \ \tilde{b}_3 = (5, 7, 10), \ \tilde{b}_4 = (3, 5, 8)$$

Considering ranking function as $R_1(a, b, c) = \frac{a+4b+c}{6}$.

In this case, based on this ranking function, the equivalent standard levels for the criteria would be as follows.

$$R_1\left(\tilde{b}_1\right) = 4.83, \ R_1\left(\tilde{b}_2\right) = 4.67, \ R_1\left(\tilde{b}_3\right) = 7.17, \ R_1\left(\tilde{b}_4\right) = 5.17$$

Considering this, by comparing each row of Table 4.3 with the values of the standard levels, we will have the following results:

Alternative A_1 is rejected because $4.67 \nleq 4$ (i.e., the second criteria does not have the minimum standard level for the second indicator. Note that this condition is also not met for the fourth criterion in this alternative.)

Alternative A_2 is rejected because $7.17 \nleq 5$ (i.e., the third criteria does not have the minimum standard level for the third indicator. Note that this condition is also not met for the fourth criterion in this alternative.)

Alternative A_3 is rejected because $4.67 \nleq 3$ (i.e., the second criteria does not have the minimum standard level for the second indicator. Note that this condition is also not met for the third and fourth criteria in this alternative.)

Alternative A_4 is rejected because $4.83 \nleq 4.67$ (i.e., the first criteria does not have the minimum standard level for the first indicator. Note that this condition is also not met for the third criterion in this alternative.)

Alternative A_5 is accepted since $5 \geq 4.83 \;\&\; 5 \geq 4.67 \;\&\; 8 \geq 7.17 \;\&\; 5 \leq 5.17$.

Alternative A_6 is rejected because $4.83 \nleq 3$ (i.e., the first criteria does not have the minimum standard level for the first indicator. Note that this condition is also not met for the second and third criteria in this alternative.)

Since out of the six alternatives, only A_5 met the standard conditions and was accepted, the final choice is alternative A_5 and the procedure concludes.

Now, let's consider the following ranking function:

$$\tilde{a} = (l, m, u),\ \tilde{b} = \left(l', m', u'\right)$$
$$\tilde{a} \leq \tilde{b} \;\; \Leftrightarrow \begin{cases} l \leq l' \\ m \leq m' \\ u \leq u' \end{cases} \tag{4.4}$$

Let's assume once again that the standard levels for each of the criteria are as follows.

$$\tilde{b}_1 = (3, 5, 6),\ \ \tilde{b}_2 = (2, 5, 6),\ \ \tilde{b}_3 = (5, 7, 10),\ \ \tilde{b}_4 = (3, 5, 8)$$

In this case, by comparing each alternative with the given standards based on the ranking function (4.4), we obtain the following results.

Alternative A_1 is rejected because

$$\begin{cases} (4, 5, 6) \geq (3, 5, 6) \\ (2, 5, 6) \nleq (2, 4, 6) \\ (5, 7, 10) \nleq (7, 8, 9) \\ (5, 8, 9) \leq (3, 5, 8) \end{cases}$$

Alternative A_2 is rejected since

$$\begin{cases} (5, 8, 10) \geq (3, 5, 6) \\ (3, 5, 7) \geq (2, 5, 6) \\ (5, 7, 10) \nleq (3, 5, 7) \\ (3, 6, 10) \nleq (3, 5, 8) \end{cases}$$

Alternative A_3 is rejected because

$$\begin{cases} (6, 8, 9) \geq (3, 5, 6) \\ (2, 5, 6) \nleq (1, 3, 5) \\ (5, 7, 10) \nleq (2, 4, 5) \\ (7, 8, 9) \nleq (3, 5, 8) \end{cases}$$

Alternative A_4 is rejected since

$$\begin{cases} (3, 5, 6) \nleq (2, 5, 6) \\ (3, 5, 7) \geq (2, 5, 6) \\ (5, 7, 10) \nleq (3, 5, 6) \\ (2, 5, 8) \leq (3, 5, 8) \end{cases}$$

Alternative A_5 is accepted because

$$\begin{cases} (3, 5, 7) \geq (3, 5, 6) \\ (4, 5, 6) \geq (2, 5, 6) \\ (6, 8, 10) \geq (5, 7, 10) \\ (3, 5, 7) \leq (3, 5, 8) \end{cases}$$

Alternative A_6 is rejected because

$$\begin{cases} (3, 5, 6) \nleq (1, 3, 5) \\ (2, 5, 6) \nleq (2, 3, 4) \\ (5, 7, 10) \nleq (4, 5, 6) \\ (2, 4, 5) \leq (3, 5, 8) \end{cases}$$

4.6 Fuzzy Disjunction Satisfying Method

Selecting an alternative in the Fuzzy Disjunction Satisfying method is based on its exceptional performance in at least one criterion. This means that if an alternative's score is either equal to a pre-determined standard or even better than the predefined standard for any criterion, then selecting that option is definite. Employing this method ensures the selection of an alternative with exceptional potential (in at least one criterion). Now, suppose a specific standard level is given for criterion "j" as $\tilde{b}_j$. If the following condition holds for alternative "i", then A_i will be accepted. Otherwise, it will be rejected. If we assume that all criteria are of a positive nature, then the acceptance conditions for alternative "i" are as follows.

$$\tilde{a}_{i1} \geq \tilde{b}_1 \ \vee \ \tilde{a}_{i2} \geq \tilde{b}_2 \ \vee \ldots \ \vee \tilde{a}_{in} \geq \tilde{b}_n \tag{4.5}$$

Note that if criterion "j" represents cost, then the relation should be $\tilde{a}_{ij} \leq \tilde{b}_j$ instead of $\tilde{a}_{ij} \geq \tilde{b}_j$.

Also, note that if the conditions (4.5) are not satisfied for at least one relation, then alternative "i" will be rejected.

Using this method, the remaining alternatives that can meet the acceptance fall into one of the following scenarios:

(a) The number of accepted alternatives based on the standard is equal to one. In this case, the final alternative is determined.
(b) The number of accepted alternatives based on the standard is greater than one. In this case, the decision-maker can choose one from the final alternatives, or in order to have only one accepted alternative, it is necessary to improve the standard of some criteria. This can be done to the extent that only one alternative is accepted.
(c) The number of accepted alternatives based on the standard is equal to zero. In this case, if the decision-maker necessarily wants to choose one of the available alternatives, some standards must be worsened. This should be done to the extent that only one alternative satisfies condition (4.5).

Example 4.5 Consider the decision matrix (4.2). Suppose the standard levels for selecting alternatives are given by the decision-maker as follows.

$$\tilde{b}_1 = (6, 8, 10)\ , \tilde{b}_2 = (5, 7, 9)\ ,\ \tilde{b}_3 = (6, 8, 10)\ , \tilde{b}_4 = (3, 4, 5)$$

Considering ranking function as $R_1(a, b, c)\ = \frac{a+4b+c}{6}$.

In this case, based on this ranking function, the equivalent standard levels for the criteria would be as follows.

$$R_1(\tilde{b}_1) = 8,\ \ R_1(\tilde{b}_2) = 7,\ \ R_1(\tilde{b}_3) = 8,\ \ R_1(\tilde{b}_4) = 4$$

Considering this, by comparing each row of Table 4.3 with the values of the standard levels, we will have the following results:

Alternative A_1 is accepted because relation $8 \geq 8$ holds for the third criterion in this alternative.

Alternative A_2 is rejected because $8 \nleq 7.83$ & $7 \nleq 5$ & $8 \nleq 5$ & $6.17 \nleq 4$.

Alternative A_3 is rejected because $8 \nleq 7.83$ & $7 \nleq 3$ & $8 \nleq 3.83$ & $8 \nleq 4$.

Alternative A_4 is rejected because $8 \nleq 4.67$ & $7 \nleq 5$ & $8 \nleq 4.83$ & $5 \nleq 4$.

Alternative A_5 is accepted because relation $8 \geq 8$ holds for the third criterion in this alternative.

Alternative A_6 is accepted because relation $3.83 \leq 8$ holds for the fourth criterion in this alternative.

Based on the results, alternatives A_1, A_5, A_6 have been accepted. Since criteria 3 and 4 have influenced the acceptance of these alternatives, the decision maker modifies the standard for criterion 3 as follows to select only one alternative.

$$\tilde{b}_3 = (7, 9, 10), \quad R_1(\tilde{b}_3) = 8.83$$

With this change, first and fifth alternatives A_1 and A_5 will be rejected. Therefore, only alternative A_6 will be the final choice.

Now, let's assume we adopt the ranking function (4.4), and suppose the standard levels are as follows:

$$\tilde{b}_1 = (6, 8, 10), \ \tilde{b}_2 = (5, 7, 9), \ \tilde{b}_3 = (6, 8, 10), \ \tilde{b}_4 = (3, 4, 5)$$

Using the ranking function (4.4), the following results are obtained.

Alternative A_1 is rejected because

$$\begin{cases} (6, 8, 10) \npreceq (4, 6, 7) \\ (5, 7, 9) \npreceq (2, 4, 6) \\ (6, 8, 10) \npreceq (7, 8, 9) \\ (5, 8, 9) \npreceq (3, 4, 5) \end{cases}$$

Alternative A_2 is rejected because

$$\begin{cases} (6, 8, 10) \npreceq (5, 8, 10) \\ (5, 7, 9) \npreceq (3, 5, 7) \\ (6, 8, 10) \npreceq (3, 5, 7) \\ (3, 6, 10) \npreceq (3, 4, 5) \end{cases}$$

Alternative A_3 is rejected because

$$\begin{cases} (6, 8, 10) \npreceq (6, 8, 9) \\ (5, 7, 9) \npreceq (1, 3, 5) \\ (6, 8, 10) \npreceq (2, 4, 5) \\ (7, 8, 9) \npreceq (3, 4, 5) \end{cases}$$

Alternative A_4 is rejected because

$$\begin{cases} (6, 8, 10) \npreceq (2, 5, 6) \\ (5, 7, 9) \npreceq (3, 5, 7) \\ (6, 8, 10) \npreceq (3, 5, 6) \\ (2, 5, 8) \npreceq (3, 4, 5) \end{cases}$$

Alternative A_5 is accepted since the following relation holds for third criterion.

$$\tilde{a}_{53} = (6, 8, 10) \geq (6, 8, 10) = \tilde{b}_3$$

Alternative A_6 is accepted since the following relation holds for the fourth criterion.

$$\tilde{a}_{64} = (2, 4, 5) \leq (3, 4, 5) = \tilde{b}_4$$

Based on the standard levels and the ranking function (4.4), alternatives A_5 and A_6 have been accepted. At this stage, in order for the decision-maker to select only one alternative, they need to improve at least one of the standard levels for criterion 3 and criterion 4. Let's assume that the standard level for criterion 4 is modified to $\tilde{b}_4 = (2, 3, 4)$. In this case, alternative A_6 will be rejected because

$$\begin{cases} (6, 8, 10) \nleq (1, 3, 5) \\ (5, 7, 9) \nleq (2, 3, 4) \\ (6, 8, 10) \nleq (4, 5, 6) \\ (2, 4, 5) \nleq (2, 3, 4) \end{cases}$$

As a result, only alternative A_5 is accepted, and it will be the final choice.

4.7 Conclusion

Decision-making in conditions of uncertainty has always been a concern for organizational managers, especially when they need to choose an alternative among multiple choices with multiple criteria. Decision-making becomes more challenging when the criteria are in conflict with each other. In this chapter, non-compensatory approaches in multi-criteria decision-making with fuzzy values are discussed. The assumption was made that all criterion values are of triangular fuzzy type. It is evident that all methods can be extended to different types of fuzzy data, such as trapezoidal, Gaussian, etc. In all methods, efforts have been made to solve examples with different fuzzy number ranking functions to demonstrate their dependence on the ranking function. Most non-compensatory methods, with fuzzy data values for criteria, were discussed. Non-compensatory methods are employed when weaknesses in one or some criteria cannot be compensated by other criteria for an alternative. These methods can be used to select one option among the alternatives when it has superiority in a criterion. It is evident that these methods cannot be used for complete ranking of the alternatives.

Acknowledgement A special thanks to the Iranian DEA society for their unwavering spiritual support and consensus in the writing of this book. Your invaluable support has been truly remarkable, and we are deeply grateful for the opportunity to collaborate with such esteemed professionals.

References

1. Allahviranloo, T., Abbasbandy, S., & Saneifard, R. (2011). An approximation approach for ranking fuzzy numbers based on weighted interval-value. *Mathematical and Computational Applications, 16*(3), 588–597.
2. Allahviranloo, T., Lotfi, F. H., Kiasary, M. K., Kiani, N. A., & Alizadeh, L. (2008). Solving fully fuzzy linear programming problem by the ranking function. *Applied mathematical sciences*, *2*(1), 19-32.
3. Armand, A., Allahviranloo, T., & Gouyandeh, Z. (2018). Some fundamental results on fuzzy calculus. *Iranian Journal of Fuzzy Systems, 15*(3), 27–46.
4. Chiu, C. Y., & Park, C. S. (1994). Fuzzy cash flow analysis using present worth criterion. *The Engineering Economist, 39*(2), 113–138.
5. Chiu, C. Y., & Park, C. S. (1998). Capital budgeting decisions with fuzzy projects. *The Engineering Economist, 43*(2), 125–150.
6. Di Nola, A., Sessa, S., Pedrycz, W., & Sanchez, E. (2013). *Fuzzy relation equations and their applications to knowledge engineering* (Vol. 3). Springer Science & Business Media.
7. Ezzati, R., Allahviranloo, T., Khezerloo, S., & Khezerloo, M. (2012). An approach for ranking of fuzzy numbers. *Expert Systems with Applications, 39*(1), 690–695.
8. Guh, Y. Y., Hon, C. C., Wang, K. M., & Lee, E. S. (1996). Fuzzy weighted average: A max-min paired elimination method. *Computers & Mathematics with Applications, 32*(8), 115–123.
9. Junior, F. R. L., Osiro, L., & Carpinetti, L. C. R. (2013). A fuzzy inference and categorization approach for supplier selection using compensatory and non-compensatory decision rules. *Applied Soft Computing, 13*(10), 4133–4147.
10. Munda, G. (2012). Intensity of preference and related uncertainty in non-compensatory aggregation rules. *Theory and Decision, 73*(4), 649–669.
11. Razavi Toosi, S. L., & Samani, J. M. V. (2014). A new integrated MADM technique combined with ANP, FTOPSIS and fuzzy max-min set method for evaluating water transfer projects. *Water Resources Management, 28*, 4257–4272.
12. Roychowdhury, S., & Pedrycz, W. (2001). A survey of defuzzification strategies. *International Journal of intelligent systems, 16*(6), 679-695.

Chapter 5
Simple Additive Weighting (SAW) Method in Fuzzy Environment

5.1 Introduction

The SAW (Simple Additive Weighting) method is a widely used and simple approach in multiple attribute decision making (MADM) methods. Its straightforward and intuitive structure has contributed to its popularity among users. This method is based on calculating the weighted average for each criterion in the decision matrix, allowing for ranking of the alternatives. Over the years, the SAW method has also been applied in the fuzzy environment, demonstrating its versatility and adaptability. We will explore some aspects and applications of the SAW method in both deterministic and non-deterministic settings [1, 4, 13].

The SAW method, originally introduced by Harsanyi in 1955 [9], is recognized as a well-established technique for comparative evaluation in MCDA. This method leverages the weights and scores assigned to each alternative to determine the best choice. Further research by Einhorn and McCutch in 1977 examined the properties of the SAW method, including its conditional monotonicity with utility and the decision-maker's risk neutrality.

The following subsections are about the application of the SAW method in different contexts. Firstly, we discuss its application in a deterministic state, specifically focusing on the selection of a hospital construction site. This practical example illustrates the SAW method's effectiveness in making informed decisions in a straightforward setting. Moving forward, we explore the utilization of the SAW method in non-deterministic environments, where uncertainties and risks are present. Real-life scenarios and case studies are presented in each subsection to provide practical insights into the proposed method [16, 20].

By examining the SAW method's various applications and its adaptability in different decision-making scenarios, we aim to provide a comprehensive understanding of this widely used MADM technique. Throughout this chapter, we will explore its strengths, limitations, and potential extensions, empowering readers to apply the SAW method effectively in their own decision-making processes.

F. Hosseinzadeh Lotfi et al., *Fuzzy Decision Analysis: Multi Attribute Decision Making Approach*, Studies in Computational Intelligence 1121,
https://doi.org/10.1007/978-3-031-44742-6_5

5.2 SAW Method

Below are the steps of using the SAW method in deterministic mode. Consider a situation where, based on a multi-attribute decision making, it is necessary to choose the best alternative among "m" alternatives based on "n" attribute (criteria).

Step 1: Constructing the decision matrix.

According to the number of criteria, the number of alternatives and the evaluation of all alternatives for different criteria, the decision matrix is formed as follows. In this method, it is assumed that all indicators are of the benefit type. If an indicator is of the cost type, it is handled by simply inverting it to represent a benefit, thus ensuring uniform treatment of all indicators.

$$X = \begin{array}{c} \quad C_1 \ldots C_n \\ \begin{bmatrix} x_{11} & \ldots & x_{1n} \\ \vdots & \ddots & \vdots \\ x_{m1} & \ldots & x_{mn} \end{bmatrix} \end{array} \tag{5.1}$$

where x_{ij} is the score of the alternative $i(i = 1, ..., m)$ related to the criterion $j(j = 1, ..., n)$.

Step 2: Constructing the Normalize decision matrix.

At this step, based on the existing methods, the decision matrix is normalized. Naturally, all cost criterion are converted into benefit criterion and they will be treated the same.

$$N = \begin{bmatrix} n_{11} & \ldots & n_{1n} \\ \vdots & \ddots & \vdots \\ n_{m1} & \ldots & n_{mn} \end{bmatrix} \tag{5.2}$$

Step 3: Determining the criteria's weights.

At this step, according to the importance of different criteria in decision-making, the weight vector of the criteria is defined as $w_j(j = 1, ..., n)$.

Step 4: Rank order all alternatives A_i according to descending score of A_i.

At this stage, the best alternative is obtained from the following relationship.

$$A^* = \max_{i=1,...,m} \{score(A_i)\} \tag{5.3}$$

where:

$$score(A_i) = \sum_{j=1}^{n} w_j n_{ij}, \quad i = 1, ..., m. \tag{5.4}$$

5.3 Choosing a Hospital Location

In the context of hospital location, various factors must be considered to ensure the effective functioning of healthcare facilities. The placement of hospitals within the city is subject to specific principles, rules, and mechanisms. While economic mechanisms and free market forces often dictate the establishment of for-profit urban elements, public and non-profit entities such as hospitals require decisions and policies based on the public interest to address market inefficiencies. The location and design of a hospital play a crucial role in its effectiveness, both during normal conditions and in critical situations such as earthquakes, fires, or conflicts. Urban designers play a significant role in addressing challenges related to hospital location and must adhere to fundamental principles that serve these centers. Factors such as connectivity to transportation arteries, including highways and public transportation services like metro, bus, and taxi, as well as proximity to fire stations and crisis management centers, are essential requirements for an ideal hospital location. Each of these elements is vital, and the absence of any of them can disrupt the efficient operation of the hospital. Moreover, the surrounding neighborhood greatly influences the hospital's location. Proximity to educational, commercial, and public centers can impact patient well-being and the treatment process. Similarly, if a hospital is situated in a residential area, congestion and construction activities can hinder air and ground relief efforts. Recent developments and the growing urban landscape have emphasized the significance of air transportation for patient transfers to medical centers. Therefore, the proximity of medical facilities to nearby buildings is crucial for providing timely aid and support. In light of these considerations, specific urban characteristics significantly influence the process of positioning any urban point, including hospitals. Each location comes with its unique conditions and features that influence the overall settlement pattern and outcomes.

Based on the aforementioned factors, it is essential to carefully select suitable land for hospital construction. Certain land uses should be avoided in close proximity to the hospital, and if necessary, a minimum distance requirement should be enforced. The following uses should be kept at a minimum distance:

I. Industrial centers (minimum distance of 244 m)—C1-benefit
II. Military barracks (except for military medical centers) (minimum distance of 1444 m)—C2-benefit
III. Airports (minimum distance of 2444 m from runways and flight paths)—C3-benefit
IV. Train, bus, and truck terminals (at least 544 m away)—C4-benfit
V. Police and law enforcement stations (minimum distance of 244 m)—C5-benfit

VI. Telecommunication, radio, and television stations and masts (minimum distance of 344 m)—C6-benefit
VII. Schools and non-related educational spaces (at least 144 m away)—C7-benefit
VIII. Stadiums (minimum distance of 244 m)—C8-benefit

Considering these criteria, the alternatives for hospital selection should be evaluated based on the minimum distance specified for each criterion. The following places, located within the city limits, are considered as potential alternatives for selection.

Step 1: Construct decision matrix (Table 5.1).

Step 2: Construct the Normalize decision matrix.

The normalized decision matrix is expressed in Table 5.2.

Step 3: Determining the criteria's weights.

According to the opinion of experts, the following weights for each of the criteria are as follows (Table 5.3):

Step 4: Rank order all alternatives A_i according to descending $score(A_i)$.

Using Eq. (5.4), the weighted sum of each alternative is as follows (Table 5.4).

So, A1 is the best place to build a hospital.

Table 5.1 Decision matrix of choosing the location of the hospital

	C1	C2	C3	C4	C5	C6	C7	C8
A1	3805	4574	8741	835	1752	412	598	3590
A2	1750	1845	7612	1710	1115	562	1423	4976
A3	854	2354	3589	4572	823	738	1578	4529
A4	902	1687	2750	5412	1985	657	809	7821
A5	1804	1927	5960	3959	2101	683	761	5361
A6	565	1836	8900	2754	714	492	843	9820
A7	1235	1580	3980	1495	389	841	480	4781

Table 5.2 The Normalized Decision Matrix of choosing the location of the hospital

	C1	C2	C3	C4	C5	C6	C7	C8
A1	0.3486	0.2894	0.2105	0.0403	0.1973	0.0940	0.0948	0.0878
A2	0.1603	0.1167	0.1833	0.0825	0.1256	0.1282	0.2256	0.1217
A3	0.0782	0.1490	0.0864	0.2205	0.0927	0.1683	0.2502	0.1108
A4	0.0826	0.1068	0.0662	0.2610	0.2236	0.1498	0.1282	0.1913
A5	0.1653	0.1219	0.1435	0.1909	0.2366	0.1558	0.1206	0.1311
A6	0.0518	0.1162	0.2143	0.1328	0.0804	0.1122	0.1336	0.2402
A7	0.1131	0.1000	0.0958	0.0721	0.0438	0.1918	0.0469	0.1170

Table 5.3 Criteria Weight of choosing the location of the hospital

Criteria	C1	C2	C3	C4	C5	C6	C7	C8
Weights	0.1667	0.1296	0.1481	0.1111	0.0741	0.1296	0.1111	0.1296

Table 5.4 Result of SAW Method for choosing the location of the hospital

	A1	A2	A3	A4	A5	A6	A7
Score (Ai)	0.1787	0.1418	0.1370	0.1397	0.1523	0.1348	0.1158
Rank	1	3	5	4	2	6	7

5.4 SAW Method in Imprecise Environments

Talking about Multiple Criteria Decision Analysis (MCDA), the Simple Additive Weighting (SAW) method has gained significant popularity due to its simplicity and ease of implementation. This method, widely utilized among users of MCDA, offers a straightforward approach to decision-making by employing a weighted average calculation on each column of a decision matrix. By assigning appropriate weights to different criteria, the SAW method enables the ranking of alternatives based on their aggregated scores. While the SAW method has been extensively applied in various decision-making scenarios, its traditional formulation assumes precise and deterministic data inputs. In recent years, researchers have extended the application of the SAW method to imprecise environments, including interval and fuzzy settings. In the imprecise environment, interval-based SAW allows decision-makers to handle interval-valued data, where the exact values of criteria or alternatives are unknown but can be bounded within certain intervals. This approach accommodates the inherent uncertainty in decision-making and provides a more realistic representation of the decision context. Furthermore, fuzzy SAW incorporates the concept of fuzzy sets, which allows for the representation of subjective and vague information. By employing fuzzy logic, decision makers can express the degree of membership or non-membership of alternatives to different criteria, capturing the inherent ambiguity in decision making processes [6, 8, 10, 11, 12].

In this section of the book, we discuss about the application of the SAW method in imprecise environments, exploring its extension to interval and fuzzy decision-making scenarios. We explain the underlying principles, methodologies, and practical examples of employing SAW in these contexts. By embracing imprecision and uncertainty, the SAW method becomes a versatile tool that empowers decision-makers to tackle complex decision problems with greater accuracy and flexibility.

5.5 Interval SAW

The mechanism of using interval data is important because, in many cases, the method of dealing with fuzzy data is based on α-cut. On the other hand, we cannot ignore the fact that in many cases, part of the data is interval. Two popular approaches of SAW-Interval are presented below. With the explanation that when part of the data are crisp, they can be considered as interval data with equal upper and lower bounds [2, 13, 22].

5.5.1 *The First Approach of Interval SAW*

This method is based on the calculation of each step while maintaining the interval characteristic of the data. In the last step, interval numbers are compared.

Step 1: Formation of interval decision matrix

$$X = \begin{array}{c} \begin{array}{ccc} C_1 & \cdots & C_n \end{array} \\ \begin{bmatrix} [x_{11}^L, x_{11}^U] & \cdots & [x_{1n}^L, x_{1n}^U] \\ \vdots & \ddots & \vdots \\ [x_{11}^L, x_{11}^U] & \cdots & [x_{mn}^L, x_{mn}^U] \end{bmatrix} \end{array} \tag{5.5}$$

where $[x_{ij}^L, x_{ij}^U]$, $(x_{ij}^L > 0)$ the score of the ith alternative $(i = 1, ..., m)$ is in relation to the jth criterion $(j = 1, ..., n)$.

Step 2: Determine the weight of each criterion.

In this step, the criterion weight is obtained either by using the opinion of experts or based on the decision matrix. These methods are introduced in the third chapter. With the explanation that if the weight of the criteria was definite, consider it as an interval with upper and lower limits equal value. Therefore, the weight of each criterion will be in the form of an interval, and without reducing the generality of the problem, it can be assumed that $0 \le w_j^L \le w_j^U \le 1$, $(j = 1, ..., n)$ and its display is as follows:

$$W = \{[w_1^L, w_1^U], \ldots, [w_n^L, w_n^U]\} \tag{5.6}$$

Step 3: Construct the normalized decision matrix.

Different methods can be extended and used to normalize decision matrix. The following relationship is for the set of Benefit and Cost criteria, which are represented by letters B and C, respectively.

$$\begin{cases} n_{ij}^{L} = \dfrac{x_{ij}^{L}}{\frac{1}{2}\sum_{i=1}^{m}(x_{ij}^{L}+x_{ij}^{U})}, & i=1,\ldots,m;\ j\in B \\ n_{ij}^{U} = \dfrac{x_{ij}^{U}}{\frac{1}{2}\sum_{i=1}^{m}(x_{ij}^{L}+x_{ij}^{U})}, & i=1,\ldots,m;\ j\in B \end{cases}$$
$$\begin{cases} n_{ij}^{L} = \dfrac{\frac{1}{x_{ij}^{U}}}{\frac{1}{2}\sum_{i=1}^{m}(\frac{1}{x_{ij}^{L}}+\frac{1}{x_{ij}^{U}})}, & i=1,\ldots,m;\ j\in C \\ n_{ij}^{U} = \dfrac{\frac{1}{x_{ij}^{L}}}{\frac{1}{2}\sum_{i=1}^{m}(\frac{1}{x_{ij}^{L}}+\frac{1}{x_{ij}^{U}})}, & i=1,\ldots,m;\ j\in C \end{cases} \tag{5.7}$$

If we consider the normalized matrix as N, it will be as follows. After applying Eq. (5.7), the terms of the normal decision matrix will be between zero and one.

$$N^{L,U} = \begin{bmatrix} [\mathrm{n}_{11}^{L}, \mathrm{n}_{11}^{U}] & \cdots & [\mathrm{n}_{1n}^{L}, \mathrm{n}_{1n}^{U}] \\ \vdots & \ddots & \vdots \\ [\mathrm{n}_{11}^{L}, \mathrm{n}_{11}^{U}] & \cdots & [\mathrm{n}_{mn}^{L}, \mathrm{n}_{mn}^{U}] \end{bmatrix} \tag{5.8}$$

Step 4: Rank order all alternatives $\mathrm{A_i}$ according to descending $score_{Interval}(A_i)$.

Using the normalized matrix shown in Eq. (5.8) and the weights obtained in the second step, multiply the weight of each criterion by its corresponding column in the normalized matrix. The weighted sum of each alternative is calculated in the following form:

$$score_{Interval}(A_i) = [p_i^{L}, p_i^{U}] = \left[\frac{\sum_{j=1}^{n} w_j^{L} n_{ij}^{L}}{\frac{1}{2}\sum_{j=1}^{n}(w_j^{L}+w_j^{U})}, \frac{\sum_{j=1}^{n} w_j^{U} n_{ij}^{U}}{\frac{1}{2}\sum_{j=1}^{n}(w_j^{L}+w_j^{U})}\right] \tag{5.9}$$

It is obvious that if there is an alternative whose lower limit of the weighted sum is greater than the upper limit of the weighted sum of the remaining alternatives, it will be the superior alternative. Otherwise, it can be done using the conventional methods of ranking intervals such as maximum regret [21].

5.5.2 The Second Approach of SAW-Interval Method

This method is based on score functions for interval data. In other words, in this method, each interval is converted into a number, and then the best weights for each alternative are selected from the weight interval of each criterion. Therefore, the steps of this method up to step 4 are the same as the first SAW-interval approach, and briefly the steps are as follows.

Step 1: Same as step 1 of the first approach in SAW-interval.

Step 2: Same as step 3 of the first approach in SAW-interval.

Step 3: Calculate the weighted sum.

In this step, we first define some scoring functions for the weighted sum in interval mode.

Consider the $w_j(j = 1, \ldots, n)$ as the weight of the jth criterion. Some examples of the weighted sum or (suitability function) for each alternative based on the matrix shown in relation (5.8) are as follows.

$$
\begin{aligned}
&(i)\gamma_1(A_i) = \sum_{j=1}^{n} w_j.(n_{ij}^L + n_{ij}^U - 1) \\
&(ii)\gamma_2(A_i) = \frac{1}{2}\sum_{j=1}^{n} w_j.(3n_{ij}^L - 1) \\
&(iii)\gamma_3(A_i) = \sum_{j=1}^{n} w_j.(\lambda n_{ij}^L + (1-\lambda)n_{ij}^U);\ \lambda \in [0, 1]
\end{aligned}
\tag{5.10}
$$

As can be seen in the above scoring functions, the weight of each criterion has been included in the calculations in a crisp manner. For this purpose and considering the interval of the weights, the following linear programming model is solved for each of the score functions.

Selecting objective functions γ_1 and γ_2 is equivalent to finding the most favorable weights for the criteria. In Eq.(511), if $n_{ij}^L + n_{ij}^U < 1$, then the model tries to assign the corresponding weight to this criterion at its lowest level. Since the coefficient of this weight in the objective function will be negative, on the other hand, if $n_{ij}^L + n_{ij}^U > 1$, the model attempts to place the corresponding weight for this criterion at its highest level.

It is worth mentioning that since after normalization, all values are within the range of [0, 1] and have real values, the occurrence probability of $n_{ij}^L + n_{ij}^U = 1$ in real data is practically zero.

$$
\begin{aligned}
Max \quad & \sum_{i=1}^{m}\sum_{j=1}^{n} w_j.(n_{ij}^L + n_{ij}^U - 1) \\
s.t. \quad & w_j^L \le w_j \le w_j^U,\ j = 1, \ldots, n, \\
& \sum_{j=1}^{n} w_j = 1.
\end{aligned}
\tag{5.11}
$$

Step 4: Rank order all alternatives A_i according to descending $\gamma(A_i)$.

After solving model (5.11), the best alternative is obtained from the following relationship and (w_1^*, ..., w_n^*) is the optimal solution model (5.11).

$$A^* = \underset{i=1,\dots,m}{Max}\{\gamma(A_i)\}; \quad \gamma(A_i) = \sum_{j=1}^{n} w_j^*.(n_{ij}^L + n_{ij}^U - 1) \tag{5.12}$$

5.5.3 Application of Interval SAW Method

The selection of machines in production and industrial systems plays a crucial role in determining the efficiency and quality of operations. A wrong choice can result in a myriad of problems, negatively impacting productivity, accuracy, and product quality. One such critical decision involves choosing the most suitable CNC lathe for a specific application.

In this scenario, we are faced with the task of selecting the optimal CNC lathe from a range of alternatives. To make an informed decision, we consider multiple criteria that capture the essential features and performance indicators of the lathe. The criteria to be evaluated are as follows:

Cost (including installation expenses in the company)—measured in thousands of dollars (C1-Cost).
Transverse axis movement of the lathe—measured in millimeters (C2-Benefit).
Longitudinal axis movement of the lathe—measured in millimeters (C3-Benefit).
Turning diameter—measured in millimeters (C4-Benefit).
Turning length (piece length in the production line)—measured in millimeters (C5-Benefit).
Failure percentage—measured in thousands (C6-Cost).

To systematically analyze the available alternatives, we employ the Simple Additive Weighting (SAW) method in an interval environment. By utilizing interval-based data, we can account for the uncertainty and imprecision associated with the criteria measurements. This approach enables us to handle data with known lower and upper bounds, providing a more realistic representation of the decision-making context.

In the following sections, we delve into the application of the SAW method in the interval environment, discussing the underlying principles, step-by-step methodology, and practical examples. By incorporating interval-based analysis into the decision-making process, we aim to enhance the accuracy and reliability of machine selection, ultimately leading to improved productivity and quality in production and industrial systems.

Step 1: Formation of interval decision matrix (Table 5.5).

Table 5.5 Decision matrix of choosing the machines

	C1	C2	C3	C4	C5	C6
A1	[4.8,5.2]	[0,500]	[0,800]	[0,200]	[0,400]	[0.1,0.2]
A2	[5.2,5.5]	[0,800]	[0,1143]	[0,220]	[0,420]	[0.1,0.2]
A3	[5.1,5.5]	[0,170]	[0,400]	[0,420]	[0,520]	[0.15,0.3]
A4	[3.1,3.9]	[0,290]	[0,1040]	[0,528]	[0,1125]	[0.15,0.3]
A5	[4.9,5.5]	[0,250]	[0,850]	[0,500]	[0,890]	[0.1,0.35]
A6	[5.5,6]	[0,505]	[0,2800]	[0,1000]	[0,2650]	[0.1,0.35]

Table 5.6 Criteria Weight of choosing the machines

Criteria	C1	C2	C3	C4	C5	C6
Weights	0.1844	0.2152	0.1139	0.1121	0.1248	0.2495

Step 2: Determine the weight of each criterion.

According to the opinion of experts, the following weights for each of the criteria are as follows (Table 5.6):

Step 3: Construct the normalize decision matrix.

Using Eq. (5.7), the normalization decision matrix will be as follows, noting that the first and sixth criteria are cost and the second, third, fourth and fifth criteria's are benefit (Table 5.7).

Step 4: Rank all alternatives A_i according to descending $score_{Interval}(A_i)$.

In this step, the weight of the criteria's is multiplied by the points, and the final score of each alternative is calculated as intervals using Eq. (5.9).

In Table 5.8, the weighted sum of each alternative is displayed as intervals, the sixth alternative has the first rank.

5.5.4 *Fuzzy SAW*

In this section, we introduce some approach for fuzzy data with the explanation that one of the most common approaches in facing fuzzy environments is the use of α-cutting. In using α-cut, the data is converted into intervals and can be used using the method mentioned above. Also, another common method is de-fuzzy and converting all values in a crisp type. It is obvious that the concept of order is also needed to compare fuzzy numbers, and therefore, in the final stages of all approaches, there must be a mechanism for sorting. For that reason, below are the approaches that are based on maintaining the characteristics of fuzzy values ([3, 19, 22].

Table 5.7 The normalize decision matrix of choosing the machines

	C1		C2		C3		C4		C5		C6	
	x_{i1}^L	x_{i1}^U	x_{i2}^L	x_{i2}^U	x_{i3}^L	x_{i3}^U	x_{i4}^L	x_{i4}^U	x_{i5}^L	x_{i5}^U	x_{i6}^L	x_{i6}^U
A1	0.1600	0.3234	0.0000	0.3976	0.0000	0.2275	0.0000	0.1395	0.0000	0.1332	0.1111	0.2143
A2	0.1620	0.3238	0.0000	0.7939	0.0000	0.3667	0.0000	0.1649	0.0000	0.1499	0.1111	0.2143
A3	0.1604	0.3239	0.0000	0.2796	0.0000	0.1572	0.0000	0.3431	0.0000	0.2006	0.1111	0.2432
A4	0.1476	0.3212	0.0000	0.5542	0.0000	0.4434	0.0000	0.5205	0.0000	0.4823	0.1111	0.2432
A5	0.1571	0.3241	0.0000	0.6605	0.0000	0.4656	0.0000	0.6661	0.0000	0.5027	0.0741	0.2751
A6	0.1594	0.3247	0.0000	1.9894	0.0000	1.9988	0.0000	1.9963	0.0000	1.9989	0.0741	0.2751

Table 5.8 Result of SAW Method for choosing the machines

	A1		A2		A3		A4		A5		A6	
	p_1^L	p_1^U	p_2^L	p_2^U	p_3^L	p_3^U	p_4^L	p_4^U	p_5^L	p_5^U	p_6^L	p_6^U
Score (Ai)	0.0572	0.2568	0.0576	0.3630	0.0573	0.2620	0.0549	0.4082	0.0474	0.4610	0.0479	1.2576
R(Ai) by wang's method	0.1996		0.3053		0.2046		0.3532		0.4135		1.2096	
Rank	6		4		5		3		2		1	

In this subsection, two approaches are discussed. The first approach is for the cases when the weight of the criteria are predetermined and the second approach is when the weight of the criteria are unknown [7, 17, 23].

5.5.5 *Fuzzy SAW Method with Predetermined Weights*

Step 1: Constructing the decision fuzzy matrix.

The decision matrix is obtained based on the rating of each index in each criterion. Sometimes the type of this rating is fuzzy, and sometimes it is the result of the transformation of linguistic variables. In any case, the decision matrix is considered as follows. It should be noted that using this method, if there are crisp values, we will convert them into a fuzzy form.

Also, we consider $z = (a, b, c)$ as $w = (a, b, b, c)$, in other words, all the values will be transformed in trapezoidal fuzzy number.

According to the above explanation, the decision matrix will be as follows:

$$X = \begin{matrix} C_1 \dots C_n \\ \begin{bmatrix} \tilde{x}_{11} & \dots & \tilde{x}_{1n} \\ \vdots & \ddots & \vdots \\ \tilde{x}_{m1} & \dots & \tilde{x}_{mn} \end{bmatrix} \end{matrix} \tag{5.13}$$

where $\tilde{x}_{ij} = (x_{ij}^L, x_{ij}^{M_1}, x_{ij}^{M_2}, x_{ij}^U), x_{ij}^L > 0$.

Step 2: Determining the weight of each criterion.

The weight of each criterion is calculated using the methods introduced in the third chapter. Therefore, we consider the weight of each criterion as:

$$\tilde{W} = [\tilde{w}_1, \dots, \tilde{w}_n]; \quad \tilde{w}_j = (w_j^L, w_j^{M_1}, w_j^{M_2}, w_j^U) \tag{5.14}$$

Step 3: Normalizing the fuzzy decision matrix.

The normal fuzzy decision matrix for profit and cost criteria is as follows.

$$\tilde{z}_{ij} = \left(z_{ij}^L, z_{ij}^{M_1}, z_{ij}^{M_2}, z_{ij}^U\right) = \left(\frac{x_{ij}^L}{x_i^{Max}}, \frac{x_{ij}^{M_1}}{x_i^{Max}}, \frac{x_{ij}^{M2}}{x_i^{Max}}, \frac{x_{ij}^U}{x_i^{Max}}\right), \quad i = 1, \dots, m; \quad j \in B$$

$$\tilde{z}_{ij} = \left(z_{ij}^L, z_{ij}^{M_1}, z_{ij}^{M_2}, z_{ij}^U\right) = \left(\frac{x_i^{Min}}{x_{ij}^U}, \frac{x_i^{Min}}{x_{ij}^{M_2}}, \frac{x_i^{Min}}{x_{ij}^{M_1}}, \frac{x_i^{Min}}{x_{ij}^L}\right), \quad i = 1, \dots, m; \quad j \in C \tag{5.15}$$

in which $x_i^{Max} = \underset{j}{Max}\{x_{ij}^U\}, (i = 1, ..., m)$ and $x_i^{Min} = \underset{j}{Min}\{x_{ij}^L\}, (i = 1, ..., m)$. Eq. (5.15), the general representation of the fuzzy normalized matrix $\tilde{Z}$ is as follows:

$$\tilde{Z} = \begin{bmatrix} \tilde{Z}_{11} & \dots & \tilde{Z}_{1n} \\ \vdots & \ddots & \vdots \\ \tilde{Z}_{m1} & \dots & \tilde{Z}_{mn} \end{bmatrix} \tag{5.16}$$

Step 4: Aggregation of weighted fuzzy scores.

In this step, using the matrix obtained in step 3 and the weights of step 2, the weighted decision matrix is as follows:

$$FWM = \tilde{W} \otimes \tilde{Z} = \begin{bmatrix} \tilde{z}_{11} & \dots & \tilde{z}_{1n} \\ \vdots & \ddots & \vdots \\ \tilde{z}_{m1} & \dots & \tilde{z}_{mn} \end{bmatrix} \times \begin{bmatrix} \tilde{w}_1 \\ \vdots \\ \tilde{w}_n \end{bmatrix} = \begin{bmatrix} \tilde{z}_{11} \otimes \tilde{w}_1 & \oplus \dots \oplus & \tilde{z}_{1n} \otimes \tilde{w}_n \\ \vdots & \ddots & \vdots \\ \tilde{z}_{m1} \otimes \tilde{w}_{m1} & \oplus \dots \oplus & \tilde{z}_{mn} \otimes \tilde{w}_n \end{bmatrix}$$
$$\cong \begin{bmatrix} \sum_{j=1}^n \left(z_{1j}^L w_j^L, z_{1j}^{M_1} w_j^{M_1}, z_{1j}^{M_2} w_j^{M_2}, z_{1j}^U w_j^U \right) \\ \vdots \\ \sum_{j=1}^n \left(z_{mj}^L w_j^L, z_{mj}^{M_1} w_j^{M_1}, z_{mj}^{M_2} w_j^{M_2}, z_{mj}^U w_j^U \right) \end{bmatrix} \tag{5.17}$$

Aggregation of weighted fuzzy scores of each alternative is obtained as follows:

$$AWFS(\widetilde{A}_i) = \sum_{j=1}^{n} \tilde{z}_{ij} \tilde{w}_j; \quad i = 1, ..., m. \tag{5.18}$$

Step 5: Ranking the alternatives.

In this step, any of the fuzzy number ranking methods can be used to sort the alternatives.

5.5.6 *Fuzzy SAW Method with Unknown Weights*

At first, there is a review of two linear programming problems that are used in the steps of the method. The fuzzy normalized matrix is known based on the Eq. (5.16). Suppose w_j the weight of the criterion is j, based on the structure of the SAW method, the score of the i-th alternative is as follows:

$$S(\tilde{A}_i) \approx \sum_{j=1}^{n} w_j \tilde{z}_{ij}; \quad i = 1, \ldots, m. \tag{5.19}$$

In other words, it can be written:

$$\begin{aligned} S(\tilde{A}_i) &\approx \left[S(A_i^L), S(A_i^{M_1}), S(A_i^{M_2}), S(A_i^U) \right] \\ &= \left[\sum_{j=1}^{n} w_j z_{ij}^L, \sum_{j=1}^{n} w_j z_{ij}^{M_1}, \sum_{j=1}^{n} w_j z_{ij}^{M_2}, \sum_{j=1}^{n} w_j z_{ij}^U \right]; i = 1, \ldots, m. \end{aligned} \tag{5.20}$$

$S(\tilde{A}_i)$ is a fuzzy number, and without loss of generality it can be assumed that $S(\tilde{A}_i) \leq \tilde{1}$ where $\tilde{1}$ is a fuzzy trapezoidal number in the form of (0.95, 1, 1, 1.05). The goal is to choose the weight in the optimistic and pessimistic state for each alternative. For each of the steps, a problem is solved with a fuzzy objective function [2, 14].

$$\begin{aligned} &\mathrm{S}_{\text{optimistic}}(\tilde{A}_i) = \mathrm{Max} \left[\sum_{j=1}^{n} w_j z_{pj}^L, \sum_{j=1}^{n} w_j z_{pj}^{M_1}, \sum_{j=1}^{n} w_j z_{pj}^{M_2}, \sum_{j=1}^{n} w_j z_{pj}^U \right] \\ &s.t. \left[\sum_{j=1}^{n} w_j z_{ij}^L, \sum_{j=1}^{n} w_j z_{ij}^{M_1}, \sum_{j=1}^{n} w_j z_{ij}^{M_2}, \sum_{j=1}^{n} w_j z_{ij}^U \right] \\ &\leq (0.95, 1, 1, 1.05); i = 1, \ldots, m; \\ &w_j > 0 \quad (j = 1, \ldots, n). \end{aligned} \tag{5.21}$$

Model (5.21) is used to score each alternative in optimistic mode. The structure of the model (5.21) is based on the optimistic policy. Considering that if the weight of each criterion becomes zero, that criterion will be ineffective in the final score, so $w_j > 0$ it has been added to the problem. That instead of $w_j > 0$ the constraint $w_j \geq \varepsilon$ is used, where ε is a small non-Archimedean number. Model (5.21) will be transformed as follows.

$$\begin{aligned} &\mathrm{S}_{\text{optimistic}}(\tilde{A}_i) = \mathrm{Max} \left[\sum_{j=1}^{n} w_j z_{pj}^L, \sum_{j=1}^{n} w_j z_{pj}^{M_1}, \sum_{j=1}^{n} w_j z_{pj}^{M_2}, \sum_{j=1}^{n} w_j z_{pj}^U \right] \\ &s.t. \left[\sum_{j=1}^{n} w_j z_{ij}^L, \sum_{j=1}^{n} w_j z_{ij}^{M_1}, \sum_{j=1}^{n} w_j z_{ij}^{M_2}, \sum_{j=1}^{n} w_j z_{ij}^U \right] \leq (0.95, 1, 1, 1.05); i = 1, \ldots, m; \\ &w_j \geq \varepsilon, \; j = 1, \ldots, n. \end{aligned} \tag{5.22}$$

Also, according to the above model, the following model can be used to score each alternative in pessimistic mode.

$$
\begin{aligned}
& S_{Pessimistic}(\tilde{A}_i) = \text{Min } [\sum_{j=1}^{n} w_j z_{pj}^{L}, \sum_{j=1}^{n} w_j z_{pj}^{M_1}, \sum_{j=1}^{n} w_j z_{pj}^{M_2}, \sum_{j=1}^{n} w_j z_{pj}^{U}] \\
& s.t. \left[\sum_{j=1}^{n} w_j z_{ij}^{L}, \sum_{j=1}^{n} w_j z_{ij}^{M_1}, \sum_{j=1}^{n} w_j z_{ij}^{M_2}, \sum_{j=1}^{n} w_j z_{ij}^{U} \right] \geq (0.95, 1, 1, 1.05); i = 1, \dots, m; \\
& w_j \geq \varepsilon, \ (j = 1, \dots, n).
\end{aligned} \tag{5.23}
$$

Models (5.22) and (5.23) can be seen as multi-objective linear programming and to solve them, the method of converting multi-objective to single-objective linear programming can be used. One of these methods is the weighted sum method. The weighted sum used below is based on a fuzzy ranking function, so the following model is solved instead of model (5.22).

$$
\begin{aligned}
& S_{optimistic}(\tilde{A}_i) = \text{Max } \frac{1}{6}(\sum_{j=1}^{n} w_j z_{pj}^{L} + 2.\sum_{j=1}^{n} w_j z_{pj}^{M_1} + 2.\sum_{j=1}^{n} w_j z_{pj}^{M_2} + \sum_{j=1}^{n} w_j z_{pj}^{U}) \\
& s.t. \sum_{j=1}^{n} w_j z_{ij}^{L} \leq 0.95, \quad \text{i} = 1,\dots,\text{m}; \\
& \sum_{j=1}^{n} w_j z_{ij}^{M_1} \leq 1, \quad \text{i} = 1,\dots,\text{m}; \\
& \sum_{j=1}^{n} w_j z_{ij}^{M_2} \leq 1, \quad \text{i} = 1,\dots,\text{m}; \\
& \sum_{j=1}^{n} w_j z_{ij}^{U} \leq 1.05, \quad \text{i} = 1,\dots,\text{m}; \\
& w_j \geq \varepsilon, \ (j = 1, \dots, n).
\end{aligned} \tag{5.24}
$$

And instead of solving model (5.23), the following model is solved.

$$\mathrm{S_{Pessimistic}}(\tilde{\mathrm{A}}_i) = \mathrm{Min}\frac{1}{6}\left(\sum_{j=1}^{n} w_j z_{pj}^{L} + 2.\sum_{j=1}^{n} w_j z_{pj}^{M_1} + 2.\sum_{j=1}^{n} w_j z_{pj}^{M_2} + \sum_{j=1}^{n} w_j z_{pj}^{U}\right)$$

$$s.t.\quad \sum_{j=1}^{n} w_j z_{ij}^{L} \geq 0.95, \quad \mathrm{i} = 1,\ldots,\mathrm{m};$$

$$\sum_{j=1}^{n} w_j z_{ij}^{M_1} \geq 1, \quad \mathrm{i} = 1,\ldots,\mathrm{m};$$

$$\sum_{j=1}^{n} w_j z_{ij}^{M_2} \geq 1, \quad \mathrm{i} = 1,\ldots,\mathrm{m};$$

$$\sum_{j=1}^{n} w_j z_{ij}^{U} \geq 1.05, \quad \mathrm{i} = 1,\ldots,\mathrm{m};$$

$$w_j \geq \varepsilon\,, \ (j = 1, \ldots, n). \tag{5.25}$$

It is worth mentioning that other methods such as ideal planning method can also be used to solve multi-objective problems. According to the issues raised above, the steps for using fuzzy SAW with unknown weights are as follows.

Step 1: Constructing the decision fuzzy matrix (Eq. (5.13)).

Step2: Normalize fuzzy decision matrix (Eq. (5.16)).

Step 3: Determining the weight of each criterion.

At this stage, the weight of each criterion for each alternative is calculated separately using models (5.24) and (5.25) and the values of $\tilde{\mathrm{S}}_{\mathrm{optimistic}}(\mathrm{A}_i)$ and $\tilde{\mathrm{S}}_{\mathrm{Pessimistic}}(\mathrm{A}_i)$ are calculated.

Step 4: Calculate the weighted sum and rank order all alternatives $\mathrm{A_i}$ according to descending $FS(A_i)$.

$$FS(\tilde{A}_i) = \frac{1}{2}(\mathrm{S_{Pessimistic}}(\tilde{\mathrm{A}}_i) + \mathrm{S_{optimistic}}(\tilde{\mathrm{A}}_i)), \quad i = 1, \ldots, m. \tag{5.26}$$

In relation to the above steps, it should be added that in the group decision-making mode and in order to aggregate the decision-making matrices, the geometric or arithmetic mean of scores can be used.

5.5.7 Fuzzy SAW Application

Airlines operate in a highly competitive market where attracting and retaining customers is a constant challenge. Initially, it was believed that price alone could serve as the sole competitive factor. However, airlines soon realized that long-term competitive advantage required more than just competitive pricing. Factors such as the regulations imposed by countries on ticket prices often lead to compromises in quality and even flight safety. Consequently, the ranking of airlines within a country is based on service quality indicators. To evaluate and rank the domestic flight services provided by Iran's airlines, a comprehensive service quality model was employed. This involved identifying and weighting service quality indicators based on expert opinions and the most comprehensive model available. The current state of services provided by each airline was then evaluated based on these indicators, ultimately resulting in a ranking of the companies. This assessment aimed to shed light on the aspects of service quality that are deemed most important by flight passengers, highlighting areas that require greater attention [10].

To gather the necessary information, a survey was conducted among airline customers, soliciting their opinions from five distinct perspectives. Each perspective represents a benefit attribute for the customers, denoted as follows:

I. C1—Tangibility: This attribute assesses the interior design, cleanliness, seat comfort, and food supply provided by the airline.
II. C2—Reliability: This attribute focuses on aspects such as flight punctuality, accuracy, and baggage delivery without delays.
III. C3—Responsiveness: The responsiveness of the airline crew, as well as their politeness and humility, are considered under this attribute.
IV. C4—Guarantee: This attribute encompasses the compensation procedures, flight safety, and the technical competence of the pilot and crew.
V. C5—Empathy: The ease of the ticketing process, handling of customer complaints, and assistance provided to customers with special needs are evaluated within this attribute.

In this section of the book, we explore the application of the Simple Additive Weighting (SAW) method in a fuzzy environment to analyze the customer survey data and assess the service quality rankings of Iran's airlines. By incorporating fuzziness into the decision-making process, we can capture the inherent uncertainty and subjective nature of customers' perceptions and preferences. The SAW method allows us to effectively aggregate and weigh the fuzzy evaluation of each service quality indicator, providing valuable insights for airline companies to enhance their competitive edge and improve overall customer satisfaction.

Table 5.9 Linguistic rating of alternatives

Linguistic terms	Fuzzy number
Very Poor (VP)	(0,0,2.5)
Poor (P)	(0,2.5,5)
Fair (F)	(2.5,5,7.5)
Good (G)	(5,7.5,10)
Very Good (VG)	(7.5,10,10)

Considering that the results of customer surveys need to be aggregated. The arithmetic mean of grades is used for evaluation. Table 5.9 is used to convert qualitative comments into quantitative data.

In this section, the approaches presented for unknown weights to select the best alternative is applied.

Step 1: Constructing the fuzzy decision matrix.

The following Table 5.10 shows the scores of 7 airline companies based on the average aggregated opinions of customers.

Step2: Normalize fuzzy decision matrix.

The normalized decision matrix is calculated based on Eq. (5.15).

Step 3: Determining the weight of each criterion.

At this stage, models (5.24) and (5.25) are solved for each alternative and the values of $S_{\text{optimistic}}(\tilde{A}_i)$ and $S_{\text{Pessimistic}}(\tilde{A}_i)$ are displayed in Table 5.11.

Step 4: Calculate the weighted sum and rank alternatives according to descending order of $FS(\tilde{A}_i)$ (Table 5.12).

According to Table 5.13, A1 airline is the best alternative. And the order of airlines is as follows: $A_1 \succ A_4 \succ A_7 \succ A_3 \succ A_5 \succ A_6 \succ A_2$.

Table 5.10 Decision matrix of choosing the airlines

	C1	C2	C3	C4	C5
A1	(3.6957,5.9053,5.9053,8.3156)	(3.6325,6.0040,6.0040,8.2065)	(1.7818,4.0454,4.0454,6.1575)	(4.5475,6.6740,6.6740,9.4150)	(2.8590,5.1031,5.1031,7.8916)
A2	(2.0340,4.9648,4.9648,7.6968)	(2.2490,4.9747,4.9747,7.2138)	(1.7924,4.5254,4.5254,7.5052)	(2.6573,4.8624,4.8624,7.0337)	(1.9445,4.2775,4.2775,7.1902)
A3	(2.1938,4.9880,4.9880,7.4039)	(2.3118,5.1156,5.1156,7.6133)	(4.3994,7.1309,7.1309,9.5517)	(2.9536,4.9637,4.9637,7.2809)	(2.5583,5.1923,5.1923,7.4391)
A4	(4.4969,7.0814,7.0814,9.5033)	(2.1826,4.8303,4.8303,7.8272)	(1.0870,3.9019,3.9019,6.5206)	(2.4201,5.4188,5.4188,8.3281)	(4.4914,6.5473,6.5473,9.1015)
A5	(1.7080,4.2340,4.2340,6.4367)	(3.6602,6.4285,6.4285,9.1896)	(2.9969,5.6576,5.6576,7.6701)	(1.9123,4.0303,4.0303,6.1695)	(3.4319,5.5136,5.5136,8.0166)
A6	(1.9115,4.3203,4.3203,7.0787)	(4.0056,6.0432,6.0432,8.6325)	(4.1481,6.8465,6.8465,9.1949)	(1.0657,3.7254,3.7254,6.4287)	(1.3779,3.7377,3.7377,6.5595)
A7	(3.1988,5.3007,5.3007,8.1245)	(4.4974,7.0275,7.0275,9.4715)	(2.8012,5.0862,5.0862,7.4558)	(1.9040,4.2061,4.2061,7.1168)	(2.5656,4.6228,4.6228,7.3151)

Table 5.11 Normal fuzzy decision matrix of choosing the airlines

	C1	C2	C3	C4	C5
A1	(0.3809,0.6086,0.6086,0.8570)	(0.3743,0.6187,0.6187,0.8457)	(0.1836,0.4169,0.4169,0.6346)	(0.4686,0.6878,0.6878,0.9703)	(0.2946,0.5259,0.5259,0.8133)
A2	(0.2096,0.5116,0.5116,0.7932)	(0.2318,0.5127,0.5127,0.7434)	(0.1847,0.4664,0.4664,0.7734)	(0.2738,0.5011,0.5011,0.7248)	(0.2004,0.4408,0.4408,0.7410)
A3	(0.2261,0.5140,0.5140,0.7630)	(0.2382,0.5272,0.5272,0.7846)	(0.4534,0.7349,0.7349,0.9843)	(0.3044,0.5115,0.5115,0.7503)	(0.2636,0.5351,0.5351,0.7666)
A4	(0.4634,0.7298,0.7298,0.9794)	(0.2249,0.4978,0.4978,0.8066)	(0.1120,0.4021,0.4021,0.6720)	(0.2494,0.5584,0.5584,0.8582)	(0.4629,0.6747,0.6747,0.9379)
A5	(0.1760,0.4363,0.4363,0.6633)	(0.3772,0.6625,0.6625,0.9470)	(0.3088,0.5830,0.5830,0.7904)	(0.1971,0.4153,0.4153,0.6358)	(0.3537,0.5682,0.5682,0.8261)
A6	(0.1970,0.4452,0.4452,0.7295)	(0.4128,0.6228,0.6228,0.8896)	(0.4275,0.7056,0.7056,0.9476)	(0.1098,0.3839,0.3839,0.6625)	(0.1420,0.3852,0.3852,0.6760)
A7	(0.3297,0.5463,0.5463,0.8373)	(0.4635,0.7242,0.7242,0.9761)	(0.2887,0.5242,0.5242,0.7683)	(0.1962,0.4335,0.4335,0.7334)	(0.2644,0.4764,0.4764,0.7539)

Table 5.12 Sum weighted values of optimistic and pessimistic

	$\tilde{S}_{\text{Pessimistic}}(A_i)$	$\tilde{S}_{\text{optimistic}}(A_i)$
A1	2.2247	0.7612
A2	1.8948	0.6308
A3	1.9617	0.7782
A4	2.0481	0.7794
A5	1.9245	0.7425
A6	1.8910	0.7680
A7	1.9785	0.7774

Table 5.13 Weighted sum of the alternatives

	$FS(A_i)$	Rank
A1	1.4930	1
A2	1.2628	7
A3	1.3700	4
A4	1.4138	2
A5	1.3335	5
A6	1.3295	6
A7	1.3780	3

5.6 Conclusion

In conclusion, the Simple Additive Weighting (SAW) method has emerged as a favored choice among decision makers, thanks to its intuitive and understandable structure. Its popularity can be attributed to the alignment between the decision making structure of individuals and the underlying principles of the SAW method. This alignment has been substantiated by numerous research studies and real world applications, further reinforcing the method's reputation. Throughout this chapter, we have explored various approaches for incorporating the SAW method in imprecise environments, specifically focusing on interval and fuzzy data. By adapting the SAW method to accommodate imprecision, decision makers gain valuable insights and guidance even in situations where data is not precise or easily quantifiable. The examples presented in this chapter have demonstrated the versatility and effectiveness of the SAW method in handling imprecise data. Decision makers can leverage these approaches to make informed choices and rank the alternatives based on multiple criteria, even when faced with uncertain or fuzzy information.

It is important to note that the SAW method extends beyond interval and fuzzy data, offering potential for further expansion into other forms of imprecision and uncertainty. As research and advancements in decision making methodologies continue, we can expect the SAW method to evolve and adapt to address new challenges and complexities that decision makers encounter. In summary, the SAW method stands as a powerful tool in decision making, particularly in imprecise environments.

Its simplicity, compatibility with human decision-making structures, and ability to handle various forms of imprecision make it a valuable asset for decision makers across diverse industries and domains. As decision making processes become increasingly complex and data uncertainty persists, the SAW method provides a reliable framework for evaluating alternatives and making informed decisions.

Acknowledgement A special thanks to the Iranian DEA society for their unwavering spiritual support and consensus in the writing of this book. Your invaluable support has been truly remarkable, and we are deeply grateful for the opportunity to collaborate with such esteemed professionals.

References

1. Afshari, A. R., Yusuff, R., & Derayatifar, A. R. (2012, May). Project manager selection by using Fuzzy Simple Additive Weighting method. In *2012 International Conference on Innovation Management and Technology Research* (pp. 412–416). IEEE.
2. Allahviranloo, T., Lotfi, F. H., Kiasary, M. K., Kiani, N. A., & Alizadeh, L. (2008). Solving fully fuzzy linear programming problem by the ranking function. *Applied mathematical sciences, 2*(1), 19–32.
3. Büyüközkan, G., & Güler, M. (2020). Smart watch evaluation with integrated hesitant fuzzy linguistic SAW-ARAS technique. *Measurement, 153*, 107353.
4. Büyüközkan, G., Karabulut, Y., & Güler, M. (2018). Strategic renewable energy source selection for turkey with hesitant fuzzy MCDM method. In: *Energy management collective and computational intelligence with theory and applications*, 229–250.
5. Chen, T. Y. (2012). Comparative analysis of SAW and TOPSIS based on interval-valued fuzzy sets: Discussions on score functions and weight constraints. *Expert Systems with Applications, 39*(2), 1848–1861.
6. Chen, L. C., & Utama, D. N. (2022). Decision support model for determining the best employee using fuzzy logic and simple additive weighting. *Journal of Computer Science, 18*(6), 530–539.
7. Chou, S. Y., Chang, Y. H., & Shen, C. Y. (2008). A fuzzy simple additive weighting system under group decision-making for facility location selection with objective/subjective attributes. *European Journal of Operational Research, 189*(1), 132–145.
8. Gandhi, N. S., Thanki, S. J., & Thakkar, J. J. (2018). Ranking of drivers for integrated lean-green manufacturing for Indian manufacturing SMEs. *Journal of Cleaner Production, 171*, 675–689.
9. Harsanyi, J. C. (1955). Cardinal welfare, individualistic ethics, and interpersonal comparisons of utility. *Journal of Political Economy, 63*(4), 309–321.
10. Hwang, C. L., Yoon, K., Hwang, C. L., & Yoon, K. (1981). Methods for multiple attribute decision making. Multiple attribute decision making: methods and applications a state-of-the-art survey, 58–191.
11. Jahanshahloo, G. R., Lotfi, F. H., & Izadikhah, M. (2006). An algorithmic method to extend TOPSIS for decision-making problems with interval data. *Applied Mathematics and Computation, 175*(2), 1375–1384.
12. Jahanshahloo, G. R., Lotfi, F. H., Balf, F. R., & Rezai, H. Z. (2008). Using Monte Carlo method for ranking interval data. *Applied mathematics and computation, 201*(1–2), 613–620.
13. Kanuganti, S., Agarwala, R., Dutta, B., Bhanegaonkar, P. N., Singh, A. P., & Sarkar, A. K. (2017). Road safety analysis using multi-criteria approach: A case study in India. *Transportation Research Procedia, 25*, 4649–4661.
14. Lotfi, F. H., Allahviranloo, T., Jondabeh, M. A., & Alizadeh, L. (2009). Solving a full fuzzy linear programming using lexicography method and fuzzy approximate solution. *Applied mathematical modelling, 33*(7), 3151–3156.

15. Mohamadinejadrashti, H., Amirteimoori, A., Kordrostami, S., & Hosseinzadeh Lotfi, F. (2021). Resource allocation and target setting based on DEA with managerial disposability: Evaluation and optimization the greenhouse gas emissions reduction in international airlines. *Journal of Decisions and Operations Research, 6*(3), 365–381.
16. Orojloo, M., Shahdany, S. M. H., & Roozbahani, A. (2018). Developing an integrated risk management framework for agricultural water conveyance and distribution systems within fuzzy decision making approaches. *Science of the Total Environment, 627*, 1363–1376.
17. Roszkowska, E., & Kacprzak, D. (2016). The fuzzy SAW and fuzzy TOPSIS procedures based on ordered fuzzy numbers. *Information Sciences, 369*, 564–584.
18. Salehi, A., & Izadikhah, M. (2014). A novel method to extend SAW for decision-making problems with interval data. *Decision Science Letters, 3*(2), 225–236.
19. Seyedmohammadi, J., Jafarzadeh, A. A., Sarmadian, F., Shahbazi, F., & Ghorbani, M. A. (2017). Application of SAW, TOPSIS and fuzzy TOPSIS models in cultivation priority planning for maize, oilseed rape, and soybean crops. *Geoderma.*
20. Wang, Y. J. (2015). A fuzzy multi-criteria decision-making model based on simple additive weighting method and relative preference relation. *Applied Soft Computing, 30*, 412–420.
21. Wang, Y. M., Greatbanks, R., & Yang, J. B. (2005). Interval efficiency assessment using data envelopment analysis. *Fuzzy Sets and Systems, 153*(3), 347–370.
22. Zamri, N., & Abdullah, L. (2015). A linear assignment method of simple additive weighting system in linear programming approach under interval type-2 fuzzy set concepts for MCDM problem. In: *Advanced Computer and Communication Engineering Technology: Proceedings of the 1st International Conference on Communication and Computer Engineering* (pp. 833–842). Springer International Publishing.
23. Zolfani, S. H., Sedaghat, M., & Zavadskas, E. K. (2012). Performance evaluating of rural ICT centers (telecenters), applying fuzzy AHP, SAW-G and TOPSIS Grey, a case study in Iran. *Technological and Economic Development of Economy, 18*(2), 364–387.
24. Żak, J., Gołda, P., Cur, K., & Zawisza, T. (2021). Assessment of airside aerodrome infrastructure by SAW method with weights from Shannon's interval entropy. *Archives of Transport* 60(4).

Chapter 6
Technique for Order Preferences by Similarity to Ideal Solutions (TOPSIS) in Uncertainty Environment

6.1 Introduction: The Essence of the TOPSIS Method and Its Application

Throughout history, humans have constantly strived for excellence and sought the ideal solutions to various challenges they encounter. Whether it is in personal achievements or professional endeavors, individuals have a natural inclination to pursue the best possible outcomes for themselves. This innate drive has led to the development of numerous decision-making methods that help individuals evaluate and select the optimal alternatives based on predefined criteria. One such prominent method used for addressing multi-attribute decision-making problems is the Technique for Order Preferences by Similarity to Ideal Solutions (TOPSIS). Introduced by Hwang and Yoon in 1981 [9], TOPSIS is a well-known approach that provides a systematic framework for decision analysis. It is based on the fundamental concept that the ideal alternative should exhibit the shortest distance to the Positive Ideal Solution (PIS) and the greatest distance from the Negative Ideal Solution (NIS). The TOPSIS method facilitates the identification of the optimal alternative by introducing two pivotal points: the positive ideal point and the negative ideal point. The positive ideal point represents the values that minimize the cost criteria and maximize the benefit criteria, while the negative ideal point represents the values that maximize the cost criteria and minimize the benefit criteria. By evaluating the proximity of each alternative to these ideal points, TOPSIS determines the optimal alternative that is closest to the positive ideal solution and farthest from the negative ideal solution. The versatility and effectiveness of the TOPSIS method have led to its widespread application in various domains. Industries such as supply chain management, engineering, manufacturing systems, business and marketing, tourism, healthcare, environmental management, and human resources have all benefited from the application of TOPSIS. Researchers and practitioners alike have explored its potential in solving complex decision-making problems across these diverse fields. To illustrate the broad scope of its application, several studies have focused on specific areas where TOPSIS

F. Hosseinzadeh Lotfi et al., *Fuzzy Decision Analysis: Multi Attribute Decision Making Approach*, Studies in Computational Intelligence 1121,
https://doi.org/10.1007/978-3-031-44742-6_6

has been successfully employed. For instance, Afshar et al. [2] utilized TOPSIS in the context of water resources, specifically examining the Karun reservoirs system. Azadeh et al. [3] applied a combination of genetic algorithms and TOPSIS to address challenges in cellular manufacturing systems. Chu et al. [6] employed the fuzzy TOPSIS method to support robot selection, while Vahdani et al. [26] proposed a fuzzy modified TOPSIS for group decision-making problems. Furthermore, Huang et al. [8] explored the application of TOPSIS in the tourism industry across various Asian countries. Behzadian et al. [4] conducted a comprehensive survey highlighting the diverse range of applications of TOPSIS. Jahanshahloo et al. [10], Lai et al. [15], Olson [19] and Roszkowska [22] also studied about TOPSIS method.

In this chapter, we delve into the details of the TOPSIS method, exploring its underlying principles, steps, and mathematical formulation. We will also examine its real-world applications and the impact it has had across different domains. By gaining a deeper understanding of this influential decision-making technique, readers will be equipped with a valuable tool to tackle complex decision problems effectively and make informed choices based on predefined criteria.

6.2 Description of the TOPSIS Method

To rank "m" alternatives $\{A_1, ..., A_m\}$ respecting "n" criteria $\{C_1, ..., C_n\}$, the TOPSIS method is summarized as following:

Step 1: Creating the Decision Matrix.

$$X = \begin{array}{c} \\ A_1 \\ A_2 \\ \vdots \\ A_m \end{array} \begin{array}{c} C_1 \; C_2 \; ... \; C_n \\ \begin{bmatrix} x_{11} & x_{12} & \cdots & x_{1n} \\ x_{21} & x_{22} & \cdots & x_{2n} \\ \vdots & \vdots & \ddots & \vdots \\ x_{m1} & x_{m2} & \cdots & x_{mn} \end{bmatrix} \end{array} \quad (6.1)$$

Step 2: Computing the Normalized Decision Matrix.

$$R = \begin{bmatrix} r_{11} & r_{12} & \cdots & r_{1n} \\ r_{21} & r_{22} & \cdots & r_{2n} \\ \vdots & \vdots & \ddots & \vdots \\ r_{m1} & r_{m2} & \cdots & r_{mn} \end{bmatrix}$$

where

$$r_{ij} = \frac{x_{ij}}{\sqrt{\sum_{i=1}^{m} x_{ij}^2}};\ j = 1, ..., n; i = 1, ..., m. \tag{6.2}$$

Step 3: Creating the Weighted Normalized Decision Matrix:

Considering $v = (v_1, ..., v_j, ..., v_n)$ as the weight vector of criteria.

$$W = \begin{bmatrix} v_1.r_{11} & v_2.r_{12} & \dots & v_n.r_{1n} \\ v_1.r_{21} & v_2.r_{22} & \dots & v_n.r_{2n} \\ \vdots & \vdots & \ddots & \vdots \\ v_1.r_{m1} & v_2.r_{m2} & \dots & v_n.r_{mn} \end{bmatrix} = \begin{bmatrix} w_{11} & w_{12} & \dots & w_{1n} \\ w_{21} & w_{22} & \dots & w_{2n} \\ \vdots & \vdots & \ddots & \vdots \\ w_{m1} & w_{m2} & \dots & w_{mn} \end{bmatrix} \tag{6.3}$$

where $w_{ij} = v_j \times r_{ij};\ i = 1, ..., m \,\&j = 1, ..., n.$

Step 4: Determining the Positive and Negative Ideal Points.

$$\begin{aligned} A^+ &= \left\{v_1^+, ..., v_j^+, ..., v_n^+\right\} \\ A^- &= \left\{v_1^-, ..., v_j^-, ..., v_n^-\right\} \end{aligned} \tag{6.4}$$

where

$$v_j^+ = \begin{cases} \max\limits_i\{v_{ij}\}, C_j \, is\, a\, benefit\, criteria. \\ \min\limits_i\{v_{ij}\}, C_j \ is \ a \ \cos t \ criteria. \end{cases};\quad j = 1, ..., n. \tag{6.5}$$

$$v_j^- = \begin{cases} \min\limits_i\{v_{ij}\}, C_j \, is\, a\, benefit\, criteria. \\ \max\limits_i\{v_{ij}\}, C_j \ is \ a \ \cos t \ criteria. \end{cases}\quad j = 1, ..., n. \tag{6.6}$$

where J^+ is the set of profit criteria and J^- is the set of cost criteria.

Step 5: Calculating the Distance of Each Alternative Form the Positive and Negative Ideal Solutions.

$$d_i^+ = \sqrt{\sum_{j=1}^{n} (v_{ij} - v_j^+)^2};\ i = 1, ..., m \tag{6.7}$$

$$d_i^- = \sqrt{\sum_{j=1}^{n} (v_{ij} - v_j^-)^2};\ i = 1, ..., m \tag{6.8}$$

That d_i^+ and d_i^- show the distance of each alternative (A_i) from the positive and negative ideal points, respectively.

Step 6: Computing the Relative Closeness Index and Ranking the Alternatives.

$$RC_i = \frac{d_i^-}{d_i^- + d_i^+}; \quad i = 1, \ldots, m \tag{6.9}$$

Obtaining the relative closeness, the alternatives will be ranked according to the descending order of RC_i. The largest RC_i shows the best alternative.

Geometrical Presentation of TOPSIS Method:

Suppose ranking six alternatives $\{A_1, \ldots, A_6\}$ considering two criteria $\{C_1, C_2\}$. Figure 6.1 illustrates the alternatives. The horizontal axis represents the first criterion (C_1) and the vertical axis represents the second criterion (C_2). Assume that the first criterion is of the cost type and the second criterion is of the profit type. Therefore, the less the first criterion the better and the more the second criterion the better. The positive ideal point (A^+) is equal to the minimum value of the cost criterion (C_1) and the maximum value of the benefit criterion (C_2) as shown in Fig. 6.1. Also, the negative ideal point (A^-) is equal to the minimum value of the benefit criterion and the maximum value of the cost criterion.

Then the distance of each alternative from the positive ideal point and the negative ideal point is calculated. The closer an alternative is to the positive ideal point and the further away from the negative ideal point, the more desirable it is. For example, comparing the two alternatives (A_1) and (A_2), the alternative A_2 is closer to the positive ideal and further away from the negative ideal point. Therefore, A_2 is preferable to A_1.

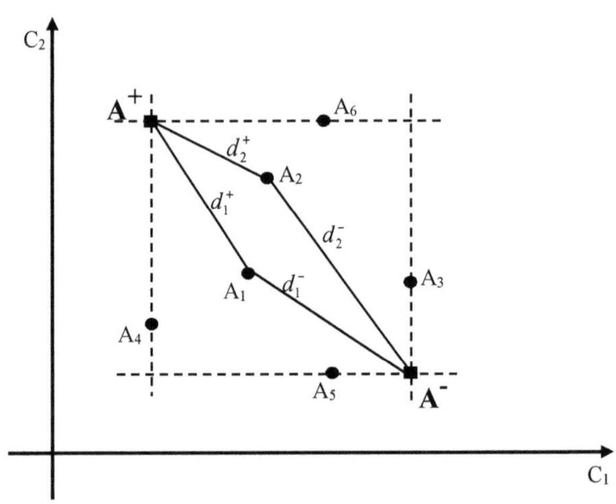

Fig. 6.1 Geometrical Presentation of TOPSIS

Some considerable advantages of the TOPSIS method include: being a logical and comprehensible method according to human judgment, low computational complexity and good computational efficiency, the ability to consider both positive and negative data.

In addition to the mentioned benefits, this method also faces challenges that affect the choice of the best alternative. Some of them include: the existence of various data normalization methods, the existence of different norms for calculating the distance to the ideal point, as well as the methods of determining the weight of criteria according to the judgment of decision makers in group decision-making and the degree of importance considered for each decision maker.

6.2.1 Case Study

In the context of waste control in hospitals, the objective is to rank five hospitals based on six criteria related to waste management. The criteria are as follows:

C1: Monthly training hours for senior managers—Training provided to hospital managers to ensure adherence to proper management practices and optimal utilization of equipment and resources.

C2: Monthly training hours for staff—Training provided to staff members to promote a culture of service delivery to patients, compliance with laws and regulations, and adherence to guidelines.

C3: Patient education—Efforts to raise awareness among patients, improve their knowledge, and modify their consumption patterns.

C4: Number of comprehensive waste management plans—The implementation of multiple waste management plans helps identify strengths, operational challenges, and weaknesses in guidelines. The greater the number of implemented plans, the better the opportunity for in-depth studies and the provision of practical solutions to enhance waste control processes.

C5: Percentage of failed waste management programs—A lower percentage of program failures indicates better performance. Analyzing and studying the reasons for program failure can contribute to the development of more practical solutions.

C6: Percentage of waste segregation at the source—The more accurate the collection and storage of waste from its initial production stage, such as radioactive waste and its proper handling, the more effective the waste control process becomes.

In this subsection, we employ the Technique for Order Preferences by Similarity to Ideal Solutions (TOPSIS) method to rank the hospitals based on their waste control performance using the aforementioned criteria. TOPSIS is a well-known decision-making technique that allows us to evaluate and compare alternatives based on their proximity to ideal solutions. By applying TOPSIS, we can determine which hospital demonstrates the best waste control practices and identify areas for improvement in others. Through this analysis, we aim to shed light on the current state of waste management in these hospitals and provide insights into effective strategies for enhancing waste control practices. By employing a systematic and quantitative approach, we can promote sustainable waste management and contribute to the overall well-being of the healthcare sector (Table 6.1).

The decision matrix is normalized using Eq. (6.2) presented in Table 6.2 Considering the weight vector $v = (v_1, v_2, v_3, v_4, v_5, v_6) = (14, 16, 11, 20, 10, 29)$ the weighted normalized decision matrix is illustrated in Table 6.3.

Positive and negative ideal solutions obtained by Eqs. (6.5) and (6.6) are presented in Table 6.4.

Table 6.5 shows the distance of each alternative from the ideal solutions attained by Eqs. (6.7) and (6.8) and the relative closeness index is calculated using Eq. (6.9). The largest RC_i shows the best alternative. Therefore, $A_3 \succ A_1 \succ A_2 \succ A_6 \succ A_5$.

Table 6.1 Decision matrix to rank five hospitals

	C1 (benefit)	C2 (benefit)	C3 (benefit)	C4 (benefit)	C5 (cost) (%)	C6 (benefit) (%)
A1	10	15	5	1	10	75
A2	8	12	4	1	15	70
A3	12	17	6	2	10	80
A4	9	7	3	0	20	60
A5	10	14	4	0	15	75

Table 6.2 Normalized decision matrix for ranking the hospitals

	C1 (benefit)	C2 (benefit)	C3 (benefit)	C4 (benefit)	C5 (cost)	C6 (benefit)
A1	0.452	0.499	0.495	0.408	0.309	0.464
A2	0.362	0.399	0.396	0.408	0.463	0.433
A3	0.543	0.566	0.594	0.816	0.309	0.495
A4	0.407	0.233	0.297	0.000	0.617	0.371
A5	0.452	0.466	0.396	0.000	0.463	0.464

Table 6.3 Weighted normalized decision matrix for ranking the hospitals

	C1 (benefit)	C2 (benefit)	C3 (benefit)	C4 (benefit)	C5 (cost)	C6 (benefit)
A1	6.331	7.987	5.446	8.165	3.086	13.450
A2	5.065	6.389	4.357	8.165	4.629	12.553
A3	7.597	9.052	6.535	16.330	3.086	14.347
A4	5.698	3.727	3.267	0.000	6.172	10.760
A5	6.331	7.454	4.357	0.000	4.629	13.450

Table 6.4 Positive and negative ideal points

	C1 (benefit)	C2 (benefit)	C3 (benefit)	C4 (benefit)	C5 (cost)	C6 (benefit)
A^+	7.597	9.052	6.535	16.330	6.172	14.347
A^-	5.065	3.727	3.267	0.000	3.086	10.760

Table 6.5 Rank based on TOPSIS method

	d_i^+	d_i^-	RC_i	Rank
A1	9.92	9.00	0.52	2
A2	8.97	9.51	0.49	3
A3	18.03	3.09	0.85	1
A4	3.15	17.95	0.15	5
A5	5.13	16.70	0.23	4

Among the five hospitals under study, based on the TOPSIS method, it has been determined that Hospital A3 ranks first and Hospital A5 ranks last in terms of waste reduction performance.

6.3 Fuzzy TOPSIS Method Using Triangle Fuzzy Numbers

In decision-making processes, it is often necessary to evaluate and rank alternatives based on multiple criteria or attributes. The Technique for Order Preference by Similarity to Ideal Solution (TOPSIS) has long been a valuable tool for addressing such multi-attribute decision-making problems. However, in many real-world scenarios, the data or criteria involved may not be precise or easily quantifiable. This is where the Fuzzy TOPSIS method comes into play. Fuzzy TOPSIS is an extension of the traditional TOPSIS method that incorporates fuzzy logic to handle uncertainty and vagueness in decision-making. Fuzzy logic allows for the representation of imprecise or subjective information by assigning degrees of membership to different categories or values. By utilizing fuzzy sets and fuzzy numbers, Fuzzy TOPSIS offers a more flexible and realistic approach to decision-making, particularly when dealing with complex and ambiguous situations. The fundamental concept behind Fuzzy TOPSIS remains the same as its crisp counterpart. The method aims to identify the most optimal alternative by considering both the positive ideal solution (the alternative that best fulfills the benefit criteria) and the negative ideal solution (the alternative that best satisfies the cost criteria). The chosen alternative should be as close as possible to the positive ideal solution and as far as possible from the negative ideal solution. In Fuzzy TOPSIS, the linguistic or fuzzy assessments provided by experts or decision-makers are transformed into fuzzy numbers, enabling the representation of subjective judgments and uncertainties. By applying appropriate fuzzy aggregation operators and fuzzy similarity measures, Fuzzy TOPSIS computes the closeness coefficient for each alternative and ranks them accordingly [7, 11, 13, 16].

The applications of Fuzzy TOPSIS are diverse and encompass various fields such as engineering, finance, environmental management, and healthcare. Its ability to handle imprecision and uncertainty makes it a valuable tool for decision-making in real-world situations where precise data may be lacking or difficult to obtain. In this section, we talk about Fuzzy TOPSIS and its application. We will examine how fuzzy sets and fuzzy numbers are employed to represent subjective judgments, and we will discuss the various fuzzy aggregation operators and similarity measures commonly used in Fuzzy TOPSIS. Furthermore, we will showcase practical examples and real-world applications of Fuzzy TOPSIS to illustrate its effectiveness in dealing with complex decision problems.

Consider that there exist "m" alternatives to be assessed based on "n" attributes. If the value of the alternatives due to each attributes are represented by linguistic term, to rank the alternatives, the TOPSIS method with Fuzzy triangular numbers is performed as follows:

Step 1: Creating the Fuzzy Decision Matrix.

If the value of ith alternative with respect to jth criteria is expressed by linguistic variables, then the following fuzzy decision matrix is obtained:

$$\tilde{X} = \begin{bmatrix} \tilde{x}_{11}\ \tilde{x}_{12} \ldots \tilde{x}_{1n} \\ \tilde{x}_{21}\ \tilde{x}_{22} \ldots \tilde{x}_{2n} \\ \vdots \quad \vdots \ \ddots \ \vdots \\ \tilde{x}_{m1}\ \tilde{x}_{m2} \ldots \tilde{x}_{mn} \end{bmatrix} \tag{6.10}$$

where each element $\tilde{x}_{ij} = (x_{ij}^l, x_{ij}^m, x_{ij}^u)$, $(x_{ij}^l > 0)$ is a triangle fuzzy number (TFN).

Step 2: Computing the Normalized Fuzzy Decision Matrix.

Studying on various criteria for decision making, some might be cost criteria and some might be profit criteria. Cost criteria need to be minimized and profit criteria should be maximized. As an example, Hiring an employee among job seekers as a multi-attribute decision making problem, job seekers are the alternatives and 'creativity', 'problem solving', 'team management', 'communication skills', 'related experience' are profit criteria and 'age', 'number of quitting job' may considered as cost criteria. The fuzzy decision matrix due to cost and profit criteria is normalizes as follows:

$$\text{For profit criteria} j \in J^+ \tilde{r}_{ij} = \left(\frac{x_{ij}^l}{x_{ij}^{u*}}, \frac{x_{ij}^m}{x_{ij}^{u*}}, \frac{x_{ij}^u}{x_{ij}^{u*}} \right) \quad where \quad x_{ij}^{u*} = \max_i \left\{ x_{ij}^u \right\} \tag{6.11}$$

$$\text{For cost criteria} j \in J^- \tilde{r}_{ij} = \left(\frac{x_{ij}^{l*}}{x_{ij}^{u}}, \frac{x_{ij}^{l*}}{x_{ij}^{m}}, \frac{x_{ij}^{l*}}{x_{ij}^{l}} \right) \quad where \quad x_{ij}^{l*} = \min_i \left\{ x_{ij}^{l} \right\} \tag{6.12}$$

$$\tilde{R} = \begin{bmatrix} \tilde{r}_{11} \ \tilde{r}_{12} \ \dots \ \tilde{r}_{1n} \\ \tilde{r}_{21} \ \tilde{r}_{22} \ \dots \ \tilde{r}_{2n} \\ \vdots \ \ \vdots \ \ \ddots \ \vdots \\ \tilde{r}_{m1} \ \tilde{r}_{m2} \ \dots \ \tilde{r}_{mn} \end{bmatrix} \text{is the normalized fuzzy decision matrix} \tag{6.13}$$

Step 3: Computing the Weighted Normalized Fuzzy Decision Matrix.

Let $\tilde{v} = (\tilde{v}_1, \ldots, \tilde{v}_j, \ldots, \tilde{v}_n)$ is the weight vector of criteria. $\tilde{v}_j = \left(v_j^l, v_j^m, v_j^u \right)$ is a fuzzy triangular number that denotes the importance of *j*th criterion. The weighted value for each element of decision matrix is equal to $\tilde{w}_{ij} = \tilde{v}_j.\tilde{r}_{ij}$ and the weighted normalized fuzzy decision matrix is as follows:

$$\tilde{W} = \begin{bmatrix} \tilde{w}_{11} & \tilde{w}_{12} & \cdots & \tilde{w}_{1n} \\ \tilde{w}_{21} & \tilde{w}_{22} & \cdots & \tilde{w}_{2n} \\ \vdots & \vdots & \ddots & \vdots \\ \tilde{w}_{m1} & \tilde{w}_{m2} & \cdots & \tilde{w}_{mn} \end{bmatrix} \tag{6.14}$$

where $\tilde{w}_{ij} = \tilde{v}_j.\tilde{r}_{ij} = \left(v_j^l, v_j^m, v_j^u \right).\left(r_{ij}^l, r_{ij}^m, r_{ij}^u \right) \simeq \left(v_j^l.r_{ij}^l, v_j^m.r_{ij}^m, v_j^u.r_{ij}^u \right) = \left(w_{ij}^l, w_{ij}^m, w_{ij}^u \right); \ j = 1, \ldots, n, \ \ i = 1, \ldots, m.$

Step 4: Obtaining the Fuzzy Positive and Negative Ideal Solutions.

Since Eq. (6.11) and Eq. (6.12) are applied to normalization, so after the normalization, cost criteria are also considered as benefits. The fuzzy positive ideal solutions and the fuzzy negative ideal solutions obtain as following:

$$\tilde{I}^+ = \left\{ \tilde{I}_1^+, \tilde{I}_2^+, \ldots, \tilde{I}_j^+, \ldots \tilde{I}_n^+ \right\}$$

$$\text{where } \tilde{I}_j^+ = \left(a^+, b^+, c^+ \right) \text{ that } a^+ = b^+ = c^+ = \max_i \left\{ w_{ij}^u \right\}; \ \ j = 1, \ldots, n \tag{6.15}$$

$$\tilde{I}^- = \left\{ \tilde{I}_1^-, \tilde{I}_2^-, \ldots, \tilde{I}_j^-, \ldots \tilde{I}_n^- \right\}$$

$$\text{where } \tilde{I}_j^- = \left(a^-, b^-, c^- \right) \text{ that } a^- = b^- = c^- = \min_i \left\{ w_{ij}^l \right\}; \ \ j = 1, \ldots, n \tag{6.16}$$

Step 5: Computing the Distance of Each Alternative from the Ideal Solutions.

Considering the distance function for TFNs presented by Chen [5] as Eq. (6.17), the distance of each alternative A_i $(i = 1, ..., m)$ from the positive ideal solution and negative ideal solutions are calculated respectively by Eqs. (6.18) and (6.19).

That the distance of two fuzzy numbers $\tilde{x}_1 = (a_1, b_1, c_1)$ and $\tilde{x}_2 = (a_2, b_2, c_2)$ is equal to

$$d(\tilde{x}_1, \tilde{x}_2) = \sqrt{\frac{1}{3}\left((a_1 - a_2)^2 + (b_1 - b_2)^2 + (c_1 - c_2)^2\right)} \tag{6.17}$$

Therefore, the distance of each alternative $\tilde{w}_{ij} = \left(w_{ij}^l, w_{ij}^m, w_{ij}^u\right)$ from the positive ideal solution $\tilde{I}_j^+ = \left(a^+, b^+, c^+\right)$ and the negative ideal solution $\tilde{I}_j^- = \left(a^-, b^-, c^-\right)$ is as follows:

$$d_i^+ = \sum_{j=1}^{n} d(\tilde{w}_{ij}, \tilde{I}_j^+) = \sum_{j=1}^{n} \sqrt{\frac{1}{3}\left(\left(w_{ij}^l - a^+\right)^2 + \left(w_{ij}^m - b^+\right)^2 + \left(w_{ij}^u - c^+\right)^2\right)}; i = 1, ..., m. \tag{6.18}$$

$$d_i^- = \sum_{j=1}^{n} d(\tilde{w}_{ij}, \tilde{I}_j^-) = \sum_{j=1}^{n} \sqrt{\frac{1}{3}\left(\left(w_{ij}^l - a^-\right)^2 + \left(w_{ij}^m - b^-\right)^2 + \left(w_{ij}^u - c^-\right)^2\right)}; i = 1, ..., m. \tag{6.19}$$

Step 6: Computing the Relative Closeness Index and Ranking the Alternatives.

The relative closeness index is computed by Eq. (6.20):

$$RC_i = \frac{d_i^-}{(d_i^- + d_i^+)}; \quad i = 1, ..., m \tag{6.20}$$

Note that the closer the alternative A_i is to positive ideal solution, the closer the value of RC_i will be to one. Therefore, the alternative with highest value of index RC_i represent the best alternative (Fig. 6.2).

6.4 Group Fuzzy TOPSIS Method

Collective memory is superior to individual memory. In group decision-making, the experiences and perspectives of multiple individuals can be harnessed at different levels. This diversity reduces the likelihood of all decision-makers holding incorrect assumptions or making the same calculation errors. As a result, group decision-making enhances the overall quality of decisions. Furthermore, one of the advantages of group decision-making is the tendency towards non-disagreement and the

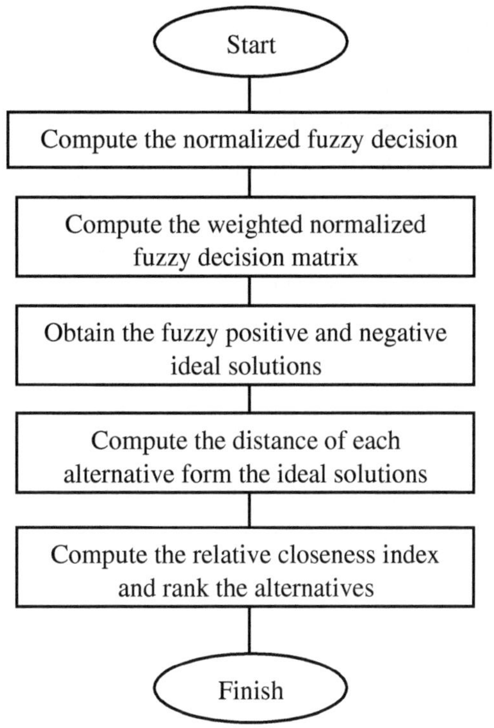

Fig. 6.2 The TOPSIS Method at a Glance

ease of accepting the final decision. When individuals contribute their ideas and insights, they are more likely to feel a sense of ownership and alignment with the chosen outcome. This fosters a cooperative and harmonious environment, which can lead to smoother implementation of decisions. Group decision-making proves to be particularly practical in addressing complex decision-making issues, such as multi-attribute decision-making problems. By leveraging the collective wisdom and diverse perspectives within the group, more comprehensive evaluations can be made, leading to the identification of a more suitable solution.

Moreover, group decision-making promotes active participation and engagement among team members, which can result in innovative ideas, creative problem-solving, and better-informed decisions. The collaborative nature of the process encourages open dialogue, brainstorming, and the exploration of alternatives. However, it is important to note that group decision-making also has its challenges. It can be time-consuming, and conflicts or power dynamics within the group may arise, potentially hindering the decision-making process. Therefore, effective facilitation and clear communication are vital to ensure productive group discussions and consensus-building.

In this section, we discuss the intricacies of group decision-making, exploring its benefits, challenges, and strategies for effective implementation. Through real-life examples and practical techniques, we aim to equip readers with the knowledge and tools to harness the power of collective decision-making and optimize the outcomes of their decision-making processes.

The procedure of group fuzzy TOPSIS method using trapezoidal fuzzy number is expressed as bellow:

Step1: Obtaining the Aggregated Fuzzy Decision Matrix.

Consider a group of decision makers with "K" members. $\tilde{x}_{ij}^{k} = (a_{ij}^{k}, b_{ij}^{k}, c_{ij}^{k}, d_{ij}^{k})$ is a trapezoidal fuzzy number which is expressed by k^{th} decision maker for ith alternative and j^{th} criterion. Hence, there exist "k" decision matrices as bellow.

$$\tilde{X}^k = \begin{bmatrix} \tilde{x}_{11}^k & \tilde{x}_{12}^k & \cdots & \tilde{x}_{1n}^k \\ \tilde{x}_{21}^k & \tilde{x}_{22}^k & \cdots & \tilde{x}_{2n}^k \\ \vdots & \vdots & \ddots & \vdots \\ \tilde{x}_{m1}^k & \tilde{x}_{m2}^k & \cdots & \tilde{x}_{mn}^k \end{bmatrix}$$

$$\tilde{X}^k = \begin{bmatrix} (a_{11}^k, b_{11}^k, c_{11}^k, d_{11}^k) & (a_{12}^k, b_{12}^k, c_{12}^k, d_{12}^k) & \cdots & (a_{1n}^k, b_{1n}^k, c_{1n}^k, d_{1n}^k) \\ (a_{21}^k, b_{21}^k, c_{21}^k, d_{21}^k) & (a_{22}^k, b_{22}^k, c_{22}^k, d_{22}^k) & \cdots & (a_{2n}^k, b_{2n}^k, c_{2n}^k, d_{2n}^k) \\ \vdots & \vdots & \ddots & \vdots \\ (a_{m1}^k, b_{m1}^k, c_{m1}^k, d_{m1}^k) & (a_{m2}^k, b_{m2}^k, c_{m2}^k, d_{m1}^k) & \cdots & (a_{mn}^k, b_{mn}^k, c_{mn}^k, d_{mn}^k) \end{bmatrix} ; k = 1, ..., K. \tag{6.21}$$

The opinion of all decision makers are aggregated as follows:

$$a_{ij} = \min_k \{a_{ij}^k\},\ b_{ij} = \frac{1}{k}\sum_{k=1}^{K} b_{ij}^k,\ c_{ij} = \frac{1}{k}\sum_{k=1}^{K} c_{ij}^k,\ d_{ij} = \max_k \{d_{ij}^k\};$$
$$i = 1, ..., m\ ,\ j = 1, ..., n \tag{6.22}$$

and $\tilde{x}_{ij} = (a_{ij}, b_{ij}, c_{ij}, d_{ij})$.

So, the aggregated fuzzy matrix is equal to:

$$\tilde{X} = \begin{bmatrix} \tilde{x}_{11} & \tilde{x}_{12} & \cdots & \tilde{x}_{1n} \\ \tilde{x}_{21} & \tilde{x}_{22} & \cdots & \tilde{x}_{2n} \\ \vdots & \vdots & \ddots & \vdots \\ \tilde{x}_{m1} & \tilde{x}_{m2} & \cdots & \tilde{x}_{mn} \end{bmatrix}$$
$$= \begin{bmatrix} (a_{11}, b_{11}, c_{11}, d_{11}) & (a_{12}, b_{12}, c_{12}, d_{12}) & \cdots & (a_{1n}, b_{1n}, c_{1n}, d_{1n}) \\ (a_{21}, b_{21}, c_{21}, d_{21}) & (a_{22}, b_{22}, c_{22}, d_{22}) & \cdots & (a_{2n}, b_{2n}, c_{2n}, d_{2n}) \\ \vdots & \vdots & \ddots & \vdots \\ (a_{m1}, b_{m1}, c_{m1}, d_{m1}) & (a_{m2}, b_{m2}, c_{m2}, d_{m2}) & \cdots & (a_{mn}, b_{mn}, c_{mn}, d_{mn}) \end{bmatrix}$$

Step 2: Computing the Normalized Fuzzy Decision Matrix.

Jahanshahloo et al. [12] applied $\alpha - cut$ to normalize triangular fuzzy numbers. Here the same procedure is applied to normalize the trapezoidal fuzzy numbers. First by $\alpha - cut$ the trapezoidal fuzzy numbers are turned into intervals and then the intervals are normalized and next the normalized intervals transformed into trapezoidal fuzzy numbers.

By using $\alpha - cut$ each trapezoidal fuzzy number $\tilde{x}_{ij} = (a_{ij}, b_{ij}, c_{ij}, d_{ij})$ turns into an interval as follows:

$$\tilde{x}_{ij}^{\alpha} = \left[[\tilde{x}_{ij}]_{\alpha}^{L}, [\tilde{x}_{ij}]_{\alpha}^{U}\right], \ \alpha \in [0, 1]$$

.

The intervals are tuned into normalized intervals as bellow:

$$[\tilde{r}_{ij}]_{\alpha}^{L} = \frac{[\tilde{x}_{ij}]_{\alpha}^{L}}{\sqrt{\sum_{i=1}^{m}\left(\left([\tilde{x}_{ij}]_{\alpha}^{L}\right)^{2} + \left([\tilde{x}_{ij}]_{\alpha}^{U}\right)^{2}\right)}}; \ i = 1, \ldots, m, \ j = 1, \ldots, n. \quad (6.23)$$

$$[\tilde{r}_{ij}]_{\alpha}^{U} = \frac{[\tilde{x}_{ij}]_{\alpha}^{U}}{\sqrt{\sum_{i=1}^{m}\left(\left([\tilde{x}_{ij}]_{\alpha}^{L}\right)^{2} + \left([\tilde{x}_{ij}]_{\alpha}^{U}\right)^{2}\right)}}; \ i = 1, \ldots, m, \ j = 1, \ldots, n. \quad (6.24)$$

So, $\tilde{x}_{ij}^{\alpha} = \left[[\tilde{x}_{ij}]_{\alpha}^{L}, [\tilde{x}_{ij}]_{\alpha}^{U}\right]$ is the normalized to $\tilde{r}_{ij}^{\alpha} = \left[[\tilde{r}_{ij}]_{\alpha}^{L}, [\tilde{r}_{ij}]_{\alpha}^{U}\right]; \ \alpha \in [0, 1]$.
To transform this interval to trapezoidal fuzzy number $\tilde{r}_{ij} = (r_{aij}, r_{bij}, r_{cij}, r_{dij})$, $\alpha = 1$ and $\alpha = 0$ is considered:

$$r_{ij}^{a} = [\tilde{r}_{ij}]_{\alpha=0}^{L}, \ r_{ij}^{b} = [\tilde{r}_{ij}]_{\alpha=1}^{L}, \ r_{ij}^{c} = [\tilde{r}_{ij}]_{\alpha=1}^{U}, \ r_{ij}^{d} = [\tilde{r}_{ij}]_{\alpha=0}^{U} \quad (6.25)$$

$\tilde{r}_{ij} = (r_{ij}^{a}, r_{ij}^{b}, r_{ij}^{c}, r_{ij}^{d})$ is the normalized trapezoidal fuzzy number and the normalize decision matrix is equal to $\tilde{R} = \left[\tilde{r}_{ij}\right]_{m\times n}$.

Step 3: Calculating the Weighted Normalized Fuzzy Decision Matrix.

If $\tilde{v}_{j}^{k} = (v_{aj}^{k}, v_{bj}^{k}, v_{cj}^{k}, v_{dj}^{k})$ denotes the weight of jth criterion represented by k^{th} decision maker then $\tilde{v}^{k} = (\tilde{v}_{1}^{k}, \ldots, \tilde{v}_{j}^{k}, \ldots, \tilde{v}_{n}^{k})$ is the weight vectors of criteria for $k = 1, \ldots, K$, and can be aggregated to weight vector $\tilde{V} = (\tilde{v}_{1}, \ldots, \tilde{v}_{j}, \ldots, \tilde{v}_{n})$ using Eq. (6.22). $\tilde{v}_{j} = \left(v_{aj}, v_{bj}, v_{cj}, v_{dj}\right)$ is a trapezoidal fuzzy number.

Multiplying each element of the aggregated weight vector by each column of the normalized decision matrix ($\tilde{w}_{ij} = \tilde{v}_j.\tilde{r}_{ij}$), the weighted normalized fuzzy decision matrix is as follows:

$$\tilde{W} = \left[\tilde{w}_{ij}\right]_{m \times n} = \begin{bmatrix} \tilde{w}_{11} & \tilde{w}_{12} & \cdots & \tilde{w}_{1n} \\ \tilde{w}_{21} & \tilde{w}_{22} & \cdots & \tilde{w}_{2n} \\ \vdots & \vdots & \ddots & \vdots \\ \tilde{w}_{m1} & \tilde{w}_{m2} & \cdots & \tilde{w}_{mn} \end{bmatrix} \tag{6.26}$$

where $\tilde{w}_{ij} = \tilde{v}_j.\tilde{r}_{ij} = \left(v_{aj}, v_{bj}, v_{cj}, v_{dj}\right).\left(r_{aij}, r_{bij}, r_{cij}, r_{dij}\right) \simeq \left(v_{aj}.r_{aij}, v_{bj}.r_{bij}, v_{cj}.r_{cij}, v_{dj}.r_{dij}\right) = \left(w_{aij}, w_{bij}, w_{cij}, w_{dij}\right)$

Step 4: Obtaining the Positive and Negative Ideal Points.

The fuzzy trapezoidal positive ideal solutions and the negative ideal solutions are as follows:

$$\tilde{I}^+ = \left\{\tilde{I}_1^+, \tilde{I}_2^+, \ldots, \tilde{I}_j^+, \ldots\tilde{I}_n^+\right\}$$

where $\tilde{I}_j^+ = \left(a^+, b^+, c^+, d^+\right)$ and $a^+ = b^+ = c^+ = d^+$

$$= \begin{cases} \max\limits_i\left\{w_{aij}\right\};\ j \in J^+ \\ \min\limits_i\left\{w_{aij}\right\};\ j \in J^- \end{cases} \tag{6.27}$$

$$\tilde{I}^- = \left\{\tilde{I}_1^-, \tilde{I}_2^-, \ldots, \tilde{I}_j^-, \ldots\tilde{I}_n^-\right\}$$

where $\tilde{I}_j^- = \left(a^-, b^-, c^-, d^-\right)$ that $a^- = b^- = c^- = d^-$

$$= \begin{cases} \min\limits_i\left\{w_{aij}\right\};\ j \in J^+ \\ \max\limits_i\left\{w_{dij}\right\};\ j \in J^- \end{cases} \tag{6.28}$$

where J^+ denotes the set of benefit criteria and J^- shows the set of cost criteria.

Step 5: Calculating the Distance of Each Alternative from the Ideal Points.

That the distance of two trapezoidal fuzzy numbers $\tilde{x}_1 = (a_1, b_1, c_1, d_1)$ and $\tilde{x}_2 = (a_2, b_2, c_2, d_2)$ calculated by

$$d(\tilde{x}_1, \tilde{x}_2) = \sqrt{\frac{1}{4}\left((a_1 - a_2)^2 + (b_1 - b_2)^2 + (c_1 - c_2)^2 + (d_1 - d_2)^2\right)} \tag{6.29}$$

The distance of alternatives from the positive ideal points, with respect to Eq. (6.29) is equal to

$$d_i^+ = \sum_{j=1}^{n} d(\tilde{w}_{ij}, \tilde{I}_j^+);\ \ i = 1, \ldots, m$$

$$d_i^+ = \sum_{j=1}^{n} \sqrt{\frac{1}{4}\left(\left(w_{aij} - a^+\right)^2 + \left(w_{bij} - b^+\right)^2 + \left(w_{cij} - c^+\right)^2 + \left(w_{dij} - d^+\right)^2\right)} \tag{6.30}$$

The distance of alternatives from the negative ideal points, with respect to Eq. (6.29) is equal to

$$d_i^- = \sum_{j=1}^{n} d(\tilde{x}_{ij}, \tilde{I}_j^-);\ \ i = 1, \ldots, m$$

$$d_i^- = \sum_{j=1}^{n} \sqrt{\frac{1}{4}\left(\left(w_{aij} - a^-\right)^2 + \left(w_{bij} - b^-\right)^2 + \left(w_{cij} - c^-\right)^2 + \left(w_{dij} - d^-\right)^2\right)} \tag{6.31}$$

Step 6: Computing the Relative Closeness Index and Ranking the Alternatives.

The relative closeness index is computed by Eq. (6.32):

$$RC_i = \frac{d_i^-}{(d_i^- + d_i^+)};\ \ i = 1, \ldots, m \tag{6.32}$$

The alternative with highest value of index RC_i represent the best alternative.

6.5 Intuitionistic Fuzzy TOPSIS Group Decision Making Method

Decisions expressed by decision makers are usually imprecise. On the other hand, intuitive fuzzy sets are suitable tools for encountering the uncertainty. Atanassov [1] proposed the theory of intuitionistic fuzzy sets in 1986. During the past decade many studies applied Intuitionistic Fuzzy TOPSIS method, from selecting the best supplier to choosing the most suitable desalination systems [20, 23, 24, 27].

In this subsection the procedure of group decision making of Intuitionistic Fuzzy TOPSIS is explained, considering $A = \{A_1, A_2, \ldots, A_m\}$ as a set of alternatives and $C = \{C_1, C_2, \ldots, C_n\}$ as a set of criteria.

Step 1: Computing the Weights of Decision Makers.

Assume K decision makers with different level of importance with respecting to their experience, skills and knowledge. The importance of each DM is considered as a linguistic terms indicated by Intuitionistic fuzzy number (IFN). Let $D^k = \left[\mu^k, v^k, \pi^k\right]$ is the IFN that denote the rank of kth decision maker. Then the weight of kth decision maker is obtained as:

$$\lambda_k = \frac{\left(\mu^k + \pi^k\left(\frac{\mu^k}{\mu^k + v^k}\right)\right)}{\sum_{k=1}^{K}\left(\mu^k + \pi^k\left(\frac{\mu^k}{\mu^k + v^k}\right)\right)};\ k = 1, \ldots, K. \tag{6.33}$$

Hence $\lambda = (\lambda_1, \lambda_2, \ldots, \lambda_K)$ is the weight vector of decision makers that $\sum_{k=1}^{K}\lambda_k = 1$ and $\lambda_k \geq 0, k = 1, \ldots, K.\lambda_k \geq 0, k = 1, \ldots, K.$

Step 2: Obtaining the Aggregated Intuitionistic Fuzzy Decision Matrix.

Let

$$\tilde{X}^k = \begin{bmatrix} \tilde{x}_{11}^k\ \tilde{x}_{12}^k \ldots \tilde{x}_{1n}^k \\ \tilde{x}_{21}^k\ \tilde{x}_{22}^k \ldots \tilde{x}_{2n}^k \\ \vdots \quad \vdots \quad \ddots \quad \vdots \\ \tilde{x}_{m1}^k\ \tilde{x}_{m2}^k \ldots \tilde{x}_{mn}^k \end{bmatrix} = \begin{bmatrix} (\mu_{11}^k, v_{11}^k, \pi_{11}^k) & (\mu_{12}^k, v_{12}^k, \pi_{12}^k) & \cdots & (\mu_{1n}^k, v_{1n}^k, \pi_{1n}^k) \\ (\mu_{21}^k, v_{21}^k, \pi_{21}^k) & (a_{22}^k, v_{22}^k, \pi_{22}^k) & \cdots & (\mu_{2n}^k, v_{2n}^k, \pi_{2n}^k) \\ \vdots & \vdots & \ddots & \vdots \\ (\mu_{m1}^k, v_{m1}^k, \pi_{m1}^k) & (\mu_{m2}^k, v_{m2}^k, \pi_{m2}^k) & \cdots & (\mu_{mn}^k, v_{mn}^k, \pi_{mn}^k) \end{bmatrix}$$

is an intuitionistic fuzzy decision matrix of kth decision maker, to aggregate the decision matrices, the Intuitionistic fuzzy weighted averaging (IFWA) operator is applied.

Suppose $\tilde{X}' = [\tilde{x}'_{ij}]_{m \times n}$ as the aggregated decision matrix that

$$\tilde{x}'_{ij} = \lambda_1\tilde{x}_{ij}^1 \oplus \lambda_2\tilde{x}_{ij}^2 \oplus \ldots \oplus \lambda_K\tilde{x}_{ij}^K \simeq \left[1 - \prod_{k=1}^{K}\left(1 - \mu_{ij}^k\right)^{\lambda K}, \prod_{k=1}^{K}\left(v_{ij}^k\right)^{\lambda K}, \prod_{k=1}^{K}\left(1 - \mu_{ij}^k\right)^{\lambda K} - \prod_{k=1}^{K}\left(v_{ij}^k\right)^{\lambda K}\right] \tag{6.34}$$

where $\tilde{x}'_{ij} = (\mu_{A_i}(x_j), v_{A_i}(x_j), \pi_{A_i}(x_j));\ i = 1, \ldots, m;\ j = 1, \ldots, n.$

The aggregated intuitionistic fuzzy decision matrix is as below:

$$\tilde{X}' = \begin{bmatrix} (\mu_{A_1}(x_1), v_{A_1}(x_1), \pi_{A_1}(x_1)) & (\mu_{A_1}(x_2), v_{A_1}(x_2), \pi_{A_1}(x_2)) & \cdots & (\mu_{A_1}(x_n), v_{A_1}(x_n), \pi_{A_1}(x_n)) \\ (\mu_{A_2}(x_1), v_{A_2}(x_1), \pi_{A_2}(x_1)) & (\mu_{A_2}(x_2), v_{A_2}(x_2), \pi_{A_2}(x_2)) & \cdots & (\mu_{A_2}(x_n), v_{A_2}(x_n), \pi_{A_2}(x_n)) \\ \vdots & \vdots & \ddots & \vdots \\ (\mu_{A_m}(x_1), v_{A_m}(x_1), \pi_{A_m}(x_1)) & (\mu_{A_m}(x_2), v_{A_m}(x_2), \pi_{A_m}(x_2)) & \cdots & (\mu_{A_m}(x_n), v_{A_m}(x_n), \pi_{A_m}(x_n)) \end{bmatrix}$$

Step 3: Determining the weights of criteria.

Let $\tilde{W}^k = (\tilde{w}_1^k, \tilde{w}_2^k, ..., \tilde{w}_j^k, ..., \tilde{w}_n^k)$ denotes the weight vector of criteria and $\tilde{w}_j^k = \left[\mu_{wj}^k, v_{wj}^k, \pi_{wj}^k\right]$ is an intuitionistic fuzzy number assigned to criterion C_j by k^{th} decision maker. The weights of criteria are computed by using IFWA operator:

$$\tilde{w}_j = \lambda_1 \tilde{w}_j^1 \oplus \lambda_2 \tilde{w}_j^2 \oplus ... \oplus \lambda_K \tilde{w}_j^K \simeq \left[1 - \prod_{k=1}^{K}\left(1-\mu_{wj}^k\right)^{\lambda K}, \prod_{k=1}^{K}\left(v_{wj}^k\right)^{\lambda K}, \prod_{k=1}^{K}\left(1-\mu_{wj}^k\right)^{\lambda K} - \prod_{k=1}^{K}\left(v_{wj}^k\right)^{\lambda K}\right] \tag{6.35}$$

The aggregated weights of criteria is equal to $\tilde{W} = (\tilde{w}_1, \tilde{w}_2, ..., \tilde{w}_n)$.

where $\tilde{w}_j = \left(\mu_w(x_j), v_w(x_j), \pi_w(x_j)\right)$; $j = 1, ..., n$.

Step 4: Calculating Aggregated Weighted Intuitionistic Fuzzy Decision Matrix.

Respecting the definition introduced by Atanassov (1986)[1]:

$$\tilde{W} \otimes \tilde{X}' \simeq \begin{pmatrix} \mu_{A_i}(x).\mu_W(x), v_{A_i}(x) + v_W(x) - v_{A_i}(x).v_W(x), \\ 1 - v_{A_i}(x) - v_W(x) - \mu_{A_i}(x).\mu_W(x) + v_{A_i}(x).v_W(x) \end{pmatrix}$$

The aggregated weighted intuitionistic fuzzy decision matrix is as follows:

$$\tilde{X} = \begin{bmatrix} (\mu_{A_1W}(x_1), v_{A_1W}(x_1), \pi_{A_1W}(x_1)) & (\mu_{A_1W}(x_2), v_{A_1W}(x_2), \pi_{A_1W}(x_2)) & \cdots & (\mu_{A_1W}(x_n), v_{A_1W}(x_n), \pi_{A_1W}(x_n)) \\ (\mu_{A_2W}(x_1), v_{A_2W}(x_1), \pi_{A_2W}(x_1)) & (\mu_{A_2W}(x_2), v_{A_2W}(x_2), \pi_{A_2W}(x_2)) & \cdots & (\mu_{A_2W}(x_n), v_{A_2W}(x_n), \pi_{A_2W}(x_n)) \\ \vdots & \vdots & \ddots & \vdots \\ (\mu_{A_mW}(x_1), v_{A_mW}(x_1), \pi_{A_mW}(x_1)) & (\mu_{A_mW}(x_2), v_{A_mW}(x_2), \pi_{A_mW}(x_2)) & \cdots & (\mu_{A_mW}(x_n), v_{A_mW}(x_n), \pi_{A_mW}(x_n)) \end{bmatrix}$$

$where \tilde{X} = [\tilde{x}_{ij}]_{m \times n}$ is the weighted matrix and $\tilde{x}_{ij} = \left(\mu_{A_iW}(x_j), v_{A_iW}(x_j), \pi_{A_iW}(x_j)\right)$. (6.36)

Step 5: Computing the Intuitionistic Fuzzy Positive and Negative Ideal Solutions.

The intuitionistic fuzzy positive ideal solution and the intuitionistic fuzzy negative ideal solution are equal to Eq. (6.37) and Eq. (6.38), respectively.

$$\tilde{A}^+ = (\mu_{A^+W}(x_j), v_{A^+W}(x_j)) \tag{6.37}$$

$$\tilde{A}^- = (\mu_{A^-W}(x_j), v_{A^-W}(x_j)) \tag{6.38}$$

where

$$\mu_{A^+W}(x_j) = \left(\left(\max_i \mu_{A_iW}(x_j) \middle| j \in J^+\right), \left(\min_i \mu_{A_iW}(x_j) \middle| j \in J^-\right)\right) \tag{6.39}$$

$$\nu_{A^+W}(x_j) = \left(\left(\min_i \nu_{A_iW}(x_j) \middle| j \in J^+\right), \left(\max_i \nu_{A_iW}(x_j) \middle| j \in J^-\right)\right) \tag{6.40}$$

$$\mu_{A^-W}(x_j) = \left(\left(\min_i \mu_{A_iW}(x_j) \middle| j \in J^+\right), \left(\max_i \mu_{A_iW}(x_j) \middle| j \in J^-\right)\right) \tag{6.41}$$

$$\nu_{A^-W}(x_j) = \left(\left(\max_i \nu_{A_iW}(x_j) \middle| j \in J^+\right), \left(\min_i \nu_{A_iW}(x_j) \middle| j \in J^-\right)\right) \tag{6.42}$$

$$\pi_{A^+W}(x_j) = \left(\begin{array}{c} \left(1 - \max_i \mu_{A_iW}(x_j) - \min_i \nu_{A_iW}(x_j) \middle| j \in J^+\right), \\ \left(1 - \min_i \mu_{A_iW}(x_j) - \max_i \nu_{A_iW}(x_j) \middle| j \in J^-\right) \end{array}\right) \tag{6.43}$$

$$\pi_{A^-W}(x_j) = \left(\begin{array}{c} \left(1 - \min_i \mu_{A_iW}(x_j) - \max_i \nu_{A_iW}(x_j) \middle| j \in J^+\right), \\ \left(1 - \max_i \mu_{A_iW}(x_j) - \min_i \nu_{A_iW}(x_j) \middle| j \in J^-\right) \end{array}\right) \tag{6.44}$$

That J^+ denotes the profit criteria and J^- shows the cost criteria.

Step 6: Computing the Separation Measure

Using the normalized Euclidean distance, Szmidt et al. [25] and Memari et al. [18] suggested the following equations to determine the measure separation between the alternatives and the ideal solutions.

$$S^+ = \sqrt{\frac{1}{2n}\sum_{j=1}^{n}\left[\left(\mu_{A_iW}(x_j) - \mu_{A^+W}(x_j)\right)^2 + \left(\nu_{A_iW}(x_j) - \nu_{A^+W}(x_j)\right)^2 + \left(\pi_{A_iW}(x_j) - \pi_{A^+W}(x_j)\right)^2\right]} \tag{6.45}$$

$$S^- = \sqrt{\frac{1}{2n}\sum_{j=1}^{n}\left[\left(\mu_{A_iW}(x_j) - \mu_{A^-W}(x_j)\right)^2 + \left(\nu_{A_iW}(x_j) - \nu_{A^-W}(x_j)\right)^2 + \left(\pi_{A_iW}(x_j) - \pi_{A^-W}(x_j)\right)^2\right]} \tag{6.46}$$

Step7: Determining the Relative Closeness Index and Ranking the Alternatives.

The relative closeness index of alternative A_i is equal to:

$$RC_i = \frac{S_i^-}{(S_i^- + S_i^+)}; \quad i = 1, \ldots, m \tag{6.47}$$

The alternatives are ranked based on descending order of the relative closeness index.

6.6 Applications

The Fuzzy TOPSIS method has found wide-ranging applications across various industries and domains. In the field of supply chain management, it has been utilized for supplier selection, determining the best logistics routes, and optimizing inventory management. In manufacturing systems, the method has been employed for production planning, equipment selection, and process optimization. In the realm of business and marketing, it has been used for market segmentation, product positioning, and customer satisfaction analysis. Furthermore, the Fuzzy TOPSIS method has been applied in healthcare for hospital performance evaluation, patient prioritization, and healthcare resource allocation. Its versatility extends to environmental management, human resources, tourism industry, and many other areas where multi-criteria decision-making is crucial for achieving optimal outcomes.

Case 1: Fuzzy and TOPSIS Methods for Selecting Optimal Sports Venues: A Comprehensive Decision-Making Approach

Cities often face challenges in establishing appropriate sports spaces that cater to the needs of the general public, ultimately promoting societal health and well-being. In this chapter, we delve into the application of Fuzzy and TOPSIS methods to address the complex decision of selecting the ideal location for a sports center among four available alternatives. The criteria considered for determining a suitable sports venue encompass a range of factors crucial to its successful establishment. These criteria include: accessibility to transportation, weather conditions, population density, safety, price, size, difficulty of acquisition, and geomorphic conditions. The first criterion, access to transportation, evaluates the ease of public transportation accessibility for different segments of society, taking into account the transportation network and road infrastructure surrounding the potential venue. Population density serves as the second criterion, considering factors such as the age distribution, the number and types of populations residing near the desired location, and the proportional relationship between the covered population and the capacity of sports facilities. To establish rankings for population density, a linguistic indicator table (Table 6.6) defines ratings ranging from "very low" (VL) to "very high" (VH). The aim is to achieve a suitable balance between population coverage and the capacity of the sports center, favoring a smaller population size to ensure proportionality. Hence, population density is classified as a cost-type indicator. The third criterion, weather conditions, encompasses factors like wind speed and direction, rainfall volume, and the number of sunny hours in the area. Similar to population density, a linguistic indicator table (Table 6.6) is used to define ratings ranging from "very inappropriate" (VL) to "very appropriate" (VH). The weather condition index is considered a profit-type indicator. Safety, the fourth criterion, focuses on assessing natural disaster risks, such as floods, tornadoes, earthquakes, and more, to ensure the safety and security of the sports venue and its users. The price index, the fifth criterion, pertains to the costs associated with land acquisition for the sports center and the initial licensing expenses. Lower costs are favored, making this criterion a cost-type indicator. Finally, the sixth criterion, geomorphic conditions, takes into account factors such as soil type, land slope, and underground water levels. These conditions play a crucial role in determining the feasibility and suitability of the location for constructing the sports venue. By employing the Fuzzy and TOPSIS methods, we aim to provide a systematic approach for decision-makers in selecting a suitable sports venue. The integration of fuzzy logic allows for the representation and handling of imprecise and uncertain information, while the TOPSIS method enables the ranking and comparison of alternatives based on their performance across multiple criteria. In the following sections of this chapter, we will delve into the implementation of the Fuzzy and TOPSIS methods, illustrating their practical application in the context of selecting the optimal sports venue among the available alternatives. By employing these techniques, decision-makers can make informed and robust decisions, ensuring the establishment of sports spaces that promote accessibility, safety, and overall well-being within the community.

In this example, linguistic variables have been transformed into triangular fuzzy numbers using the Table 6.6.Using the Triangular Fuzzy TOPSIS method, the four available alternatives have been ranked based on the six mentioned criteria (Tables 6.7 and 6.8).

Equation (6.11) applied to normalize the benefit criteria C1,C3,C4 and C5. Also, Eq. (6.12) applied to normalize the cost criteria C2 and C5. The result are shown in Table 6.9. According to step3, the weighted normalize decision matrix obtained and presented in Table 6.11 (Using linguistic variables, Table 6.10 shows the Fuzzy weight of each criterion).

Applying Eqs. (6.15) and (6.16) the positive and the negative ideal solutions is attained and displayed in Table 6.12. Table 6.13 shows that A2 ranks as the best alternative and $A_2 \succ A_4 \succ A_1 \succ A_3$.

Table 6.6 Linguistic variables for evaluating the alternatives

Linguistic terms	Fuzzy number
Very low (VL)	(0,1,2)
Low (L)	(1,2,3)
Medium (M)	(2,3,4)
High (H)	(3,4,5)
Very High (VH)	(4,5,6)

Table 6.7 Decision matrix using linguistic variable

	C1 (benefit)	C2 (cost)	C3 (benefit)	C4 (benefit)	C5 (cost)	C6 (benefit)
A1	H	M	L	L	VH	L
A2	M	L	H	VH	L	VH
A3	L	H	VL	M	H	M
A4	VH	L	M	H	M	M

Table 6.8 Fuzzy decision matrix for selecting optimal sport venues

	C1 (benefit)	C2 (cost)	C3 (benefit)	C4 (benefit)	C5 (cost)	C6 (benefit)
A1	(3,4,5)	(2,3,4)	(1,2,3)	(1,2,3)	(4,5,6)	(1,2,3)
A2	(2,3,4)	(1,2,3)	(3,4,5)	(4,5,6)	(1,2,3)	(4,5,6)
A3	(1,2,3)	(3,4,5)	(0,1,2)	(2,3,4)	(3,4,5)	(2,3,4)
A4	(4,5,6)	(1,2,3)	(2,3,4)	(3,4,5)	(2,3,4)	(2,3,4)

Case 2: Fuzzy TOPSIS in Human Resources Evaluation and Employee Ranking

In the field of human resource management, one of the critical tasks is selecting the most suitable candidate for a job position from a pool of applicants. This process involves evaluating and ranking the candidates based on specific criteria that align with the desired qualifications and attributes. In this scenario, three interviewers, who are managers at the same level, collectively participate in the decision-making process. The evaluation and selection of candidates consider several indicators that provide insights into their suitability for the position. These indicators include:

C1: Personality Traits: This criterion assesses personal discipline, honesty of expression, and good behavior, which are essential qualities for successful integration into the company's culture.

C2: Work Experience: Work experience is a crucial factor that encompasses the number of years of service and specialized abilities acquired over time. It highlights the candidate's expertise and competency in their field.

C3: Analytical Power and Problem-Solving Ability: This criterion evaluates the candidate's cognitive skills, critical thinking abilities, and aptitude for analyzing complex situations and finding effective solutions.

C4: Leadership and Management Skills: The ability to lead and manage a group is an important attribute, particularly for positions that involve supervisory or managerial responsibilities.

C5: Job Stability: The rate of job changes over the past five years is considered in this criterion. If a candidate has frequently switched jobs within a short span, it raises concerns about their commitment and potential longevity within the company. Therefore, this criterion is deemed as a negative factor.

To facilitate the evaluation process, the linguistic variables corresponding to the criteria have been transformed into trapezoidal fuzzy numbers using Table 6.14. This allows for the representation of imprecise and uncertain information associated with subjective assessments.

To rank the four job applicants, the Group Trapezoidal Fuzzy TOPSIS method is employed. This method takes into account the collective opinions of the three interviewers and considers the overall performance of each candidate across the five aforementioned criteria. By utilizing the Fuzzy TOPSIS approach in human resources evaluation, organizations can enhance their decision-making process and ensure the

Table 6.9 Normalized fuzzy decision matrix for selecting optimal sport venues

	C1	C2	C3	C4	C5	C6
A1	(0.50,0.67,0.83)	(0.25,0.33,0.50)	(0.20,0.40,0.60)	(0.17,0.33,0.50)	(0.17,0.20,0.25)	(0.17,0.33,0.50)
A2	(0.33,0.50,0.67)	(0.33,0.50,1.00)	(0.60,0.80,1.00)	(0.67,0.83,1.00)	(0.33,0.50,1.00)	(0.67,0.83,1.00)
A3	(0.17,0.33,0.50)	(0.20,0.25,0.33)	(0,0.20,0.40)	(0.33,0.50,0.67)	(0.20,0.25,0.33)	(0.33,0.50,0.67)
A4	(0.67,0.83,1.00)	(0.33,0.50,1.00)	(0.40,0.60,0.80)	(0.50,0.67,0.83)	(0.20,0.25,0.33)	(0.33,0.50,0.67)

Table 6.10 Fuzzy weight vector

	C1	C2	C3	C4	C5	C6
$\tilde{v}$	(0.3,0.4,0.5)	(0.1,0.2,0.3)	(0.1,0.2,0.3)	(0.2,0.3,0.4)	(0.4,0.5,0.6)	(0.1,0.2,0.3)
Normalized $\tilde{v}$	(0.13,0.17,0.21)	(0.04,0.08,0.13)	(0.04,0.08,0.13)	(0.08,0.13,0.17)	(0.17,0.21,0.25)	(0.04,0.08,0.13)

Table 6.11 Weighted normalized fuzzy decision matrix for selecting optimal sport venues

	C1	C2	C3	C4	C5	C6
A1	(0.063,0.083,0.104)	(0.010,0.014,0.021)	(0.008,0.017,0.0258)	(0.014,0.028,0.042)	(0.028,0.033,0.042)	(0.007,0.014,0.021)
A2	(0.042,0.063,0.083)	(0.014,0.021,0.042)	(0.025,0.033,0.042)	(0.056,0.069,0.083)	(0.056,0.083,0.167)	(0.028,0.035,0.042)
A3	(0.021,0.042,0.063)	(0.008,0.010,0.014)	(0,0.008,0.017)	(0.028,0.042,0.056)	(0.003,0.042,0.056)	(0.014,0.021,0.028)
A4	(0.083,0.104,0.125)	(0.014,0.021,0.042)	(0.017,0.025,0.033)	(0.042,0.056,0.069)	(0.042,0.056,0.083)	(0.014,0.021,0.028)

Table 6.12 Positive and negative ideal solutions

	C1	C2	C3	C4	C5	C6
A^+	(0.125,0.125,0.125)	(0.042,0.042,0.042)	(0.042,0.042,0.042)	(0.083,0.083,0.083)	(0.167,0.167,0.167)	(0.042,0.042,0.042)
A^-	(0.021,0.021,0.021)	(0.008,0.008,0.008)	(0,0,0)	(0.014,0.014,0.014)	(0.028,0.028,0.028)	(0.007,0.007,0.007)

Table 6.13 Rank based on Fuzzy TOPSIS method

	d_i^+	d_i^-	RC_i	Rank
A1	0.315	0.126	0.286	3
A2	0.203	0.273	0.574	1
A3	0.338	0.104	0.236	4
A4	0.224	0.227	0.503	2

Table 6.14 Linguistic variables for evaluating the alternatives

Linguistic terms	Trapezoidal fuzzy number
Very Poor (VP)	(0,0,1,2)
Poor (P)	(1,2,2,3)
Medium Poor (MP)	(2,3,4,5)
Fair (F)	(4,5,5,6)
Medium Good (MG)	(5,6,7,8)
Good (G)	(7,8,8,9)
Very Good (VG)	(8,9,10,10)

Table 6.15 Decision matrix using linguistic variable

		C1 (benefit)	C2 (benefit)	C3 (benefit)	C4 (benefit)	C5 (cost)
A1	DM1	G	G	P	G	VP
	DM2	MG	G	MP	F	P
	DM3	F	F	MP	MG	MP
A2	DM1	VG	G	VG	F	F
	DM2	G	F	G	G	F
	DM3	MG	MG	VG	MP	MG
A3	DM1	VP	G	G	P	P
	DM2	P	VG	MG	P	F
	DM3	VP	MG	VG	F	F
A4	DM1	G	F	G	VG	MP
	DM2	MG	MG	G	G	F
	DM3	G	G	F	VG	MP

selection of candidates who align with the desired qualifications and contribute to the company's success. In following, we will delve deeper into the implementation of the Fuzzy TOPSIS method, discussing the steps involved and showcasing its practical application in the context of employee ranking and recruitment (Table 6.15).

Table 6.16 shows the opinion of all three decision makers. Applying Eq. (6.22) three matrices are aggregated in Table 6.17. Table 6.18 demonstrates the normalized

Table 6.16 Trapezoidal fuzzy Decision matrix

		C1 (benefit)	C2 (benefit)	C3 (benefit)	C4 (benefit)	C5 (cost)
A1	DM1	(7,8,8,9)	(7,8,8,9)	(1,2,2,3)	(7,8,8,9)	(0,0,1,2)
	DM2	(5,6,7,8)	(7,8,8,9)	(2,3,4,5)	(4,5,5,6)	(1,2,2,3)
	DM3	(4,5,5,6)	(4,5,5,6)	(2,3,4,5)	(5,6,7,8)	(2,3,4,5)
A2	DM1	(8,9,10,10)	(7,8,8,9)	(8,9,10,10)	(4,5,5,6)	(4,5,5,6)
	DM2	(7,8,8,9)	(4,5,5,6)	(7,8,8,9)	(7,8,8,9)	(4,5,5,6)
	DM3	(5,6,7,8)	(5,6,7,8)	(8,9,10,10)	(2,3,4,5)	(5,6,7,8)
A3	DM1	(0,0,1,2)	(7,8,8,9)	(7,8,8,9)	(1,2,2,3)	(1,2,2,3)
	DM2	(1,2,2,3)	(8,9,10,10)	(5,6,7,8)	(1,2,2,3)	(4,5,5,6)
	DM3	(0,0,1,2)	(5,6,7,8)	(8,9,10,10)	(4,5,5,6)	(4,5,5,6)
A4	DM1	(7,8,8,9)	(4,5,5,6)	(7,8,8,9)	(8,9,10,10)	(2,3,4,5)
	DM2	(5,6,7,8)	(5,6,7,8)	(7,8,8,9)	(7,8,8,9)	(4,5,5,6)
	DM3	(7,8,8,9)	(7,8,8,9)	(4,5,5,6)	(8,9,10,10)	(2,3,4,5)

Table 6.17 Aggregated fuzzy decision matrix

	C1 (benefit)	C2 (benefit)	C3 (benefit)	C4 (benefit)	C5 (cost)
A1	(4,6.33,6.67,9)	(4,7,7,9)	(1,2.67,3.33,5)	(4,6.33,6.67,9)	(0,1.67,2.33,5)
A2	(5,7.67,8.33,10)	(4,6.33,6.67,9)	(7,8.67,9.33,10)	(2,5.33,5.67,9)	(4,5.33,5.67,8)
A3	(0,0.67,1.33,3)	(5,7.67,8.33,10)	(5,7.67,8.33,10)	(1,3,3,6)	(1,4,4,6)
A4	(5,7.33,7.67,9)	(4,6.33,6.67,9)	(4,7,7,9)	(7,8.67,9.33,10)	(2,3.67,4.33,6)

Trapezoidal fuzzy numbers using $\alpha - cut$ and turning into intervals by Eqs. (6.23), (6.24) and (6.25).

First row of Table 6.19 shows the aggregated weight vector $\tilde{v} = (\tilde{v}_1, \tilde{v}_2, \tilde{v}_3, \tilde{v}_4, \tilde{v}_5)$ presented by decision makers and the second row shows the normalized weight vector.

Note that considering $\tilde{v}_j = \left(v_j^l, v_j^{m1}, v_j^{m2}, v_j^u\right)$ as a trapezoidal fuzzy number, it is normalized by using $\left(\frac{v_j^l}{c}, \frac{v_j^{m1}}{c}, \frac{v_j^{m2}}{c}, \frac{v_j^u}{c}\right)$, where $c = \sum_{j=1}^{n} v_j^u$.

The weighted normalized decision matrix is illustrated in Table 6.20. Positive and negative ideal solutions obtained by Eqs. (6.27) and (6.28) that is presented in Table 6.21.

Ranking the alternatives with respect to the relative closeness index, A4 is presented as the best alternative and A3 is the last one. As shown in Table 6.22 the ranking result is as follows: $A_4 \succ A_2 \succ A_1 \succ A_3$.

Table 6.18 Normalized fuzzy decision matrix

	C1 (benefit)	C2 (benefit)	C3 (benefit)	C4 (benefit)	C5 (cost)
A1	(0.22,0.35,0.37,0.49)	(0.20,0.35,0.35,0.44)	(0.05,0.13,0.17,0.25)	(0.21,0.35,0.37,0.47)	(0,0.14,0.20,0.37)
A2	(0.27,0.42,0.46,0.54)	(0.20,0.32,0.34,0.44)	(0.35,0.43,0.46,0.50)	(0.10,0.30,0.31,0.47)	(0.30,0.46,0.49,0.59)
A3	(0,0.04,0.07,016)	(0.25,0.39,0.42,0.49)	(0.25,0.38,0.41,0.50)	(0.05,0.17,0.17,0.31)	(0.07,0.35,0.35,0.44)
A4	(0.27,0.41,0.42,0.49)	(0.20,0.32,0.34,0.44)	(0.20,0.35,0.35,0.45)	(0.36,0.48,0.52,0.52)	(0.15,0.32,0.38,0.44)

Table 6.19 Weight vector

	$\tilde{v}_1$	$\tilde{v}_2$	$\tilde{v}_3$	$\tilde{v}_4$	$\tilde{v}_5$
$\tilde{v}$	(0.4,0.5,0.5,0.6)	(0.7,0.8,0.8,0.9)	(0.4,0.5,0.5,0.6)	(0.9,0.9,1,1)	(0.1,0.2,0.2,0.3)
Normalized	(0.12,0.15,0.15,0.18)	(0.21,0.24,0.24,0.26)	(0.12,0.15,0.15,0.18)	(0.26,0.26,0.29,0.29)	(0.03,0.06,0.06,0.09)

Table 6.20 Weighted normalized fuzzy decision matrix

	C1 (benefit)	C2 (benefit)	C3 (benefit)	C4 (benefit)	C5 (cost)
A1	(0.026,0.041,0.043,0.058)	(0.040,0.072,0.072,0.091)	(0.006,0.016,0.019,0.030)	(0.055,0.093,0.098,0.124)	(0,0.004,0.006,0.011)
A2	(0.032,0.050,0.054,0.064)	(0.040,0.066,0.069,0.091)	(0.041,0.051,0.054,0.059)	(0.028,0.078,0.083,0.124)	(0.009,0.014,0.014,0.017)
A3	(0,0.004,0.009,0.019)	(0.050,0.079,0.086,0.101)	(0.030,0.045,0.049,0.059)	(0.014,0.044,0.044,0.083)	(0.002,0.010,0.010,0.013)
A4	(0.032,0.048,0.050,0.058)	(0.040,0.066,0.069,0.091)	(0.024,0.041,0.041,0.053)	(0.097,0.127,0.137,0.138)	(0.004,0.009,0.011,0.013)

Table 6.21 Positive and negative ideal solutions

	C1 (benefit)	C2 (benefit)	C3 (benefit)	C4 (benefit)	C5 (cost)
A^+	(0.064,0.064,0.064,0.064)	(0.101,0.101,0.101,0.101)	(0.059,0.059,0.059,0.059)	(0.138,0.138,0.138,0.138)	(0,0,0,0)
A^-	(0,0,0,0)	(0.040,0.040,0.040,0.040)	(0.006,0.006,0.006,0.006)	(0.014,0.014,0.014,0.014)	(0.017,0.017,0.017,0.017)

Table 6.22 Rank based on group Fuzzy TOPSIS method

	d_i^+	d_i^-	RC_i	Rank
A1	0.162	0.187	0.536	3
A2	0.150	0.207	0.580	2
A3	0.207	0.145	0.412	4
A4	0.112	0.235	0.678	1

6.7 Fuzzy DEA-TOPSIS

Data Envelopment Analysis (DEA) is a widely used method for evaluating decision-making units (DMU) by calculating the relative efficiency of each DMU based on its inputs and outputs. It offers a non-parametric framework that does not require explicit assumptions about the functional forms of production or cost functions, making it suitable for various industries and sectors. DEA has proven to be particularly valuable in benchmarking studies, performance evaluation, resource allocation, and decision-making processes, providing valuable insights for improving productivity and optimizing resource utilization.

The CCR model introduced by Charnes-Cooper-Rhodes is the first mathematical model proposed in Data Envelopment Analysis (DEA). It allows decision-makers to evaluate DMUs based on their relative efficiency scores. The efficiency score falls within the range of [0,1]. A DMU with an efficiency score of one is considered efficient. The decision-making units can be ranked based on their efficiency scores. However, it is possible to have more than one efficient DMU. In such cases, ranking the efficient DMU solely based on their efficiency scores is not possible. Multiple methods have been proposed in Data Envelopment Analysis to rank the efficient DMUs. One of these methods involves first calculating the distance of each efficient DMU from the positive and negative ideal DMUs. Then, DMUs are ranked based on the distance from the positive and negative ideals. Which is the same as the TOPSIS perspective. Lotfi et al. [17] applied TOPSIS method to rank efficient Decision Making Units in Data Envelopment Analysis. Jahanshahloo et al. [14] and Rakhshan [21] studied about ranking DMUs by positive and negative ideal points with interval and crisp data in DEA. Ranking efficient DMUs by TOPSIS method with fuzzy data is explained in detail below.

Step 1: Generating the Decision Making Matrix

In general, the efficiency of a Decision Making Unit (DMU) improves when it generates higher values for its outputs while having lower values for its inputs. To reflect this, we consider inputs as criteria with negative attributes and outputs as criteria with positive attributes. This approach involves creating a matrix, known as the Decision Matrix, which incorporates all the available information about the inputs and outputs of the DMUs. Each row of the matrix represents a DMU, while each column, represents any input as cost criterion and any output as benefit criterion.

Note that that since the goal is to rank the efficient DMUs, only the efficient decision-making units are considered as alternatives. therefore, efficient DMUs are included in each row.

Also note that inputs and outputs are fuzzy numbers. The decision matrix is as follows, where $(C_1, ..., C_t)$ are inputs as cost criteria and $(C_{t+1}, ..., C_n)$ are outputs as benefit criteria and $(A_1, ..., A_m)$ are efficient DMUs as the alternatives.

$$\tilde{X} = \begin{array}{c} \\ A_1 \\ A_2 \\ \vdots \\ A_m \end{array} \begin{array}{c} \begin{array}{cccccc} C_1 & \cdots & C_t & C_{t+1} & \cdots & C_n \end{array} \\ \begin{bmatrix} \tilde{x}_{11} & \cdots & \tilde{x}_{1t} & \tilde{x}_{1t+1} & \cdots & \tilde{x}_{1n} \\ \tilde{x}_{21} & \cdots & \tilde{x}_{2t} & \tilde{x}_{2t+1} & \cdots & \tilde{x}_{2n} \\ \vdots & \ddots & \vdots & \vdots & \ddots & \vdots \\ \tilde{x}_{m1} & \tilde{x}_{mt} & \tilde{x}_{mt} & \tilde{x}_{mt+1} & \cdots & \tilde{x}_{mn} \end{bmatrix} \end{array} \tag{6.48}$$

where each element $\tilde{x}_{ij} = (x_{ij}^l, x_{ij}^m, x_{ij}^u)$ is a triangle fuzzy number.

Step2: Introducing the Weight Vector

To determine weight vector different methods which mentioned in chapter3 can be applied, such as Shannon entropy method. Also, the weight of all attributes can be considered equal to one. Moreover, since benefit and cost criteria are inputs and outputs, Multiplier form of CCR model may applied to determine weights as explained bellow.

First, using the fuzzy CCR model (6.50), all decision-making units are evaluated, and the efficient DMUs are determined. Note that for crisp values if the efficiency score equals to one the evaluated DMU is considered efficient. Meanwhile, in the context of fuzzy model (6.50) if the upper bound of efficiency score is equal to one, that DMU is considered efficient. In other words, let the efficiency score of DMU_j is equal to $Eff = (a, b, c)$ if $c = 1$ then DMU_j is efficient.

After obtaining the efficient units, model (6.51) is applied to evaluate the efficient DMUs resulting in the coefficients (U,V). The weights needed for this stage are obtained by averaging the coefficients (U,V).

$$\begin{aligned} Max \quad & \sum_{j=t+1}^{n} \tilde{u}_j \tilde{x}_{jp} \\ s.t. \quad & \sum_{j=1}^{t} \tilde{v}_j \tilde{x}_{jp} = (1, 1, 1) \\ & \sum_{j=t+1}^{n} \tilde{u}_j \tilde{x}_{ji} - \sum_{j=1}^{t} \tilde{v}_j \tilde{x}_{ji} \le (0, 0, 0); \; i = 1, ..., m \\ & \tilde{v}_j \ge (0, 0, 0) \qquad , j = 1, ..., t \\ & \tilde{u}_j \ge (0, 0, 0) \qquad , j = t + 1, ..., n. \end{aligned} \tag{6.49}$$

$$
\begin{aligned}
& Max \sum_{j=t+1}^{n} \left(u_j^l x_{jp}^l, u_j^m x_{jp}^m, u_j^u x_{jp}^u\right) \\
& s.t. \sum_{j=1}^{t} v_j^l x_{jp}^l \leq 1 \\
& \sum_{j=1}^{t} v_j^m x_{jp}^m \leq 1 \\
& \sum_{j=1}^{t} v_j^u x_{jp}^u \leq 1 \\
& \sum_{j=t+1}^{n} u_j^l x_{ji}^l - \sum_{j=1}^{t} v_j^l x_{ji}^l \leq 0;\ i = 1, \ldots, m \\
& \sum_{j=t+1}^{n} u_j^m x_{ji}^m - \sum_{j=1}^{t} v_j^m x_{ji}^m \leq 0;\ i = 1, \ldots, m \\
& \sum_{j=t+1}^{n} u_j^u x_{ji}^u - \sum_{j=1}^{t} v_j^u x_{ji}^u \leq 0;\ i = 1, \ldots, m \\
& v_j^l \leq v_j^m \leq v_j^u;\ j = 1, \ldots, t, \\
& u_j^l \leq u_j^m \leq u_j^u;\ j = t+1, \ldots, n, \\
& (v_j^l, v_j^m, v_j^u) \geq (0, 0, 0);\ j = 1, \ldots, t, \\
& (u_j^l, u_j^m, u_j^u) \geq (0, 0, 0);\ j = t+1, \ldots, n.
\end{aligned} \tag{6.50}
$$

$$
\begin{aligned}
& Max \sum_{j=t+1}^{n} \left(d_j^1 + d_j^2 + d_j^3\right) \\
& s.t. \sum_{j=1}^{t} \left(v_j^l + v_j^m + v_j^u\right) + \sum_{j=t+1}^{n} \left(u_j^l + u_j^m + u_j^u\right) = 1 \\
& \sum_{j=t+1}^{n} u_j^l x_{ji}^l - \sum_{j=1}^{t} v_j^l x_{ji}^l + d_j^1 = 0;\ i = 1, \ldots, m \\
& \sum_{j=t+1}^{n} u_j^m x_{ji}^m - \sum_{j=1}^{t} v_j^m x_{ji}^m + d_j^2 = 0;\ i = 1, \ldots, m \\
& \sum_{j=t+1}^{n} u_j^u x_{ji}^u - \sum_{j=1}^{t} v_j^u x_{ji}^u + d_j^3 = 0;\ i = 1, \ldots, m \\
& 0 \leq v_j^l \leq v_j^m \leq v_j^u;\ j = 1, \ldots, t \\
& 0 \leq u_j^l \leq u_j^m \leq u_j^u;\ j = t+1, \ldots, n
\end{aligned} \tag{6.51}
$$

Let, $\tilde{V} = (\tilde{v}_1, ..., \tilde{v}_t, \tilde{u}_{t+1}, ..., \tilde{u}_n)$ is the weight vector of criteria attained by model (6.51). That $(\tilde{v}_1, ..., \tilde{v}_t)$ are weights of inputs and $(\tilde{u}_{t+1}, ..., \tilde{u}_n)$ are weights of outputs obtained by the above mentioned model. Also,$\tilde{v}_j = \left(v_j^l, v_j^m, v_j^u\right); j = 1, ..., t$ and $\tilde{u}_j = \left(u_j^l, u_j^m, u_j^u\right); j = t + 1, ..., n$ are a fuzzy triangular number that denote the importance of jth criterion.

Step 3: Calculating the Weighted Decision Matrix

The weighted decision matrix obtains as bellow:

$$\tilde{W} = \begin{bmatrix} \tilde{w}_{11} & \tilde{w}_{12} & \cdots & \tilde{w}_{1n} \\ \tilde{w}_{21} & \tilde{w}_{22} & \cdots & \tilde{w}_{2n} \\ \vdots & \vdots & \ddots & \vdots \\ \tilde{w}_{m1} & \tilde{w}_{m2} & \cdots & \tilde{w}_{mn} \end{bmatrix} \tag{6.52}$$

where,

$$\tilde{w}_{ij} = \left(w_{ij}^l, w_{ij}^m, w_{ij}^u\right) = \begin{cases} \tilde{v}_j.\tilde{x}_{ij} \simeq \left(v_j^l, v_j^m, v_j^u\right).\left(x_{ij}^l, x_{ij}^m, x_{ij}^u\right) \simeq \left(v_j^l.x_{ij}^l, v_j^m.x_{ij}^m, v_j^u.x_{ij}^u\right);\ j = 1, ..., t \\ \tilde{u}_j.\tilde{x}_{ij} \simeq \left(u_j^l, u_j^m, u_j^u\right).\left(x_{ij}^l, x_{ij}^m, x_{ij}^u\right) \simeq \left(u_j^l.x_{ij}^l, u_j^m.x_{ij}^m, u_j^u.x_{ij}^u\right);\ j = t+1, ..., n \end{cases}$$

Step 4: Obtaining Positive and Negative Ideal Points

In data envelopment analysis, a DMU with lower input consumption and higher desired output production is considered favorable. Equations (6.53) and (6.54) are used to determine the positive and negative ideals, respectively.

$$\tilde{A}^+ = \left\{\tilde{w}_1^+, ..., \tilde{w}_t^+, \tilde{w}_{t+1}^+, ..., \tilde{w}_n^+\right\}$$

$$\tilde{w}_j^+ = \left(w_j^{+l}, w_j^{+m}, w_j^{+u}\right) \ that\ w_j^{+l} = w_j^{+m} = w_j^{+u} = \begin{cases} \min\limits_i(\tilde{w}_{ij}^l);\ j = 1, ..., t \\ \max\limits_i(\tilde{w}_{ij}^u);\ j = t+1, ..., n \end{cases} \tag{6.53}$$

$$\tilde{A}^- = \left\{\tilde{w}_1^-, ..., \tilde{w}_t^-, \tilde{w}_{t+1}^-, ..., \tilde{w}_n^-\right\}$$

$$\tilde{w}_j^- = \left(w_j^{-l}, w_j^{-m}, w_j^{-u}\right) \ that\ w_j^{-l} = w_j^{-m} = w_j^{-u} = \begin{cases} \max\limits_i(\tilde{w}_{ij}^u);\ j = 1, ..., t \\ \min\limits_i(\tilde{w}_{ij}^l);\ j = t+1, ..., n \end{cases} \tag{6.54}$$

Step 5: Computing the Distance of Each Alternative from the Ideal Solutions.

The distance of each alternative $\tilde{w}_{ij} = \left(w_{ij}^l, w_{ij}^m, w_{ij}^u\right)$ from the positive ideal point $\tilde{w}_j^+ = \left(w_j^{+l}, w_j^{+m}, w_j^{+u}\right)$ and negative ideal point $\tilde{w}_j^- = \left(w_j^{-l}, w_j^{-m}, w_j^{-u}\right)$ are calculated respectively by Eqs. (6.55) and (6.56).

$$d_i^+ = \sum_{j=1}^{n} d(\tilde{w}_{ij}, \tilde{w}_j^+) = \sum_{j=1}^{n} \sqrt{\frac{1}{3}\left(\left(w_{ij}^l - w_j^{+l}\right)^2 + \left(w_{ij}^m - w_j^{+m}\right)^2 + \left(w_{ij}^u - w_j^{+u}\right)^2\right)}; \; i = 1, \ldots, m \tag{6.55}$$

$$d_i^- = \sum_{j=1}^{n} d(\tilde{w}_{ij}, \tilde{w}_j^-) = \sum_{j=1}^{n} \sqrt{\frac{1}{3}\left(\left(w_{ij}^l - w_j^{-l}\right)^2 + \left(w_{ij}^m - w_j^{-m}\right)^2 + \left(w_{ij}^u - w_j^{-u}\right)^2\right)}; \; i = 1, \ldots, m \tag{6.56}$$

Step 6: Computing the Relative Closeness Index and Ranking the Alternatives.

The relative closeness index is computed by Eq. (6.57).

$$RC_i = \frac{d_i^-}{(d_i^- + d_i^+)}; \quad i = 1, \ldots, m \tag{6.57}$$

The closer the alternative A_i is to positive ideal solution, the closer the value of RC_i will be to one. Since the aim is to rank the efficient DMUs, any efficient decision making unit with highest value of index RC_i represent the best efficient DMU.

6.8 Conclusion

In this chapter, we have explored the TOPSIS method in the fuzzy environment and its application in various contexts. We have discussed different approaches and presented several case studies where the TOPSIS method has been successfully implemented. By incorporating fuzzy logic, the method enables the handling of imprecise and uncertain information, allowing decision-makers to navigate complex and ambiguous scenarios. One notable application of the TOPSIS method discussed in this chapter was the selection of suitable locations for sports centers. We examined the criteria involved in the decision-making process, such as accessibility to transportation, weather conditions, population density, safety, price, size, difficulty of acquisition, and geomorphic conditions. By employing the TOPSIS method, decision-makers can systematically evaluate and rank potential alternatives based on their performance across these criteria, facilitating the establishment of sports spaces that promote accessibility, safety, and overall well-being within the community. Additionally, we explored the application of the TOPSIS method in the evaluation and ranking of job applicants in the context of human resources management. By considering criteria such as personality traits, work experience, analytical power, problem-solving ability, leadership and management skills, and job stability, organizations can make informed decisions when selecting candidates for employment. The integration of fuzzy logic allows for a more comprehensive assessment of candidates, considering subjective assessments and capturing the inherent uncertainties present in the evaluation process. The idea of TOPSIS has been discussed and examined for evaluating efficient decision-making units with triangular fuzzy data in data envelopment analysis. Overall, the TOPSIS method in the fuzzy environment has proven to be a valuable tool for decision-makers across various domains. Its ability to handle imprecise and uncertain information, combined with its practicality and effectiveness, makes it a suitable choice for complex decision-making problems. As we conclude this chapter, it is important to note that the TOPSIS method is not limited to the applications discussed here. Its versatility allows for its utilization in a wide range of scenarios, from strategic planning and project selection to supplier evaluation and investment decision-making. By incorporating the TOPSIS method into their decision-making processes, organizations and decision-makers can enhance their ability to make informed and robust decisions, taking into account multiple criteria and navigating the complexities of the fuzzy environment. Furthermore, the application of TOPSIS in fuzzy data envelopment analysis has been discussed.

Acknowledgement A special thanks to the Iranian DEA society for their unwavering spiritual support and consensus in the writing of this book. Your invaluable support has been truly remarkable, and we are deeply grateful for the opportunity to collaborate with such esteemed professionals.

References

1. Atanassov, K.T.: Intuitionistic fuzzy sets. Physica-Verlag HD, pp. 1–137 (1999)
2. Afshar, A., Mariño, M. A., Saadatpour, M., & Afshar, A. (2011). Fuzzy TOPSIS multi-criteria decision analysis applied to Karun reservoirs system. *Water Resources Management, 25*, 545–563.
3. Azadeh, A., Kor, H., & Hatefi, S. M. (2011). A hybrid genetic algorithm-TOPSIS-computer simulation approach for optimum operator assignment in cellular manufacturing systems. *Journal of the Chinese Institute of Engineers, 34*(1), 57–74.
4. Behzadian, M., Otaghsara, S. K., Yazdani, M., & Ignatius, J. (2012). A state-of the-art survey of TOPSIS applications. *Expert Systems with Applications, 39*(17), 13051–13069.
5. Chen, C. T. (2000). Extensions of the TOPSIS for group decision-making under fuzzy environment. *Fuzzy Sets and Systems, 114*(1), 1–9.
6. Chu, T. C., & Lin, Y. C. (2003). A fuzzy TOPSIS method for robot selection. *The International Journal of Advanced Manufacturing Technology, 21*, 284–290.
7. Gupta, P., Mehlawat, M. K., Grover, N., & Pedrycz, W. (2018). Multi-attribute group decision making based on extended TOPSIS method under interval-valued intuitionistic fuzzy environment. *Applied Soft Computing, 69*, 554–567.
8. Huang, J. H., & Peng, K. H. (2012). Fuzzy Rasch model in TOPSIS: A new approach for generating fuzzy numbers to assess the competitiveness of the tourism industries in Asian countries. *Tourism Management, 33*(2), 456–465.
9. Hwang, C. L., Yoon, K., Hwang, C. L., & Yoon, K. (1981). Methods for multiple attribute decision making. *Multiple attribute decision making: methods and applications a state-of-the-art survey*, 58–191.

10. Jahanshahloo, G. R., Khodabakhshi, M., Lotfi, F. H., & Goudarzi, M. M. (2011). A cross-efficiency model based on super-efficiency for ranking units through the TOPSIS approach and its extension to the interval case. *Mathematical and Computer Modelling, 53*(9–10), 1946–1955.
11. Jahanshahloo, G. R., Lotfi, F. H., & Davoodi, A. R. (2009). Extension of TOPSIS for decision-making problems with interval data: Interval efficiency. *Mathematical and Computer Modelling, 49*(5–6), 1137–1142.
12. Jahanshahloo, G. R., Lotfi, F. H., & Izadikhah, M. (2006). Extension of the TOPSIS method for decision-making problems with fuzzy data. *Applied Mathematics and Computation, 181*(2), 1544–1551.
13. Jahanshahloo, G. R., Lotfi, F. H., & Izadikhah, M. (2006). An algorithmic method to extend TOPSIS for decision-making problems with interval data. *Applied Mathematics and Computation, 175*(2), 1375–1384.
14. Jahanshahloo, G. R., Lotfi, F. H., Rezaie, V., & Khanmohammadi, M. (2011). Ranking DMUs by ideal points with interval data in DEA. *Applied Mathematical Modelling, 35*(1), 218–229.
15. Lai, Y. J., Liu, T. Y., & Hwang, C. L. (1994). TOPSIS for MODM. *European journal of Operational Research, 76*(3), 486–500.
16. Lotfi, F. H., Allahviranloo, T., Jondabeh, M. A., & Kiani, N. A. (2007). A new method for complex decision making based on TOPSIS for complex decision making problems with fuzzy data. *Applied Mathematical Sciences, 1*(60), 2981–2987.
17. Lotfi, F. H., Fallahnejad, R., & Navidi, N. (2011). Ranking efficient units in DEA by using TOPSIS method. *Applied Mathematical Sciences, 5*(17), 805–815.
18. Memari, A., Dargi, A., Jokar, M. R. A., Ahmad, R., & Rahim, A. R. A. (2019). Sustainable supplier selection: A multi-criteria intuitionistic fuzzy TOPSIS method. *Journal of Manufacturing Systems, 50*, 9–24.
19. Olson, D. L. (2004). Comparison of weights in TOPSIS models. *Mathematical and Computer Modelling, 40*(7–8), 721–727.
20. Ouyang, Y., & Pedrycz, W. (2016). A new model for intuitionistic fuzzy multi-attributes decision making. *European Journal of Operational Research, 249*(2), 677-682.
21. Rakhshan, S. A. (2017). Efficiency ranking of decision making units in data envelopment analysis by using TOPSIS-DEA method. *Journal of the Operational Research Society, 68*, 906–918.
22. Roszkowska, E. (2011). Multi-criteria decision making models by applying the TOPSIS method to crisp and interval data. *Multiple Criteria Decision Making/University of Economics in Katowice, 6*(1), 200–230.
23. Shakouri, B., Abbasi Shureshjani, R., Daneshian, B., & Hosseinzadeh Lotfi, F. (2020). A parametric method for ranking intuitionistic fuzzy numbers and its application to solve intuitionistic fuzzy network data envelopment analysis models. *Complexity, 2020*, 1-25.
24. Shariatmadari Serkani, E., Hosseinzadeh Lotfi, F., Najafi, E., & Ahadzadeh Namin, M. (2022). Efficiency measurement for hierarchical network systems using network DEA and intuitionistic fuzzy ANP. *Scientia Iranica, 29*(4), 2252-2269.
25. Szmidt, E., & Kacprzyk, J. (2000). Distances between intuitionistic fuzzy sets. *Fuzzy Sets and Systems, 114*(3), 505–518.
26. Vahdani, B., Mousavi, S. M., & Tavakkoli-Moghaddam, R. (2011). Group decision making based on novel fuzzy modified TOPSIS method. *Applied Mathematical Modelling, 35*(9), 4257–4269.
27. Xie, D., Xiao, F., & Pedrycz, W. (2022). Information quality for intuitionistic fuzzy values with its application in decision making. *Engineering Applications of Artificial Intelligence, 109*, 104568.

Chapter 7
Elimination Choice Translating Reality (ELECTRE) in Uncertainty Environment

7.1 Introducing Different Versions of the ELECTRE

The ELECTRE (ELimination and Choice Translating REality) family of methods has been widely used in multi-criteria decision analysis. Over the years, various versions and extensions of ELECTRE have been developed, each offering unique features and improvements to address different decision-making scenarios. This chapter provides a comprehensive overview of the different versions of ELECTRE methods, highlighting their advancements, and applications. The ELECTRE proposed by Roy in 1991. Since then, researchers have made significant contributions to enhance the methodology, address its limitations, and adapt it to diverse decision contexts. This chapter talk about the evolution of ELECTRE methods, ranging from ELECTRE I to ELECTRE III, as well as their fuzzy extensions, such as ELECTRE with fuzzy sets and hesitant fuzzy sets. By exploring the distinct characteristics of each version, readers will gain insights into the strengths, weaknesses, and practical considerations associated with different ELECTRE variants. Furthermore, this chapter sheds light on recent extensions and variations of ELECTRE, including integration with utility theory, handling interval-valued intuitionistic fuzzy information, and incorporating linguistic term sets. Understanding the evolution and capabilities of ELECTRE methods is crucial for decision-makers, analysts, and researchers involved in multi-criteria decision analysis. This chapter serves as a valuable resource to grasp the foundations, advancements, and practical applications of ELECTRE methods, enabling informed decision-making in complex and uncertain decision environments [5, 6].

F. Hosseinzadeh Lotfi et al., *Fuzzy Decision Analysis: Multi Attribute Decision Making Approach*, Studies in Computational Intelligence 1121,
https://doi.org/10.1007/978-3-031-44742-6_7

7.2 Electre I

ELECTRE I is the pioneering method in multi-criteria decision-making. With its qualitative approach, it revolutionized decision analysis by incorporating preferences and judgments, handling uncertainty, and offering computational efficiency. From engineering to finance, its practical applications have made it a go-to tool for informed decision-making in various domains. In a multi-criteria decision making problem where the decision maker is supposed to choose the best alternative among "m" alternative based on "n" criteria [2, 14, 16].

Step 1. Construct the decision matrix.

According to the number of criteria, the number of alternatives and the evaluation of all alternative for different criteria, the decision matrix is formed as follows.

$$X = \begin{matrix} & \begin{matrix} C_1 & \dots & C_n \end{matrix} \\ & \begin{bmatrix} x_{11} & \dots & x_{1n} \\ \vdots & \ddots & \vdots \\ x_{m1} & \dots & x_{mn} \end{bmatrix} \end{matrix} \tag{7.1}$$

where x_{ij} is the score of the alternative $i(i = 1, ..., m)$ related to the criterion $j(j = 1, ..., n)$.

Step 2. Construct the normalize decision matrix.

At this step, based on existing methods, the decision matrix is normalized. Naturally, all cost criterion are converted into benefit criterion and they will be treated the same.

$$\begin{cases} n_{ij} = \dfrac{x_{ij}}{\sum_{i=1}^{m} x_{ij}}, & i = 1, ..., m;\ j \in B \\ n_{ij} = \dfrac{\frac{1}{x_{ij}}}{\sum_{i=1}^{m} \frac{1}{x_{ij}}}, & i = 1, ..., m;\ j \in C \end{cases} \tag{7.2}$$

Using Eq. (7.2), the normalized matrix will be as follows.

$$N = \begin{bmatrix} n_{11} & \dots & n_{1n} \\ \vdots & \ddots & \vdots \\ n_{m1} & \dots & n_{mn} \end{bmatrix} \tag{7.3}$$

Step 3: Determining the criteria's weights.

At this step, according to the importance of different criteria in decision-making, the weight vector of the criteria is defined as $w_j (j = 1, ..., n)$.

$$W = [w_1, w_2, ..., w_n]$$

Step 4. Determining the weighted decision matrix.

The weighted decision matrix is obtained by multiplying the normalized matrix by the criteria's weight vector and its elements are as follows.

$$v_{ij} = w_j n_{ij}, i = 1, ..., m; \; j = 1, ..., n. \tag{7.4}$$

Step 5. Determining the concordance and discordance sets.

For both alternative t and b, the set of criteria is divided into two groups: concordance and discordance. The concordance set (S_{tb}) is the set of criteria where alternative t outperformed (preferred) b. And the complementary set is discordance set (I_{tb}). The set of concordance criteria are determined as follows.

$$S_{tb} = \{j | v_{tj} \geq v_{bj}\}, (t, b = 1, ..., m). \tag{7.5}$$

$$I_{tb} = \{j | v_{tj} < v_{bj}\}, (t, b = 1, ..., m). \tag{7.6}$$

It is important to note that in the method chosen to normalize the points, the behavior of all of them is beneficial. If another method of normalization is used, the above definition should be revised.

Step 6. Construct the concordance matrix.

The concordance matrix is a square matrix whose dimension is the number of alternatives. Each element of this matrix is called concordance index between two alternatives. Which is calculated as follows.

$$c_{tb} = \frac{\sum\limits_{j \in S_{tb}} w_j}{\sum\limits_{j=1}^{n} w_j}, (t, b = 1, ..., m). \tag{7.7}$$

The concordance index indicates the superiority of alternative t over alternative b and its value is between zero and one. In general, this matrix is not symmetric.

$$C = \begin{bmatrix} - & c_{12} & \dots & c_{1m} \\ c_{21} & - & \dots & c_{2m} \\ \vdots & & \ddots & \vdots \\ c_{m1} & & \dots \ c_{m(m-1)} & - \end{bmatrix} \tag{7.8}$$

Step 7. Construct the discordance matrix

The discordance matrix is a square matrix whose dimension is the number of alternatives. Each element of this matrix is called discordance index between two alternatives. The value of this index is obtained from the following relationship.

$$d_{tb} = \frac{\underset{j\in I_{tb}}{Max} |v_{tj} - v_{bj}|}{\underset{j\in j}{Max} |v_{tj} - v_{bj}|}, (t, b = 1, ..., m). \tag{7.9}$$

By calculating the discordance index for all pairs of alternatives, the discordance matrix can be defined as follows. In general, this matrix is not symmetric.

$$D = \begin{bmatrix} - & d_{12} \dots & & d_{1m} \\ d_{21} & - & \dots & d_{2m} \\ \vdots & & \ddots & \vdots \\ d_{m1} \dots & d_{m(m-1)} & & - \end{bmatrix} \tag{7.10}$$

Step 8. Determine the Boolean concordance matrix.

In the sixth step, how to calculate the (c_{tb}) index was explained. In this step, a certain value for the concordance index is determined, which is called the concordance threshold and is represented by $\overline{c}$. The concordance index is obtained by averaging the concordance indices (elements of the concordance matrix) or in other words it is calculated as follows.

$$\overline{c} = \sum_{t=1}^{m} \sum_{\substack{b=1 \\ b \neq t}}^{m} \frac{c_{tb}}{m(m-1)}$$

The Boolean concordance matrix (F) is formed according to the concordance threshold value. If c_{tb} is greater than $\overline{c}$, the superiority of alternative t over b is acceptable, otherwise, alternative t is not superior to b. Therefore, the terms of the Boolean matrix are determined from the following equation.

$$f_{tb} = \begin{cases} 1, & c_{tb} \geq \overline{c} \\ 0, & c_{tb} < \overline{c} \end{cases} \tag{7.11}$$

Step 9. Determine the Boolean discordance matrix.

As in the previous step, the Boolean discordance matrix (G) can be formed. For this purpose, the discordance threshold $(\overline{d})$ should be calculated from the averaging of the discordance indices (elements of the discordance matrix).

$$\overline{d} = \sum_{t=1}^{m} \sum_{\substack{b=1 \\ b \neq t}}^{m} \frac{d_{tb}}{m(m-1)} \tag{7.12}$$

As stated in the seventh step, the lower the (d_{tb}) value, the better. So the elements of Boolean discordance matrix are calculated as follows.

$$g_{tb} = \begin{cases} 0 & , g_{tb} > \overline{d} \\ 1 & , g_{tb} \leq \overline{d} \end{cases} \tag{7.13}$$

Step 10. Construct the global matrix H.

This matrix is obtained by multiplying the elements of Boolean concordance matrix and Boolean discordance matrix.

$$h_{tb} = f_{tb} \cdot g_{tb}, (t, b = 1, \ldots, m). \tag{7.14}$$

Step 10. Choosing the best alternative.

The matrix H expresses the partial preferences of the alternatives. For example, if the value h_{tb} is equal to one, it means that the superiority of the alternative t over the alternative b is acceptable in both concordance and discordance states. An alternative should be chosen that dominates rather than defeats. From this point of view, the alternative can be ranked.

7.3 Electre II

ELECTRE II, an evolution of the ELECTRE family, takes multi-criteria decision-making to new heights. Building upon the foundations of its predecessor, ELECTRE I, this method introduces enhanced ranking techniques and a refined treatment of imprecision. By incorporating advanced decision rules, ELECTRE II provides decision-makers with a robust framework to navigate complex decision landscapes and make well-informed choices efficiently.

The ELECTRE II method is the closest method from the ELECTRE method family to ELECTRE I. Its difference with the ELECTRE I method is in the definition of two prioritization relations in this method instead of one relation of the ELECTRE I method. In fact, the expression of the intensity of preference is clearly expressed in the ELECTRE II method. The steps of the ELECTRE II method are completely similar to the ELECTRE I method in several steps, and the only difference in this method is in defining the thresholds and identifying the advantages and preferences of the alternatives, and the steps of this method are as follows [10].

Steps 1 to 7 are identical to those of the ELECTRE I method.

Step 8. Concordance threshold and discordance threshold.

This step marks the starting point of the difference between the ELECTRE I and ELECTRE II methods. In this stage, four threshold values need to be determined as follows:

C^*	Threshold of strong concordance
D^*	Threshold of weak discordance
C^-	Threshold of weak concordance
D^-	Threshold of weak discordance

These four thresholds are typically determined by group decision-making methods. However, in some situations where access to experts is not feasible, the decision-maker can set these thresholds themselves. It is important to note that the condition of $0 < C^- < C^* < 1$ and $0 < D^- < D^* < 1$ must always be satisfied.

Step 9: Formation of the final dominance comparison matrix.

This matrix encompasses the relationships between all pairs of alternatives in terms of dominance. The alternatives are compared pairwise to construct this matrix. Based on the concordance matrix (C) and the discordance matrix (D), the following two dominance conditions are evaluated for each pair of alternatives, denoted as alternative a and alternative b:

I. If ($c_{tb} \geq c_{bt}$, $d_{tb} \leq D^*$, $c_{tb} \geq C^*$) alternative "t" strongly dominates alternative "b," the symbol S^F is assigned to the corresponding element in the final dominance matrix.
II. ($c_{tb} \geq c_{bt}$, $d_{tb} \leq D^-$, $c_{tb} \geq C^-$) If alternative "t" weakly dominates alternative "b," the symbol S^f is assigned to the corresponding element in the matrix.

Elements that do not meet the above conditions remain empty.

Step 10. Performing the Ranking Process.

After constructing the final dominance matrix for the alternatives based on their mastery comparison, the alternatives are ranked in two ways in terms of dominance.

I. Decreasing Order: In this approach, the alternatives are ranked from the best to the worst based on the final dominance matrix. The best alternative is the one that exhibits higher and stronger dominance compared to other alternatives. In other words, the alternative with more S^F and S^f symbols in the rows of the final dominance matrix will have a higher degree of dominance.

II. Increasing Order: In this approach, the alternatives are ranked from the worst to the best, starting with the least favorable alternative. The worst alternative is the one that is more dominated by other alternatives. In other words, the alternative with more S^F and S^f symbols in the columns of the final dominance matrix indicates a higher degree of being dominated.

7.4 Electre III

The ELECTRE III method is considered the strongest among the ELECTRE methods due to its sensitivity to preference thresholds and its consideration of various types of relationships between alternatives. It is worth mentioning that implementing this method requires more time compared to other methods, but the high accuracy and capability of this approach justify the challenges it presents. The steps of this method are as follows [9].

Step 1: Construct the decision matrix, similar to Step 1 of the ELECTRE I method.

Step 2: Construct the normalized matrix, similar to Step 2 of the ELECTRE I method.

Step 3: Define thresholds and assign weights to criteria.

The names and abbreviations used for the thresholds are as follows:

Indifference threshold, preference threshold, and veto threshold. Each of these thresholds is defined separately for each criterion. For example, in criterion "k" the indifference, preference, and veto thresholds are denoted by p_k, q_k and τ_k respectively. When comparing two alternatives, "t" and "b" with respect to criterion "k" five possible cases can occur.

I. If $-q_k < n_{tk} - n_{bk} < q_k$, then alternative "t" is approximately equal to alternative "b" and is denoted by tI_kb.

II. If $q_k < n_{tk} - n_{bk} < p_k$, then alternative "t" prefers alternative "b" and is denoted by tQ_kb.

III. If $p_k < n_{tk} - n_{bk}$, then alternative "t" completely rejects alternative "b" and is denoted by tP_kb.

IV. If $-p_k < n_{tk} - n_{bk} < q_k$, then alternative "b" prefers alternative "t" and is denoted by bQ_kt.

V. If $n_{tk} - n_{bk} < -p_k$, then alternative "b" completely rejects alternative "t" and is denoted by tP_kb.

Step 4: Formation of the concordance matrix for each criterion.

In this step, using the decision matrix formed in Step 1 and the thresholds determined in Step 2, a concordance matrix is constructed for each criterion based on the relationships between alternatives.

$$C_k(t,b) = \begin{cases} \frac{g_{tk}+p_{bk}-g_{bk}}{p_{bk}-q_{bk}} & , q_{bk} < g_{bk} - g_{tk} < p_{bk}, \\ 1 & , g_{bk} - g_{tk} \le q_{bk} \\ 0 & , p_{bk} < g_{tk} - q_{bk} \end{cases} \tag{7.15}$$

Step 5: Construction of the overall concordance matrix.

Based on the matrices constructed for each criterion, the overall concordance matrix is formed by weighted averaging.

$$C(t,b) = \frac{\sum\limits_{k=1}^{n} w_k C_k(t,b)}{\sum\limits_{k=1}^{n} w_k} \tag{7.16}$$

where w_k is the weight of the kth criterion.

Step 6. Construction of the discordance matrix for each criterion.

In this step, similar to Step 4, the discordance matrices for the alternatives are constructed for each criterion using the decision matrix and the determined thresholds.

$$D_k(t,b) = \begin{cases} \dfrac{g_{tk}+p_{bk}-g_{bk}}{\tau_{bk}-p_{bk}} & , p_{bk} < g_{bk} - g_{tk} < \tau_{bk}; \\ 1 & , \tau_{bk} < q_{bk} - g_{tk}; \\ 0 & , g_{bk} - g_{tk} \le p_{bk}. \end{cases} \tag{7.17}$$

Step 7. Construction of the overall discordance matrix.

Based on the matrices constructed for each criterion, the overall discordance matrix is created by applying weighted averaging.

$$D(t,b) = \frac{\sum\limits_{k=1}^{n} w_k D_k(t,b)}{\sum\limits_{k=1}^{n} w_k} \tag{7.18}$$

where w_k is the weight of the kth criterion.

Step 8. Formation of the credibility matrix.

In this step, the credibility matrix (S) is formed between different alternatives, and its elements are calculated as follows.

$$S(t,b) = \begin{cases} C(t,b) & , D(t,b) \leq C(t,b) \\ C(t,b)\prod\limits_{k} \frac{1-D_k(t,b)}{1-C(t,b)}, & otherwise \end{cases} \tag{7.19}$$

Step 9. Formation of the final comparison matrix.

Based on the credibility matrix (S), the λ, $s(\lambda)$ is calculated as follow.

$$\lambda = \max(S) \tag{7.20}$$

$$S(\lambda) = 0.3 + 0.15\lambda \tag{7.21}$$

Then, the final comparison matrix (T) is formed as follows:

$$T(\mathrm{t,b}) = \begin{cases} 1 \text{ , } S(\mathrm{t,b}) > \lambda - \mathrm{S}(\lambda) \\ 0 \text{ , } otherwise \end{cases} \tag{7.22}$$

Step 10. Performing the Ranking Process.

After the formation of the final matrix, the alternatives are ranked to prioritize them for further consideration. The alternatives are sorted once from the best state to the worst state, and once from the worst state to the best state. Finally, by comparing the descending and ascending rankings, the final ranking of the alternatives is obtained.

7.5 Electre IV

ELECTRE IV method is one of the less commonly used versions of the ELECTRE method. This method is typically employed in priority ranking problems with criteria of equal importance. In this approach, alternatives are compared, preferred, and sometimes excluded from each other in seven possible cases. Then, the weak and strong priorities among the alternatives are determined. The results obtained from the dominance relationships can be represented as a matrix, which bears a strong resemblance to the dominance matrix in ELECTRE III. In this method, to prioritize the alternatives, they are sorted once from the best state to the worst state and once from the worst state to the best state. Finally, by comparing the descending and ascending rankings, the final ranking of the alternatives is determined [1].

This subsection presents an overview of the primary limitations or disadvantages associated with the utilization of ELECTRE methods. This is particularly evident when the quantitative nature of the criteria necessitates the adoption of an alternative approach, when assigning a score to each action becomes necessary, when ensuring independence from irrelevant actions and maintaining the stability of the action set are essential requirements, or when the occurrence of intransitivities poses potential challenges. In certain scenarios, it becomes essential to attribute a score to each individual action. In cases where decision-makers mandate the assignment of scores to all actions, the ELECTRE methods are not suitable for fulfilling this objective. Instead, it is advisable to employ scoring methods. However, decision-makers should be aware that the use of scoring methods prevents the provision of information that may lead to instances of intransitivities or incomparabilities between certain pairs of actions. Furthermore, it should be noted that the score itself is highly delicate. When all the criteria within a family possess a quantitative nature, it is considered preferable to employ alternative methods. Nevertheless, if the intention is to incorporate a noncompensatory approach, either partially or completely, along with considering the supporting and opposing reasons, even in situations where all the criteria are quantitative, ELECTRE methods should be utilized. This is applicable even when dealing with imperfect knowledge regarding at least one criterion.

One of the topics that should be considered in the ELECTRE method is the phenomenon of rank reversal. which is one of the main factors of this incident, The presence of discriminating thresholds and the appropriate values that should have been assigned to them.

7.6 ELECTRE I for Prioritizing Parks

Parks are an essential part of urban living and contribute to the quality of life in cities. However, the prioritization of parks can vary depending on the factors considered. In this section, we will explore the importance of various indicators such as the quality of equipment and facilities, the quality of amenities and infrastructure, safety, distance to the city center, greenery, and area size in prioritizing parks.

Quality of Equipment and Facilities: The quality of equipment and facilities is a crucial factor in the prioritization of parks. Adequate and well-maintained equipment and facilities ensure that visitors have an enjoyable experience. Parks with high-quality equipment and facilities are likely to attract more visitors, leading to increased usage and benefits for the community. Therefore, the quality of equipment and facilities should be considered when prioritizing parks. (C1-Benefit).

Quality of Amenities and Infrastructure: The quality of amenities and infrastructure is another essential factor in prioritizing parks. Well-maintained and accessible amenities such as restrooms, picnic areas, and walking paths are necessary for a pleasant park experience. Additionally, infrastructure such as parking areas and transportation alternatives play a vital role in park accessibility. Parks that are easily accessible to the public are more likely to be used, providing greater benefits to the community. Therefore, the quality of amenities and infrastructure should also be considered when prioritizing parks. (C2-Benefit).

Safety: Safety is a critical factor in the prioritization of parks. Parks that are safe and secure are more likely to be used by the community. Measures such as adequate lighting, security personnel, and surveillance systems can improve the safety of parks. Prioritizing safety can increase park usage and improve the quality of life in the surrounding community. (C3-Benefit).

Distance to the City Center: The distance of a park from the city center is another important factor to consider when prioritizing parks. Parks that are located closer to the city center are more accessible to the public and tend to be more frequently used. These parks also have the potential to provide a more significant impact on the community, as they are often located in densely populated areas. Therefore, parks that are closer to the city center should be prioritized when considering the location of parks. (C4-Cost).

Greenery and Beauty: The amount of greenery and beauty of a park is also an essential factor in prioritizing parks. Parks with a significant amount of greenery and scenic beauty offer a serene and peaceful environment for visitors. Additionally, these parks can contribute to the overall aesthetic of the community, providing an attractive destination for residents and visitors. Therefore, parks with abundant greenery and natural beauty should be considered when prioritizing parks. (C5-Benefit).

Area Size: The size of a park is also an important factor to consider when prioritizing parks. Larger parks have more space for a variety of activities and amenities, allowing for greater usage and more significant benefits to the community. However, smaller parks can be suitable for more localized activities and may be more accessible to the public. Therefore, the area size of parks should be considered when prioritizing parks (C6-Benefit).

In conclusion, the prioritization of parks should consider various indicators such as the quality of equipment and facilities, the quality of amenities and infrastructure, safety, distance to the city center, greenery, and area size. These factors contribute to the accessibility and overall quality of parks, providing benefits to the community. Parks that meet these indicators should be prioritized, ensuring that they provide maximum benefits to the surrounding community.

The prioritization of parks has been performed step-by-step using the ELECTRE I method.

Table 7.1 Decision matrix of parks evlaution

	C1	C2	C3	C4	C5	C6
A1	80	80	7	3400	8	12000
A2	25	20	8	485	3	1890
A3	40	55	5	3790	5	4570
A4	20	60	7	4565	4	6140
A5	30	35	9	2875	6	2280

Table 7.2 Normalizing the decision matrix of parks evaluation

	C1	C2	C3	C4	C5	C6
A1	0.4103	0.32	0.1944	0.2249	0.0923	0.4464
A2	0.1282	0.08	0.2222	0.0321	0.647	0.0703
A3	0.2051	0.22	0.1389	0.2507	0.0828	0.17
A4	0.1026	0.24	0.1944	0.302	0.0687	0.2284
A5	0.1538	0.14	0.25	0.1902	0.1091	0.0848

Table 7.3 Weights of criteria

Criteria	C1	C2	C3	C4	C5	C6
Weights	0.1487	0.1629	0.1935	0.169	0.1731	0.1527

Step 1: Constructing the Decision Matrix.

In this step, a decision matrix is created to represent the evaluation of parks. Each row of the matrix corresponds to a park, and each column represents a criterion. The values in the matrix indicate the performance or score of each park on each criterion (Table 7.1).

Step 2: Normalizing the decision matrix.

In this step, the decision matrix is normalized using Eq. (7.2), taking into account that indicator C4 represents cost (Table 7.2).

Step 3: Determining the weights of criteria (Table 7.3)

The weights of each criterion are determined based on expert opinions as shown in Table 7.3:

Step 4. Determining the Weighted Normalized Matrix.

By multiplying the weight of each criterion with its corresponding column, the weighted normalized matrix is obtained as follows (Table 7.4).

Step 5. Formation of Concordance and Discordance Sets (Table 7.5).

Step 6: Formation of the Concordance Matrix (Table 7.6).

Step 7: Formation of the Discordance Matrix (Table 7.7).

Step 8: Formation of the Concordance Boolean Dominance Matrix.

Table 7.4 The weighted normalized matrix of parks evaluation

	C1	C2	C3	C4	C5	C6
A1	0.061	0.0521	0.0376	0.038	0.016	0.0682
A2	0.0191	0.013	0.043	0.0054	0.112	0.0107
A3	0.0305	0.0358	0.0269	0.0424	0.0143	0.026
A4	0.0153	0.0391	0.0376	0.051	0.0119	0.0349
A5	0.0229	0.0228	0.0484	0.0321	0.0189	0.013

Table 7.5 Concordance and discordance sets of parks evaluation

Pairwise Comparisons	Concordance Set S	Discordance Set I
A1 compared to A2	C1, C2, C4, C6	C3, C5
A1 compared to A3	C1, C2, C3, C5, C6	C4
A1 compared to A4	C1, C2, C3, C5, C6	C4
A1 compared to A5	C1, C2, C4, C6	C3, C5
A2 compared to A1	C3, C5	C1, C2, C4, C6
A2 compared to A3	C3, C5	C1, C2, C4, C6
A2 compared to A4	C1, C3, C5	C2, C4, C6
A2 compared to A5	C5	C1, C2, C3, C4, C6
A3 compared to A1	C4	C1, C2, C3, C5, C6
A3 compared to A2	C1, C2, C4, C6	C3, C5
A3 compared to A4	C1, C5	C2, C3, C4, C6
A3 compared to A5	C1, C2, C4	C3, C5, C6
A4 compared to A1	C4	C1, C2, C3, C5, C6
A4 compared to A2	C2, C4, C6	C1, C3, C5
A4 compared to A3	C2, C3, C4, C6	C1, C5
A4 compared to A5	C2, C4, C6	C1, C3, C5
A5 compared to A1	C3, C5	C1, C2, C4, C6
A5 compared to A2	C1, C2, C3, C4, C6	C5
A5 compared to A3	C3, C5, C6	C1, C2, C4
A5 compared to A4	C1, C3, C5	C2, C4, C6

In this step, based on the average of the Concordance matrix, $\overline{c} = 0.5$ is calculated, resulting in the formation of the Concordance Boolean dominance matrix (Table 7.8).

Step 9. Formation of the Discordant Boolean Dominance Matrix $\left(\overline{d} = 0.7385\right)$ (Table 7.9).

Step 10. Formation of the Final Dominance Matrix

Table 7.6 Concordance Matrix of parks evaluation

	A1	A2	A3	A4	A5
A1	–	0.6334	0.8310	0.8310	0.6334
A2	0.3666	–	0.3666	0.5153	0.1731
A3	0.1690	0.6334	–	0.3218	0.4807
A4	0.1690	0.4847	0.6782	–	0.4847
A5	0.3666	0.8269	0.5193	0.5153	–

Table 7.7 Discordance matrix of parks evaluation

	A1	A2	A3	A4	A5
A1	–	1.0000	0.1033	0.0854	0.5519
A2	0.5982	–	0.3784	0.4558	0.2870
A3	1.0000	1.0000	–	0.7049	1.0000
A4	1.0000	1.0000	1.0000	–	1.0000
A5	1.0000	1.0000	0.6061	1.0000	–

Table 7.8 The concordance boolean dominance matrix of parks evaluation

	A1	A2	A3	A4	A5
A1	–	1	1	1	1
A2	0	–	0	1	0
A3	0	1	–	0	0
A4	0	0	1	–	0
A5	0	1	1	1	–

Table 7.9 Discordant boolean dominance matrix of parks evaluation

	A1	A2	A3	A4	A5
A1	–	0	1	1	1
A2	1	–	1	1	1
A3	0	0	–	1	0
A4	0	0	0	–	0
A5	0	0	1	0	–

Table 7.10 Final Dominance Matrix of parks evaluation

	A1	A2	A3	A4	A5
A1	–	0	1	1	1
A2	0	–	0	1	0
A3	0	0	–	0	0
A4	0	0	0	–	0
A5	0	0	1	0	–

Table 7.11 Ranking the alternatives of parks evaluation

	Number of times dominated	Number of defeats	Difference
A1	3	0	3
A2	1	0	1
A3	0	2	−2
A4	0	2	−2
A5	1	1	0

In this step, the final dominance matrix is formed by multiplying element-wise the components of the final dominance matrix, resulting in the formation of the Boolean dominance matrix (Table 7.10).

Step 11. Selection of the Best alternative

Based on the final dominance matrix, the number of times each alternative dominates and the number of times it is defeats are calculated and presented in the following (Table 7.11).

Based on the above results, the priority of alternatives is as follows:

$$A_1 \succ A_2 \succ A_5 \succ A_3 = A_4$$

Clearly, using the introduced versions of the ELECTRE method, the above results can be calculated based on each of them.

7.7 Fuzzy ELECTRE Method

The fuzzy ELECTRE method is a multi-criteria decision-making technique that combines the principles of fuzzy logic and the ELECTRE method. In contrast to traditional decision-making methods, the fuzzy ELECTRE method allows for the consideration of imprecise and uncertain data, which is common in real-world decision-making situations. By incorporating fuzzy logic, the method can handle the ambiguity and vagueness in the decision-making process and provide more accurate and realistic results. The fuzzy ELECTRE method is widely used in fields such as engineering, management, finance, and environmental sciences, where decision-making involves complex, multi-dimensional problems [15].

Also, group decision-making can be a complex process, particularly when there are multiple experts involved with varying opinions and expertise. However, using the expertise of these individuals can lead to better decision outcomes. One way to incorporate the opinions of experts in the decision-making process is through the application of the ELECTRE method. The ELECTRE method is a multi-criteria decision analysis method that uses a set of weighted criteria to evaluate different alternatives and determine the best solution. In group decision-making using the ELECTRE method, each expert provides their opinions on the criteria and weights, which are then combined to form a group decision. One advantage of using the ELECTRE method in group decision-making is that it allows for the consideration of multiple perspectives and the incorporation of both quantitative and qualitative data. Additionally, the use of a systematic approach helps to ensure that the decision is based on a rigorous evaluation of the alternatives and criteria. To apply the ELECTRE method in group decision-making, it is important to have a clear understanding of the criteria being evaluated and the weight assigned to each criterion. The experts should be chosen based on their expertise and knowledge in the area being evaluated, and their opinions should be considered equally in the decision-making process. Overall, group decision-making using experts' opinions and the ELECTRE method can lead to better decision outcomes and ensure that all perspectives are taken into account in the decision-making process [2, 6, 8, 9].

Below, some methods for different fuzzy situations have been presented

7.7.1 ELECTRE—Fuzzy Trapezoidal Form

The decision matrix is obtained based on the rating of each index for each criterion. Sometimes, the type of rating used is fuzzy, and at other times, it is the result of transforming linguistic variables. In either case, the decision matrix is determined as follows. It should be noted that when using this method, if certain values exist, they must be converted into a fuzzy form. Additionally, we will consider z = (a, b, c) as z = (a, b, b, c). In other words; all values will be transformed into trapezoidal fuzzy form.

Consider a scenario where the decision maker (DM) has defined a weight vector and a decision matrix for a multi-criteria decision problem, where there are m alternatives and m criteria. In this scenario, some values are expressed qualitatively and linguistic variables are used. The weights and ratings (h = 1,…,k) of DM_h are aggregated as follows, where the weights and ratings for DM1 and DM2 are $\tilde{V}_{ij}^{h} = (v_{ij}^{h,L}, v_{ij}^{h,M_1}, v_{ij}^{h,M_2}, v_{ij}^{h,U}), (v_{ij}^{h,L} > 0)$ and $\tilde{Y}_j^h = (y_j^{h,L}, y_j^{h,M_1}, y_j^{h,M_2}, y_j^{h,U}), (y_j^{h,L} > 0)$, respectively [8].

$$\tilde{w}_j = (w_j^L, w_j^{M_1}, w_j^{M_2}, w_j^U);\ j = 1, \ldots, n.$$

$$w_j^L = \min_{h=1,\ldots,k} \{y_j^{h,L}\},\ w_j^{M_1} = \frac{1}{k}\sum_{h=1}^{k} y_j^{h,M_1},\ w_j^{M_2} = \frac{1}{k}\sum_{h=1}^{k} y_j^{h,M_2},\ w_j^U = \max_{h=1,\ldots,k} \{y_j^{h,U}\}$$

$$\tilde{W} = (\tilde{w}_1, \tilde{w}_2, \ldots, \tilde{w}_n) \tag{7.23}$$

Furthermore, the fuzzy decision matrix $\tilde{X}$ is aggregated and constructed as follows.

$$\tilde{x}_{ij} = (x_{ij}^{L}, x_{ij}^{M_1}, x_{ij}^{M_2}, x_{ij}^{U});\ j = 1, ..., n.$$

$$x_{ij}^{L} = \underset{h=1,...,k}{\mathrm{Min}}\{v_{ij}^{h,L}\},\ x_{ij}^{M_1} = \frac{1}{k}\sum_{h=1}^{k} v_{ij}^{h,M_1},\ x_{ij}^{M_2} = \frac{1}{k}\sum_{h=1}^{k} v_{ij}^{h,M_2},\ x_{ij}^{U} = \underset{h=1,...,k}{\mathrm{Max}}\{v_{ij}^{h,U}\}$$

$$\tilde{X} = \begin{bmatrix} \tilde{x}_{11} & \dots & \tilde{x}_{1n} \\ \vdots & \ddots & \dots \\ \tilde{x}_{m1} & \dots & \tilde{x}_{mn} \end{bmatrix} \quad (7.24)$$

Normalizing the fuzzy decision matrix would result in the following representation, which is displayed using $\tilde{N}$:

$$\tilde{n}_{ij} = (n_{ij}^{L}, n_{ij}^{M_1}, n_{ij}^{M_2}, n_{ij}^{U}) = (\frac{x_{ij}^{L}}{x_j^{Max}}, \frac{x_{ij}^{M_1}}{x_j^{Max}}, \frac{x_{ij}^{M2}}{x_j^{Max}}, \frac{x_{ij}^{U}}{x_j^{Max}}),$$

$$i = 1, ..., m;\ j \in B;\ \mathrm{B} = \{\mathrm{j}|\mathrm{C_j}\ \text{is benefit criteria}\}$$

$$\tilde{n}_{ij} = (n_{ij}^{L}, n_{ij}^{M_1}, n_{ij}^{M_2}, n_{ij}^{U}) = (\frac{x_j^{Min}}{x_{ij}^{L}}, \frac{x_j^{Min}}{x_{ij}^{M_1}}, \frac{x_j^{Min}}{x_{ij}^{M_2}}, \frac{x_j^{Min}}{x_{ij}^{U}}),$$

$$i = 1, ..., m\ ;\ j \in C;\ \mathrm{C} = \{\mathrm{j}|\mathrm{C_j}\ \text{is cost criteria}\}$$

$$\tilde{N} = \begin{bmatrix} \tilde{n}_{11} & \dots & \tilde{n}_{1n} \\ \vdots & \ddots & \dots \\ \tilde{n}_{m1} & \dots & \tilde{n}_{mn} \end{bmatrix} \quad (7.25)$$

in which $x_j^{Max} = \underset{i}{Max}\{x_{ij}^{U}\},\ (j = 1, ..., n)$ and $x_j^{Min} = \underset{i}{Min}\{x_{ij}^{L}\},\ (j = 1, ..., n)$

The weighted normalized fuzzy decision matrix is obtained by multiplying each normalized rating by its corresponding weight.

$$\tilde{z}_{ij} \simeq (w_j^{L} n_{ij}^{L}, w_j^{M_1} n_{ij}^{M_1}, w_j^{M_2} n_{ij}^{M_2}, w_j^{U} n_{ij}^{U})$$

$$\tilde{Z} = \begin{bmatrix} \tilde{z}_{11} & \dots & \tilde{z}z_{1n} \\ \vdots & \ddots & \dots \\ \tilde{z}_{m1} & \dots & \tilde{z}z_{mn} \end{bmatrix} \quad (7.26)$$

The concordance and discordance sets are determined using the Hamming distance.[1] $\tilde{z}_{tj} \geq \tilde{z}_{bj}$ If and only if $d(max(\tilde{z}_{tj}, \tilde{z}_{bj}), \tilde{z}_{tj}) \geq d(max(\tilde{z}_{tj}, \tilde{z}_{bj}), \tilde{z}_{bj})$.Based on $J_{tb}^{C} = \{j|\tilde{z}_{tj} \geq \tilde{z}_{bj}\}$, $J_{tb}^{D} = \{j|\tilde{z}_{tj} \leq \tilde{z}_{bj}\}$, the concordance and discordance matrices are determined as follows:

$$\tilde{c}_{tb} = (c_{tb}^{L}, c_{tb}^{M_1}, c_{tb}^{M_2}, c_{tb}^{U}) = (\sum_{j \in J_{tb}^{C}} w_j^{L}, \sum_{j \in J_{tb}^{C}} w_j^{M_1}, \sum_{j \in J_{tb}^{C}} w_j^{M_2}, \sum_{j \in J_{tb}^{C}} w_j^{U})$$

---

[1] $d(\tilde{A}, \tilde{B}) = \int_{R} |\mu_{\tilde{A}}(x) - \mu_{\tilde{B}}(x)| dx$

$$\tilde{C} = \begin{bmatrix} - & \tilde{c}_{12} \ldots & & \tilde{c}_{1m} \\ \tilde{c}_{21} & - & \ldots & \tilde{c}_{2m} \\ \vdots & & \ddots & \vdots \\ \tilde{c}_{m1} & \ldots & \tilde{c}_{m(m-1)} & - \end{bmatrix} \tag{7.27}$$

Also, the discordance matrix is determined as follows.

$$d_{\text{tb}} = \frac{\max\limits_{j \in J_{tb}^{D}} \{|\tilde{z}_{tj} - \tilde{z}_{bj}|\}}{\max\limits_{j \in \{1,\ldots,n\}} \{|\tilde{z}_{tj} - \tilde{z}_{bj}|\}} = \frac{\max\limits_{j \in J_{tb}^{D}} \{d(max(\tilde{z}_{tj}, \tilde{z}_{bj}), \tilde{z}_{bj})\}}{\max\limits_{j \in \{1,\ldots,n\}} \{d(max(\tilde{z}_{tj}, \tilde{z}_{bj}), \tilde{z}_{bj})\}}$$

$$D = \begin{bmatrix} - & d_{12} \ldots & & d_{1m} \\ d_{21} & - & \ldots & d_{2m} \\ \vdots & & \ddots & \vdots \\ d_{m1} & \ldots & d_{m(m-1)} & - \end{bmatrix} \tag{7.28}$$

In order to compare the elements of the concordance matrix with the concordance level, this value must be determined. that the average of the matrix members is used for this purpose. Therefore, the concordance level will be calculated as follows.

$$\tilde{\bar{c}} = (\bar{c}^{L}, \bar{c}^{M_1}, \bar{c}^{M_2}, \bar{c}^{U}) = \Bigg(\sum_{\substack{t=1 \\ t \neq b}}^{m} \sum_{\substack{b=1 \\ b \neq t}}^{m} \frac{c_{tb}^{L}}{m(m-1)}, \sum_{\substack{t=1 \\ t \neq b}}^{m} \sum_{\substack{b=1 \\ b \neq t}}^{m} \frac{c_{tb}^{M_1}}{m(m-1)}, \sum_{\substack{t=1 \\ t \neq b}}^{m} \sum_{\substack{b=1 \\ b \neq t}}^{m} \frac{c_{tb}^{M_2}}{m(m-1)}, \sum_{\substack{t=1 \\ t \neq b}}^{m} \sum_{\substack{b=1 \\ b \neq t}}^{m} \frac{c_{tb}^{U}}{m(m-1)}\Bigg) \tag{7.29}$$

Using the concordance level, the Boolean concordance matrix is determined as follows.

$$f_{\text{tb}} = \begin{cases} 1 \text{ , } \tilde{c}_{\text{tb}} \geq \tilde{\bar{c}} \\ 0 \text{ , otherwise} \end{cases}$$

$$F = \begin{bmatrix} - & f_{12} \ldots & & f_{1m} \\ f_{21} & - & \ldots & f_{2m} \\ \vdots & & \ddots & \vdots \\ f_{m1} & \ldots & f_{m(m-1)} & - \end{bmatrix} \tag{7.30}$$

Similarly, the discordance level and Boolean discordance matrix are determined as follows.

$$\overline{d} = \sum_{t=1}^{m} \sum_{\substack{b=1 \\ b \neq t}}^{m} \frac{d_{tb}}{m(m-1)} \tag{7.31}$$

$$g_{\text{tb}} = \begin{cases} 1 \text{ , } d_{\text{tb}} < \overline{d} \\ 0 \text{ , otherwise} \end{cases}$$

$$G = \begin{bmatrix} - & g_{12} & \cdots & g_{1m} \\ g_{21} & - & \cdots & g_{2m} \\ \vdots & & \ddots & \vdots \\ g_{m1} & & \cdots & g_{m(m-1)} - \end{bmatrix} \tag{7.32}$$

Using the elements of Boolean concordance and discordance matrices, the global matrix will be formed as follows.

$$h_{\text{tb}} = f_{\text{tb}} \cdot g_{\text{tb}}$$

$$H = \begin{bmatrix} - & h_{12} \cdots & & h_{1m} \\ h_{21} & - & \cdots & h_{2m} \\ \vdots & & \ddots & \vdots \\ h_{m1} \cdots & h_{m(m-1)} & & - \end{bmatrix} \tag{7.33}$$

Based on the above hierarchy, the steps of the fuzzy triangular ELECTRE I method in group decision-making are as follows:

Step 1: The decision-makers' opinions are aggregated and a fuzzy weight vector is formed using Eq. (7.23).

Step 2: The fuzzy decision matrix is constructed using Eq. (7.24).

Step 3: The normalized fuzzy decision matrix is obtained using Eq. (7.25).

Step 4: The normalized fuzzy decision matrix is weighted, similar to Eq. (7.26).

Step 5: The concordance matrix is determined using Eq. (7.27).

Step 6: The discordance matrix is determined using Eq. (7.28).

Step 7: The Boolean concordance matrix is determined using Eq. (7.30).

Step 8: The Boolean discordance matrix is determined using Eq. (7.32).

Step 9: The global matrix is determined using Eq. (7.33).

Step 10: The best alternative is chosen based on the difference in the number of times the alternative has been dominant or dominated.

7.7.2 *Manager Selection: Employing ELECTRE Method and Fuzzy Linguistic Variables*

This subsection employs the ELECTRE (Elimination Choice Translating Reality) method to facilitate the selection of managers through multi-criteria decision making. The objective is to identify and assess a range of indicators that can effectively evaluate candidates' suitability for managerial positions. These indicators encompass objective factors, management skills, knowledge and expertise, personality characteristics, attitude and vision, as well as professional and public credibility. By leveraging the ELECTRE method, organizations can make well-informed decisions in selecting competent and qualified individuals for managerial roles.

I. Objective Factors: (C1)
Objective factors play a crucial role in assessing managerial candidates. The following factors are considered: past years' secret score, service history, completed educational courses, and educational certificates. These objective measures provide a foundation for evaluating the qualifications and achievements of potential managers.

II. Management Skills: (C1)
Effective managers possess essential management skills that contribute to organizational success. Communication skills, decision-making abilities, and creativity-related skills are key indicators used to assess a candidate's managerial competency in this regard.

III. Knowledge and Expertise: (C2)
Knowledge and expertise are fundamental attributes for managers. Evaluating a candidate's work expertise, familiarity with management concepts and theories, as well as knowledge of laws and regulations, helps gauge their ability to navigate complex managerial challenges.

IV. Personality Characteristics: (C3)
Personality characteristics greatly influence managerial effectiveness. Assessing justice and fairness, high competitive spirit, responsibility-optimism, criticism, and stability of personality assists in identifying individuals who possess the requisite traits for effective leadership.

V. Attitude and Vision: (C4)
Managers need to demonstrate appropriate attitudes and vision for the role. Factors such as respecting subordinates, a desire to gain power, and a genuine interest in managing people provide insights into a candidate's aptitude for leadership and team-building.

The selection of managers requires a comprehensive evaluation of various indicators. By employing multi-attribute decision making methods, organizations can systematically assess the qualifications, skills, knowledge, personality traits, attitudes, and credibility of managerial candidates. This subsection serves as a guide for organizations seeking to make informed decisions in selecting competent managers who can effectively contribute to the success of the organization.

We assume there are five managers and three decision-makers. Each decision-maker has evaluated the managers using linguistic variables and assigned weights to them. The results of their evaluations are as follows. Before proceeding, let's define the criteria for converting linguistic variables to fuzzy numbers, as shown in Tables 7.12, 7.13.

Decision makers have expressed their opinions for each criterion, which are listed in (Table 7.14).

The opinions of decision makers regarding the performance of each alternative on each criterion are as Table 7.15:

Step 1: Using Eq. (7.23) and the results from Table 7.14, the weights of each criterion are as Table 7.16.

Step 2: The decision matrix using relation (7.24) and Table 7.15 is as Table 7.17.

Step 3: The normalized decision matrix is as Table 7.18.

Step 4: The weighted normalized decision matrix is as follows (Table 7.19).

Step 5: The concordance matrix is determined using Eq. (7.27) (Table 7.20).

Step 6: The discordance matrix is determined using Eq. (7.28) (Table 7.21)

Step 7: The Boolean concordance matrix is determined using Eq. (7.31) (Table 7.22).

Step 8: The Boolean discordance matrix is determined using Eq. (7.32) (Table 7.23).

Step 9: The global matrix is determined using Eq. (7.34) (Table 7.24).

Step 10: The best alternative is chosen based on the difference in the number of times the alternative has been dominant or dominated.

Based on the results of Table 7.25, the order of alternatives is as follows: $A_2 = A_3 \succ A_1 = A_4 \succ A_5$.

Table 7.12 Transformation of linguistic variables in determining the importance of criteria weight

Linguistic terms	Fuzzy number
Very low importance (VL)	(0,0,0.1,0.2)
Low importance (L)	(0.1,0.2,0.2,0.3)
Medium low importance (ML)	(0.2,0.3,0.4,0.5)
Medium importance (M)	(0.4,0.5,0.5,0.6)
Medium high importance (MH)	(0.5,0.6,0.7,0.8)
High importance (H)	(0.7,0.8,0.8,0.9)
Very High importance (VH)	(0.9,0.9,1,1)

Table 7.13 Transformation of linguistic variables in determining the performance of alternatives

Linguistic terms	Fuzzy number
Very Poor (VP)	(0,0,1,2)
Poor (P)	(1,2,2,3)
Medium Poor (MP)	(2,3,4,5)
Fair (F)	(4,5,5,6)
Medium Good (MG)	(5,6,7,8)
Good (G)	(7,8,8,9)
Very Good (VG)	(8,9,10,10)

Table 7.14 Opinions of decision makers on the importance of each criterion

Criteria	Opinions of decision maker
C1	H,H,H
C2	VH,VH,VH
C3	VH,VH,H
C4	H,H,H
C5	H,H,H

Table 7.15 Decision makers' opinions on the performance of each alternative on each criterion

Alternative	Criteria				
	C1	C2	C3	C4	C5
A1	MG,MG,MG	MG,MG,VG	G,G,G	G,G,G	G,G,G
A2	G,G,G	VG,VG,VG	VG,VG,VG	G,VG,VG	VG,VG,VG
A3	VG,VG,G	VG,G,G	VG,VG,G	VG,VG,VG	G,VG,G
A4	G,G,G	G,G,MG	MG,MG,G	G,G,G	G,G,VG
A5	MG,MG,MG	MG,G,G	MG,MG,MG	MG,MG,G	MG,MG,MG

Table 7.16 Weight of each criterion based on expert's average opinion

Criteria	Weights of each criterion
C1	(0.7,0.8,0.8,0.9)
C2	(0.8,0.9,1,1)
C3	(0.7,0.87,0.93,1)
C4	(0.7,0.8,0.8,0.9)
C5	(0.7,0.8,0.8,0.9)

Table 7.17 Fuzzy decision matrix

Alternative	C1	C2	C3	C4	C5
A1	(5,6,7,8)	(5,7,8,10)	(7,8,8,9)	(7,8,8,9)	(7,8,8,9)
A2	(7,8,8,9)	(8,9,10,10)	(8,9,10,10)	(7,8.7,9.3,10)	(8,9,10,10)
A3	(7,8.7,9.3,10)	(7,8.3,8.7,10)	(7,8.7,9.3,10)	(8,9,10,10)	(7,8.3,8.7,10)
A4	(7,8,8,9)	(5,7.3,7.7,9)	(5,6.7,7.3,9)	(7,8,8,9)	(7,8.3,8.7,10)
A5	(5,6,7,8)	(5,7.3,7.7,9)	(5,6,7,8)	(5,6.7,7.3,9)	(5,6,7,8)

7.8 The ELECTRE III Method and Interval-Valued Intuitionistic Fuzzy Sets

This section focuses on two powerful tools that aid decision-makers in tackling the complexities of evaluating alternatives: the ELECTRE III method and interval-valued intuitionistic fuzzy sets. The ELECTRE III method, an evolution of the ELECTRE family, stands as a prominent multi-criteria decision-making approach. Its foundations lie in outranking relations, providing a structured framework to rank alternatives based on a set of criteria. By capturing preference relationships and incorporating positive and negative criteria, ELECTRE III offers a robust and comprehensive analysis that assists decision-makers in making well-informed choices. With its ability to handle conflicting objectives, ELECTRE III proves invaluable in domains where trade-offs and compromises are inevitable. Complementing the ELECTRE III method, we turn to the field of interval-valued intuitionistic fuzzy sets (IVIFS). Traditional fuzzy sets extend the boundaries of crisp sets by accommodating uncertainty, but IVIFS takes this concept even further by embracing the notions of hesitation and non-membership. This innovative framework allows decision-makers to express their preferences in a more nuanced manner, acknowledging the imprecision and vagueness that often accompany decision-making processes. By capturing the degrees of membership, non-membership, and hesitation, IVIFS provides a rich and realistic representation of decision-makers' opinions. Within this section, we aim to equip readers with a solid understanding of ELECTRE III Method and Interval-Valued Intuitionistic Fuzzy Sets and their utility in the real world decision problems. Examples and case studies illustrate the application of the ELECTRE III method and interval-valued intuitionistic fuzzy sets in diverse domains, showcasing their strengths and limitations in different decision-making contexts. Whether you are a researcher, practitioner, or student seeking to expand your knowledge in multi-criteria decision-making or explore cutting-edge methodologies, this section serves as a valuable resource. By illuminating the intricacies and practicalities of the ELECTRE III method and interval-valued intuitionistic fuzzy sets, we empower readers to effectively leverage these tools in their decision-making endeavors. By incorporating these methodologies into your decision-making toolkit, you can enhance your ability to navigate complex decision problems, make informed choices, and embrace the challenges of uncertainty and ambiguity with confidence [3, 4].

Table 7.18 Fuzzy normalized decision matrix

Alternative	C1	C2	C3	C4	C5
A1	(0.5,0.6,0.7,0.8)	(0.5,0.7,0.8,1)	(0.7,0.8,0.8,0.9)	(0.7,0.8,0.8,0.9)	(0.7,0.8,0.,8,0.9)
A2	(0.7,0.8,0.8,0.9)	(0.8,0.9,1,1)	(0.8,0.9,1,1)	(0.7,0.87,0.93,1)	(0.8,0.9,1,1)
A3	(0.7,0.87,0.93,1)	(0.7,0.83,0.87,1)	(0.7,0.87,0.93,1)	(0.8,0.9,1,1)	(0.7,0.83,0.87,1)
A4	(0.7,0.8,0.8,0.9)	(0.5,0.73,0.77,0.9)	(0.5,0.67,0.73,0.9)	(0.7,0.8,0.8,0.9)	(0.7,0.83,0.87,1)
A5	(0.5,0.6,0.7,0.8)	(0.5,0.73,0.77,0.9)	(0.5,0.6,0.7,0.8)	(0.5,0.67,0.73,0.9)	(0.5,0.6,0.7,0.8)

Table 7.19 Fuzzy weighted normalized decision matrix

Alternative	C1	C2	C3	C4	C5
A1	(0.35,0.48,0.56,0.72)	(0.4,0.63,0.8,1)	(0.49,0.7,0.74,0.9)	(0.49,0.64,0.64,0.81)	(0.49,0.64,0.64,0.81)
A2	(0.49,0.64,0.64,0.81)	(0.64,0.81,1,1)	(0.56,0.78,0.93,1)	(0.49,0.7,0.74,0.9)	(0.56,0.72,0.8,0.9)
A3	(0.49,0.7,0.74,0.9)	(0.56,0.75,0.87,1)	(0.49,0.76,0.86,1)	(0.56,0.72,0.8,0.9)	(0.49,0.66,0.7,0.9)
A4	(0.49,0.64,0.64,0.81)	(0.4,0.66,0.77,0.9)	(0.35,0.58,0.68,0.9)	(0.49,0.64,0.64,0.81)	(0.49,0.66,0.7,0.9)
A5	(0.35,0.48,0.56,0.72)	(0.4,0.66,0.77,0.9)	(0.35,0.52,0.65,0.8)	(0.35,0.54,0.58,0.81)	(0.35,0.48,0.56,0.72)

Table 7.20 The concordance matrix

Alternative	A1	A2	A3	A4	A5
A1	–	(0,0,0,0)	(0.7,0.8,0.8,0.9)	(2.2,2.57,2.73,2.9)	(3.6,4.17,4.33,4.7)
A2	(3.6,4.17,4.33,4.7)	–	(2.2,2.57,2.73,2.9)	(3.6,4.17,4.33,4.7)	(3.6,4.17,4.33,4.7)
A3	(3.6,4.17,4.33,4.7)	(1.4,1.6,1.6,1.8)	–	(3.6,4.17,4.33,4.7)	(3.6,4.17,4.33,4.7)
A4	(2.1,2.4,2.4,2.7)	(0.7,0.8,0.80,0.9)	(0.7,0.8,0.8,0.9)	–	(3.6,4.17,4.33,4.7)
A5	(0.7,0.8,0.8,0.9)	(0,0,0,0)	(1.4,1.6,1.6,1.8)	(1.5,1.7,1.8,1.9)	–

Table 7.21 The discordance matrix

Alternative	A1	A2	A3	A4	A5
A1	–	1	1	0.65	0
A2	0	–	0.65	0	0
A3	0	1	–	0	0
A4	1	1	1	–	0
A5	1	1	1	1	–

Table 7.22 The Boolean concordance matrix

Alternative	A1	A2	A3	A4	A5
A1	–	0	0	1	1
A2	1	–	1	1	1
A3	1	0	–	1	1
A4	0	0	0	–	1
A5	0	0	0	0	–

Table 7.23 The Boolean discordance matrix

Alternative	A1	A2	A3	A4	A5
A1	–	0	0	0	1
A2	1	–	0	1	1
A3	1	0	–	1	1
A4	0	0	0	–	1
A5	0	0	0	0	–

Table 7.24 The global matrix

Alternative	A1	A2	A3	A4	A5
A1	–	0	0	0	1
A2	1	–	0	1	1
A3	1	0	–	1	1
A4	0	0	0	–	1
A5	0	0	0	0	–

Table 7.25 Final dominance

	Number of times dominated	Number of defeats	Difference
A1	1	2	−1
A2	3	0	3
A3	3	0	3
A4	1	2	−1
A5	0	4	−4

The degree of net credibility serves as a measure of value, wherein a higher degree indicates a greater level of attractiveness for the alternative xi. Consequently, the net credibility degree allows for a complete ranking of all alternatives. Furthermore, in order to address practical challenges encountered in fuzzy multi-attribute decision making (FMADM), it is imperative to involve evaluators and decision makers who encompass diverse stakeholders and interest groups. These groups consist of individuals with distinct backgrounds and positions, leading to a significant variance in subjective judgments. For instance, the aforementioned thresholds (concordance, discordance, and veto) could be expressed using fuzzy data. This illustrates that ELECTRE III and IV offer more suitable approaches for evaluating real-world problems [7, 13].

In group decision making, it is initially necessary to aggregate the opinions of decision makers (DMs) regarding the criteria and the performance of alternatives, and then proceed with the steps of each method. For this purpose, the steps of the method are as follows:

Step 1: The weights of each decision maker need to be determined, which we represent here as $\lambda_t (t = 1, ..., k)$. It is important to note that the sum of weights should be equal to 1.

Step 2: In this step, the rating of each decision maker for the performance of each alternative in each criterion is represented as $\tilde{x}_{ij}^t (i = 1, ..., m;\ j = 1, ..., \text{n})$. This rating is represented as IVIFN, and the matrix elements for each decision maker are represented as $\tilde{x}_{ij}^t = ([\alpha_{Lij}^t, \alpha_{Uij}^t], [\beta_{Lij}^t, \beta_{Uij}^t])$. The aggregated decision matrix is represented as $\tilde{x}_{ij}$.

$$\tilde{x}_{ij} = ([\alpha_{Lij}, \alpha_{Uij}], [\beta_{Lij}, \beta_{Uij}]) = \left(\left[\left(1 - \prod_{t=1}^{k} (1 - \alpha_{Lij}^t)^{\lambda_t}\right), \left(1 - \prod_{t=1}^{k} (1 - \alpha_{Uij}^t)^{\lambda_t}\right)\right], \left[\left(\prod_{t=1}^{k} (\beta_{Lij}^t)^{\lambda_t}, \prod_{t=1}^{k} (\beta_{Uij}^t)^{\lambda_t}\right)\right]\right); \quad j = 1, ..., n. \tag{7.34}$$

Step 3: In this step, the weight vector is determined by aggregating the opinions of decision makers regarding each criterion. The opinion of each decision maker for each criterion is represented as $\tilde{w}_j^k = ([\theta_{Lj}^k, \theta_{Uj}^k], [\xi_{Lj}^k, \xi_{Uj}^k])$. The aggregated weight vector is represented as $\tilde{W} = [\tilde{w}_1, \tilde{w}_2, ..., \tilde{w}_n]$.

$$\tilde{w}_{ij} = ([\theta_{Lij}, \theta_{Uij}], [\xi_{Lij}, \xi_{Uij}]) = \left(\left[\left(1 - \prod_{t=1}^{k} (1 - \theta_{Lj}^t)^{\lambda_t}\right), \left(1 - \prod_{t=1}^{k} (1 - \theta_{Uj}^t)^{\lambda_t}\right)\right], \left[\left(\prod_{t=1}^{k} (\xi_{Lj}^t)^{\lambda_t}, \prod_{t=1}^{k} (\xi_{Uj}^t)^{\lambda_t}\right)\right]\right); \quad j = 1, ..., n. \tag{7.35}$$

Step 4: Determine the concordance matrix

Build the concordance matrix by first establishing the thresholds, denoted as P for strong preference, Q for weak preference, and I for indifference. When determining the model's thresholds, we consider that the performance of alternatives can be assessed with a certain level of accuracy. We also account for the inherent uncertainty in evaluations. Multiple techniques can be employed in this regard. Some techniques directly define a threshold, while others utilize a dispersion threshold concept. The dispersion threshold incorporates the idea of probable values and the distinction between optimistic and pessimistic values. It represents the plausible range of difference caused by overestimation or underestimation, influencing the evaluation of consequences or performance levels. It is important to acknowledge that there are no absolute values for the thresholds. Instead, the selected values for thresholds are chosen to best capture the imperfect nature of our knowledge [11, 12].

Next, it is necessary to compare a pair of alternatives and determine the concordance matrix using the following procedure:

If the value of $s(\tilde{A}_t)$ is found to be less than the value of $s(\tilde{A}_b)$, it can be concluded that $\tilde{A}_t$ is less preferred than $\tilde{A}_b$. Here, the value of $s(\tilde{A}_i) = s([a, b], [c, d])$ is computed using $\frac{1}{2}(a - c + b - d)$.

$$c_{tb} = \frac{\sum_{j=1}^{n} w_j c_j(\mathrm{t, b})}{\sum_{j=1}^{n} w_j}, (t, b = 1, ...m)$$

$$c_j(\mathrm{t, b}) = \begin{cases} 1 & , \mathrm{s}(\tilde{x}_{\mathrm{tj}}) - \mathrm{s}(\tilde{x}_{\mathrm{bj}}) \le \mathrm{s}(\tilde{p}_i) \\ 0 & , \mathrm{s}(\tilde{x}_{\mathrm{tj}}) - \mathrm{s}(\tilde{x}_{\mathrm{bj}}) > s(\tilde{q}_i) \\ \frac{\mathrm{s}(\tilde{x}_{\mathrm{tj}}) - \mathrm{s}(\tilde{x}_{\mathrm{bj}})}{\mathrm{s}(\tilde{p}_i) - s(\tilde{q}_i)}, & \text{otherwise} \end{cases} \tag{7.36}$$

Step 5: Determine the discordance matrix

$$d_{tb}^j = \begin{cases} 1 & , \mathrm{s}(\tilde{x}_{bj}) - \mathrm{s}(\tilde{x}_{tj}) \le \mathrm{s}(\tilde{p}_j) \\ 0 & , \mathrm{s}(\tilde{x}_{bj}) - \mathrm{s}(\tilde{x}_{tj}) > s(\tilde{v}_j) \\ \frac{\mathrm{s}(\tilde{x}_{bj}) - \mathrm{s}(\tilde{x}_{tj}) - \mathrm{s}(\tilde{p}_j)}{\mathrm{s}(\tilde{v}_j) - s(\tilde{p}_j)}, & \text{otherwise} \end{cases} \tag{7.37}$$

Step 6: Determining the degree of outranking between two alternatives

$$\mathrm{S}_{tb} = \begin{cases} C_{tb} & , \mathrm{d}_{tb}^j \le c_{tb}(\mathrm{j} = 1, ..., \mathrm{n}) \\ C_{tb} \prod_{j \in J_{tb}} \frac{1 - d_{tb}^j}{1 - c_{tb}} & , \text{otherwise} \end{cases} \tag{7.38}$$

where, $J_{tb} = \{j | d_{tb}^j > c_{tb}\}$

Step 7: Ranking the alternatives

The arrangement of alternatives can be determined through an evaluation of their concordance credibility degree, discordance credibility degree, and net credibility degree. These values can be computed using the following calculations.

The concordance credibility degree:

$$CCD(A_i) = \sum_{t=1}^{m} S_{it}, i = 1, ..., m. \tag{7.39}$$

The discordance credibility degree:

$$DCD(A_i) = \sum_{t=1}^{m} S_{ti}, i = 1, ..., m. \tag{7.40}$$

The net credibility degree:

$$NCD(A_i) = CCD(A_i) - DCD(A_i), i = 1, ..., m. \tag{7.41}$$

The degree of net credibility serves as a measure of value, wherein a higher degree indicates a greater level of attractiveness for the alternative xi. Consequently, the net credibility degree allows for a complete ranking of all alternatives. Furthermore, in order to address practical challenges encountered in fuzzy multi-attribute decision making (FMADM), it is imperative to involve evaluators and decision makers who encompass diverse stakeholders and interest groups. These groups consist of individuals with distinct backgrounds and positions, leading to a significant variance in subjective judgments. For instance, the aforementioned thresholds (concordance, discordance, and veto) could be expressed using fuzzy data. This illustrates that ELECTRE III and IV offer more suitable approaches for evaluating real-world problems.

7.9 Unraveling Employee Commitment: Key Factors for Ranking and Evaluation Using the Interval-Valued Intuitionistic Fuzzy Number

In this section of the book, we discuss about the factors that influence the work commitment of employees within an organization. These factors can be broadly categorized into four distinct groups, each playing a crucial role in shaping the level of commitment exhibited by individuals.

The first category focuses on the organizational factors that significantly impact employee commitment. These include the presence of a cooperative management system, an evaluation system based on performance, fair treatment and adherence to organizational justice, the size of the organization, leadership styles, and management behavior towards employee appreciation, the effectiveness of the promotion system, and the alignment of organizational identity, values, and beliefs with employee values. Additionally, the level of organizational support extended to employees plays a vital role in fostering commitment.

Moving on to the second category, we explore the factors related to job interest and satisfaction. This encompasses aspects such as career advancement opportunities, challenging job roles, the perceived value of the work, quality of working life, job security, the absence of job tension and stress, availability of training programs, and the presence or absence of conflict and role ambiguity. Job satisfaction also contributes significantly to the commitment levels of employees.

The third category present the influence of the organization's image within society. Factors like feeling safe and comfortable in the work environment, favorable physical conditions at the workplace, positive relationships between employees and superiors, consideration of environmental and physical factors in the work environment, an interesting and attractive physical work environment, and the provision of social services and facilities contribute to the overall commitment of employees.

Finally, individual factors play a role in shaping employee commitment. These factors include the level of education, social class, work experience and history, age, expertise, adherence to moral principles, and the economic base of employees.

Understanding and analyzing these factors is crucial when attempting to rank employees based on expert language judgments. By recognizing the various elements that influence work commitment, organizations can better assess and evaluate employee performance, leading to more effective decision-making processes and improved overall outcomes.

The objective was to rank four employees of the organization who had been evaluated by five experts through interviews and organizational data. The four aforementioned criteria were ultimately used to rank these employees. It should be noted that the weight assigned to each expert was equal. Initially, a table of linguistic variable conversions for the importance level of each criterion was provided, followed by the conversion of linguistic variables for the performance of each employee.

Opinions from the five experts were gathered concerning the criteria, with the results presented in Table 7.27. The conversion process utilized Table 7.26 and Eq. (7.35). Furthermore, Table 7.28 and Table 7.29 provides the linguistic judgment of the experts regarding the employees, compiled using relation (7.34).

Threshold values for each criterion are specified below (Table 7.32).

Furthermore, the matrix of concordance is presented accordingly Table 7.33.

Following that, the matrix of discordance is computed, and its outcomes are displayed in Table 7.34. Subsequently, a comparison is made between the matrices of agreement and discordance. The outcomes of this comparison and the corresponding values are presented in Tables 7.35 and 7.36, respectively.

7.10 Conclusion

This chapter discussed about the application of the ELECTRE approach and fuzzy ELECTRE, along with fuzzy intuitionistic methods, in the field of multi-criteria decision-making (MCDM). The ELECTRE approach was introduced as a powerful family of methods that can greatly support decision-making processes. While the

Table 7.26 Linguistic variable transformation in determining the performance of alternatives

Linguistic terms	IVIFNs
Extremely good (EG)/extremely high (EH)	([1, 1], [0, 0])
Very very good (VVG)/very very high (VVH)	([0.9, 0.9], [0.1, 0.1])
Very good (VG)/very high (VH)	([0.7333, 0.825], [0, 0.125])
Good (G)/high (H)	([0.6333, 0.725], [0.1, 0.225])
Medium good (MG)/medium high (MH)	([0.5333, 0.625], [0.2, 0.325])
Fair (F)/medium (M)	([0.4333, 0.525], [0.3, 0.425])
Medium bad (MB)/medium low (ML)	([0.3333, 0.425], [0.4, 0.525])
Bad (B)/low (L)	([0.15, 0.2875], [0.45, 0.6375])
Very bad (VB)/very low (VL)	([0, 0.1375], [0.6, 0.7875])
Very very bad (VVL)/very very low (VVL)	([0.1, 0.1], [0.9, 0.9])

Table 7.27 Linguistic variable transformation in determining criterion weights

Linguistic terms	IVIFNs
Very important (VI)	([0.9, 0.9], [0.1, 0.1])
Important (I)	([0.4, 0.7625], [0, 0.2115])
Medium (M)	([0.15, 0.5125], [0.25, 0.4625])
Unimportant (U)	([0, 0.3625], [0.4, 0.6125])
Very unimportant (VU)	([0.1, 0.1], [0.9, 0.9]

Table 7.28 Expert opinions on each indicator

Criteria	Expert opinions
C1	VI,I,I,M,VI
C2	VI,VI,M,U,M
C3	M,I,U,U,U
C4	VI,M,VI,M,I

various variants of the ELECTRE approach share similarities, they also differ in terms of their approaches, parameters, and roles within the decision-making process. It is crucial to select the most appropriate variant based on the specific problem under consideration.

Table 7.29 Expert opinions on performance of each employee

Alternative	Criteria			
	C1	C2	C3	C4
A1	H,M,H,H,M	M,M,H,VH,M	VH,M,M,H,H	H,ML,VH,ML,VH
A2	H,M,M,M,M	ML,H,VH,VH,H	VL,L,M,VL,ML	H,H,M,VH,M
A3	M,M,L,VL,L	VVL,L,M,M,M	M,VVL,L,M,VL	VH,M,H,M,M
A4	M,VH,H,M,VH	VH,M,MH,H,VH	H,MH,M,H,VH	M,H,VVH,VH,M

Table 7.30 Weight of each criterion

Criteria	Weight	$S(w_j)$
C1	([0.7648, 0.8712], [0, 0.1768])	0.729590194
C2	([0.7084, 0.8027], [0.1257, 0.2307])	0.577357189
C3	([0.1549, 0.5838], [0, 0.3872])	0.175780762
C4	([0.7434, 0.8458], [0, 0.3872])	0.601035037

Furthermore, the chapter explored the practical application of fuzzy ELECTRE and fuzzy intuitionistic methods. These approaches offer valuable tools for decision-making in real-world scenarios where uncertainty and ambiguity prevail. By incorporating fuzzy logic and intuitionistic reasoning, these methods enable decision-makers to effectively handle imprecise and subjective information. The practical examples provided demonstrated the versatility and effectiveness of these methods across diverse domains, showcasing their ability to generate nuanced and reliable decision outcomes.

Overall, the integration of fuzzy ELECTRE and fuzzy intuitionistic methods enhances the decision-making process, enabling more informed and robust choices in complex decision environments. By embracing these methods, decision-makers can effectively navigate uncertainties and make well-informed decisions, ultimately leading to improved outcomes.

Table 7.31 Decision matrix with Interval-Valued intuitionistic fuzzy numbers

Alternative	Criteria			
	C1	C2	C3	C4
A1	([0.645,0.738],[0.97,0.212])	([0.634,0.732],[0,0.215])	([0.672,0.766],[0,0.183])	([0.671,0.770],[0,0.176])
A2	([0.559,0.656],[0.168,0.292])	([0.699,0.798],[0,0.149])	([0.514,0.638],[0,0.293])	([0.672,0.766],[0,0.183])
A3	([0.200,0.349],[0.384,0.572])	([0.389,0.488],[0.323,0.458])	([0.259,0.405],[0.384,0.534])	([0.634,0.732],[0,0.215])
A4	([0.697,0.791],[0,0.158])	([0.688,0.779],[0,0.171])	([0.688,0.779],[0,0.171])	([0.763,0.818],[0,0.150])

Table 7.32 Threshold values

Criteria	C1	C2	C3	C4
Q	0.116	−0.325	0.116	−0.325
P	0.516	0.316	0.316	−0.083
V	0.716	0.516	0.516	0.116

Table 7.33 Concordance matrix

Alternative	A1	A2	A3	A4
A1	1	0.528	0.711	0.529
A2	0.542	1	0.627	0.389
A3	0	0	1	0.0843
A4	0.602	0.565	0.779	1

Table 7.34 Discordance matrix

A1				A2				A3				A4			
1	0	0	0	1	0	0	0	1	0	0	0	1	0.38	0.13	0.83
0	1	0	0	0	1	0	0	0	1	0	0	0.44	1	0.15	0.85
1	0.32	1	1	1	1	1	1	1	1	1	1	0.7	0.67	1	1.11
0	0	0	1	0	0	0	1	0	0	0	1	0.002	0	0	1

Table 7.35 The credibility matrix

Alternative	A1	A2	A3	A4
A1	1	0.528	0.711	3.44
A2	0.542	1	0.627	1.02
A3	0	0	1	0
A4	0.602	0.656	0.779	1

Table 7.36 Ranking of the alternatives

Alternative	A1	A2	A3	A4
score	3.545	1.095	−2.118	−2.522
Rank	1	2	3	4

Acknowledgement A special thanks to the Iranian DEA society for their unwavering spiritual support and consensus in the writing of this book. Your invaluable support has been truly remarkable, and we are deeply grateful for the opportunity to collaborate with such esteemed professionals.

References

1. Brito, A. J., de Almeida, A. T., & Mota, C. M. (2010). A multicriteria model for risk sorting of natural gas pipelines based on ELECTRE TRI integrating Utility Theory. *European Journal of Operational Research, 200*(3), 812–821.
2. Chen, N., Xu, Z., & Xia, M. (2015). The ELECTRE I multi-criteria decision-making method based on hesitant fuzzy sets. *International Journal of Information Technology & Decision Making, 14*(03), 621–657.
3. Chen, Z. S., Zhang, X., Pedrycz, W., Wang, X. J., & Skibniewski, M. J. (2020). Bid evaluation in civil construction under uncertainty: A two-stage LSP-ELECTRE III-based approach. *Engineering Applications of Artificial Intelligence, 94*, 103835.
4. Chen, Z. S., Zhang, X., Rodríguez, R. M., Pedrycz, W., & Martínez, L. (2021). Expertise-based bid evaluation for construction-contractor selection with generalized comparative linguistic ELECTRE III. *Automation in Construction, 125*, 103578.
5. Figueira, J., & Roy, B. (2002). Determining the weights of criteria in the ELECTRE type methods with a revised Simos' procedure. *European journal of operational research, 139*(2), 317–326.
6. Figueira, J. R., Greco, S., Roy, B., & Słowiński, R. (2013). An overview of ELECTRE methods and their recent extensions. *Journal of Multi-Criteria Decision Analysis, 20*(1–2), 61–85.
7. Hashemi, S. S., Hajiagha, S. H. R., Zavadskas, E. K., & Mahdiraji, H. A. (2016). Multicriteria group decision making with ELECTRE III method based on interval-valued intuitionistic fuzzy information. *Applied mathematical modelling, 40*(2), 1554–1564.
8. Hatami-Marbini, A., & Tavana, M. (2011). An extension of the Electre I method for group decision-making under a fuzzy environment. *Omega, 39*(4), 373–386.
9. Leyva-Lopez, J. C., & Fernandez-Gonzalez, E. (2003). A new method for group decision support based on ELECTRE III methodology. *European journal of operational research, 148*(1), 14–27.
10. Liao, H., Yang, L., & Xu, Z. (2018). Two new approaches based on ELECTRE II to solve the multiple criteria decision making problems with hesitant fuzzy linguistic term sets. *Applied Soft Computing, 63*, 223–234.
11. Lupo, T. (2015). Fuzzy ServPerf model combined with ELECTRE III to comparatively evaluate service quality of international airports in Sicily. *Journal of air transport management, 42*, 249–259.
12. Mohamadghasemi, A., Hadi-Vencheh, A., Lotfi, F. H., & Khalilzadeh, M. (2020). An integrated group FWA-ELECTRE III approach based on interval type-2 fuzzy sets for solving the MCDM problems using limit distance mean. *Complex & Intelligent Systems, 6*, 355–389.
13. Mohamadghasemi, A., Hadi-Vencheh, A., & Lotfi, F. H. (2021). The interval Type-2 fuzzy ELECTRE III method to prioritize machines for preventive maintenance. *International Journal of Industrial Engineering & Production Research, 32*(4), 1–19.
14. Roy, B. (1991). The outranking approach and the foundations of ELECTRE methods. *Theory and decision, 31*, 49–73.
15. Sevkli, M. (2010). An application of the fuzzy ELECTRE method for supplier selection. *International Journal of Production Research, 48*(12), 3393–3405.
16. Wang, X., & Triantaphyllou, E. (2008). Ranking irregularities when evaluating alternatives by using some ELECTRE methods. *Omega, 36*(1), 45–63.

Chapter 8
Analytical Hierarchy Process (AHP) in Fuzzy Environment

Abstract This chapter provides an overview of hierarchical decision structures, focusing on the AHP and ANP methods. It explains how these structures break down complex problems and organize criteria, sub-criteria, and alternatives in a hierarchical manner. The Analytic Hierarchy Process (AHP) is explored in detail, including its step-by-step implementation process and practical examples. The chapter then introduces the Analytic Network Process (ANP), which extends AHP to decision problems with interdependencies and feedback loops. The construction of networks, calculation of priority weights, and interpretation of results within the ANP framework are discussed. Finally, Fuzzy AHP is introduced as a method that addresses the limitations of AHP by incorporating fuzzy sets theory, supported by real-world case studies. Fuzzy AHP allows decision-makers to express preferences in a flexible and nuanced manner, capturing uncertainties and ambiguity present in real-world scenarios. Its advantage lies in handling vague and imprecise information systematically, enabling the effective incorporation of subjective assessments and expert opinions

8.1 Hierarchical Decision Structure (Threats and Opportunities)

In the ever-evolving landscape of decision-making, organizations face complex challenges and opportunities that require careful evaluation and analysis. To make informed choices, decision-makers often rely on structured approaches that systematically assess different factors and prioritize alternatives. One such approach is the hierarchical decision structure, which provides a framework for organizing and prioritizing information in a hierarchical manner. In this chapter, we explore the concepts and methodologies associated with hierarchical decision structures, with a specific focus on two prominent methods: the Analytic Hierarchy Process (AHP) and the Analytic Network Process (ANP). Hierarchical decision structures offer a systematic approach to decision-making by breaking down complex problems into manageable components. The structure consists of a hierarchy of criteria, sub-criteria, and

F. Hosseinzadeh Lotfi et al., *Fuzzy Decision Analysis: Multi Attribute Decision Making Approach*, Studies in Computational Intelligence 1121,
https://doi.org/10.1007/978-3-031-44742-6_8

alternatives, organized in a tree-like format. This hierarchical arrangement allows decision-makers to consider multiple dimensions of a problem and evaluate the relative importance and relationships between different elements. By structuring decisions hierarchically, organizations can navigate through intricate decision landscapes more effectively and improve the quality of their choices. The Analytic Hierarchy Process (AHP) is a widely adopted decision-making technique that enables decision-makers to quantitatively evaluate and prioritize alternatives based on a set of criteria. AHP employs pairwise comparisons to establish the relative importance of criteria and alternatives, assigning numerical values to indicate preferences. By systematically comparing elements within each level of the hierarchy, AHP provides decision-makers with a mathematical framework for making consistent and rational choices. Furthermore, AHP incorporates a consistency check to ensure the reliability of the decision-making process [14, 15, 24].

The Analytic Network Process (ANP), extends the principles of AHP to decision problems with interdependencies and feedback loops. ANP acknowledges that decisions are often interconnected and that the relationships between criteria and alternatives can be complex. ANP allows decision-makers to model the interactions between elements in a network-like structure, facilitating a more comprehensive and accurate evaluation. By capturing the interdependencies and feedback within a decision framework, ANP provides a robust methodology to address complex decision scenarios. Threats and opportunities are inherent to any decision-making process, and the hierarchical decision structure, along with AHP and ANP methodologies, can be instrumental in effectively managing them. By using these methods, decision-makers can assess potential threats and opportunities, identify their relative significance, and make informed decisions accordingly. AHP and ANP enable a systematic evaluation of different factors, considering both qualitative and quantitative aspects, thus providing a comprehensive framework to analyze risks, prioritize actions, and exploit opportunities. In the upcoming sections, we will delve deeper into the theoretical foundations of the hierarchical decision structure, exploring the key principles, techniques, and considerations involved. We will discuss the step-by-step process of implementing AHP and ANP methodologies, providing practical guidance and examples to illustrate their applications. Additionally, we will highlight real-world case studies where these methods have been successfully utilized to address threats and opportunities in diverse domains. By the end of this chapter, readers will have gained a solid understanding of hierarchical decision structures and will be equipped to leverage AHP and ANP to make effective decisions in the face of threats and opportunities [22].

Hierarchical decision structures offer a powerful framework for navigating complex decision landscapes. The AHP and ANP methods, in particular, provide decision-makers with quantitative and qualitative tools to evaluate criteria, alternatives, and their relationships systematically. By adopting these methodologies, organizations can enhance their decision-making processes, mitigate risks posed by threats, and capitalize on opportunities that arise.

8.2 Analytical Hierarchy Process (AHP)

Hierarchical Analysis Method is one of the powerful techniques for multi-criteria decision-making and, despite some criticisms, it has been well-received by the scientific community. This method, which reflects natural behavior and human thinking, enables the decision-maker to interact between different criteria in complex and unstructured situations. This technique facilitates decision-making by organizing emotions, perceptions, aspirations, and judgments, and identifies the influential forces on the decision. The process of hierarchical analysis is one of the most comprehensive processes designed for decision-making with multiple criteria, as it allows for formulating the problem in a hierarchical form. It also allows for considering various quantitative and qualitative criteria in the problem. In this process, different alternatives intervene in the decision-making, and there is the possibility of sensitivity analysis on the criteria and sub-criteria. Another excellent advantage of this multi-criteria decision-making method is determining the level of compatibility and incompatibility of the decision. This method has a strong theoretical basis and is based on intuitive principles. The hierarchical analysis process decomposes complex and problematic issues and simplifies them to solve them. In fact, hierarchical analysis is a method to assist decision-makers in prioritizing and classifying their goals and solutions in a complex, unstructured, and opaque environment.

In this method, the decision-making problem is divided into different levels of objectives, criteria, sub-criteria, and alternatives so that the decision-maker can easily focus on the smallest decision. To build a decision model, at the highest level, the objective is defined, and at intermediate levels, criteria are identified, and at the lowest level, possible alternatives are placed [12].

In the following, the mentioned levels are briefly introduced:

Objective: The main research question or problem to be solved is referred to as the objective. The objective is the highest level in the hierarchical diagram and has only one parameter, which is chosen as the task of the highest level of project decision-making.

Criterion: Criteria are essentially the attributes of the objective or the means of measuring it. The more criteria cover the components of the objective and better represent the objective, the higher the likelihood of obtaining more accurate results. Criteria are the second level in the hierarchical diagram after the objective. At this level, the required number of criteria can be drawn and adjusted horizontally based on the necessity. Criteria can be divided into sub-criteria, and sub-criteria can be further divided into subsequent sub-criteria. This situation can increase to n sub-criteria in vertical and horizontal levels, depending on the necessity.

Alternative: Alternatives are actually the intended targets or destinations in the hierarchical diagram, and the answer to the objective is derived from the drawn alternatives. Alternatives are the last level in the hierarchical diagram and depend on how the hierarchical analysis method is used. In cases where this method is used for selection or prioritization purposes, the determination of alternatives is usually done by the researcher. It is the researcher who decides which alternatives should be chosen or prioritized among the available ones.

The most important capability of the hierarchical analysis method is its ability to transform the hierarchical structure of a complex multi-criteria problem into an extended structure to better understand the decision-maker's perception of the decision problem. This method uses pairwise comparisons of decision elements, taking into account the criteria or alternatives, to determine their relative importance.

The Analytic Hierarchy Process (AHP) is based on dividing the decision problem into smaller parts, guiding decision-makers towards pairwise comparisons between influential criteria in the decision problem. This process aims to express the importance of each criterion and the extent of their impact through a hierarchical structure, as assessed by experts. In this method, numerical values are assigned to the importance of each criterion relative to other criteria based on mental judgments. Ultimately, the criteria with the highest importance are identified, and the priority order of the criteria is determined. The overall score for each possible alternative is obtained by multiplying the relative score of each path by the decision alternative and then adding this score to each decision alternative.

The following four principles are stated as the principles of the Analytic Hierarchy Process, and all calculations, rules, and regulations are built upon these principles:

I. Reciprocal Condition: If the preference of element A over element B is equal to n, then the preference of element B over element A will be equal to 1/n.
II. Consistency: Element A and element B must be consistent and comparable. In other words, the superiority of element A over element B cannot be infinite or zero.
III. Dependence: Each hierarchical element can depend on its higher-level element, and this dependence can continue linearly up to the highest level.
IV. Expectations: Whenever there is a change in the hierarchical structure, the evaluation process must be repeated.

These principles form the basis of the Analytic Hierarchy Process and serve as the foundation for all computations and decision-making rules in this method.

The steps of implementing the AHP method are as follows:

Step 1. Constructing a Hierarchy Chart

The first step in the Analytic Hierarchy Process is to construct a hierarchy chart, which typically displays the problem's hierarchy in the order of the goal, criteria (and sub-criteria if applicable), and alternatives. The hierarchy chart illustrates the decision's objective, the factors being compared, and the competing alternatives being evaluated. Creating a graphical representation of the problem that showcases the goal, criteria, and alternatives is essential for this purpose.

In order to create an effective hierarchy chart, the following steps can be followed:

Identify the Goal: Clearly define the main objective or question that the decision-making process aims to address. This should represent the highest level in the hierarchy.

Determine Criteria: Identify the criteria or factors that are relevant and influential in evaluating the alternatives. These criteria should be directly related to achieving the goal. Arrange the criteria in a logical order beneath the goal, indicating their hierarchical relationship.

Establish Sub-criteria (if applicable): If necessary, further break down the criteria into sub-criteria. These sub-criteria should contribute to the evaluation of their parent criteria. Arrange them hierarchically under their respective parent criteria.

Specify the Alternatives: Identify the available alternatives or alternatives that will be compared and evaluated against the established criteria. List them below the criteria or sub-criteria they relate to.

Visualize the Hierarchy: Create a graphical representation of the hierarchy chart, commonly using a tree-like structure. The goal should be at the top, followed by the criteria and sub-criteria, and finally, the alternatives at the bottom.

By constructing a hierarchy chart, decision-makers can visually understand the relationships between the goal, criteria, and alternatives. This chart serves as a foundation for conducting pairwise comparisons and assigning relative weights to the criteria during the subsequent steps of the Analytic Hierarchy Process.

Figure 8.1 shows a hierarchical structure that the objective is to select among 4 cars according to the four criteria of security, style, capacity and cost. The "cost" criterion has four sub-criteria and also "capacity" is the parent of two sub-criteria "cargo capacity" and "passenger capacity".

Step 2. Formation of Pairwise Comparison Matrices

In this step, the elements within each level are compared pairwise with other relevant elements from the higher level, and pairwise comparison matrices are formed. Numerical values are assigned to represent the relative importance of comparing two alternatives or two criteria.

To construct the pairwise comparison matrices, the following steps can be followed:

Define the Scale: Determine a scale that represents the relative importance or preference between elements. A common scale is a numerical scale from 1 to 9, where

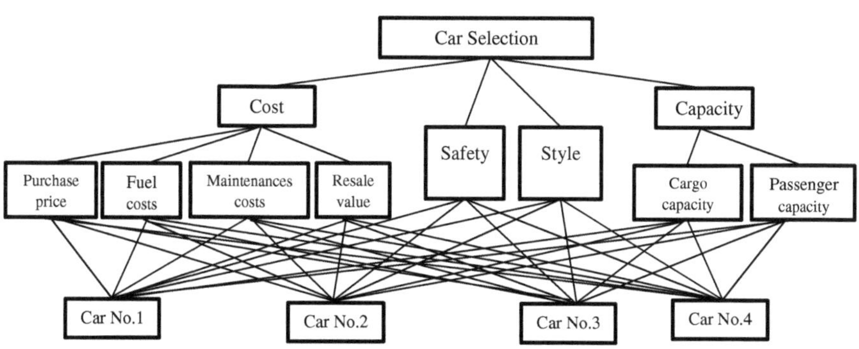

Fig. 8.1 Hierarchical structure for car selection

1 indicates equal importance or preference, and higher values represent increasing importance or preference.

Compare Pairwise: Compare each element within a level with every other element from the higher level, one pair at a time. Assess the relative importance or preference of one element over the other based on the defined scale. Record the numerical value representing the degree of importance.

Complete the Matrix: Fill in the pairwise comparison matrix by placing the numerical values in the corresponding cells. The matrix should be square, with the same elements on both rows and columns. The diagonal elements should have a value of 1, indicating that an element is equally important or preferred compared to itself. Reciprocal Property: Ensure the reciprocal property by taking the reciprocal of each value outside the diagonal and placing it in the corresponding cell of the opposite pair. For example, if the importance of element A over element B is assigned as 5, the importance of element B over element A should be assigned as 1/5.

Repeat for Each Pair: Repeat the pairwise comparison process for all relevant pairs of elements within each level of the hierarchy. Create separate matrices for each pairwise comparison.

By constructing pairwise comparison matrices, decision-makers can quantify the relative importance or preference between elements. These matrices provide a structured framework for comparing and prioritizing elements within the hierarchy. The numerical values within the matrices will be used in subsequent calculations to determine the weights or priorities of the elements in the decision-making process.

The provided pairwise comparison matrix is represented as follows:

$$A = \left[a_{ij}\right]_{n \times n}$$

Where, a_{ij} represents the preference of element "i" over element "j".

According to the reciprocal property, in pairwise comparisons of criteria, the relationship is as follows:

$$a_{ij} = \frac{1}{a_{ji}}; i, j = 1, ..., n.$$

The pairwise comparison matrix is a $n \times n$ matrix, where "n" represents the number of elements being compared.

Note that, in most cases there is more than one decision maker and a group of decision makers give opinions regarding the formation of the decision matrix. Considering "K" decision makers, the judgments can be aggregated by arithmetic mean. Let a_{ijk} denotes the preference of kth decision maker, then to aggregate $A^k = \left[a_{ijk}\right]$; $k = 1, ..., K$ matrices into a unit decision matrix $A = \left[a_{ij}\right]$, apply $a_{ij} = \frac{1}{K}\sum_{k=1}^{K} a_{ijk}$.

Step 3. Calculating Weights

In this step, the weights of each element are determined using the pairwise comparison matrix and applying the methods introduced in Chap. 3 of this book.

Step 4. Calculating Inconsistency Ratio

Calculating the inconsistency ratio is highly important in the AHP method. It generally depends on the decision-maker, and it is typically considered as 0.1. If the level of inconsistency exceeds 0.1, it is advisable to reconsider the judgments [7].

To calculate the inconsistency rate, first subtract λ_{Max} from equation $A \times W = \lambda_{Max} \times W$, and then calculate the inconsistency index (I.I) as follows.

$$I.I = \frac{\lambda_{Max} - n}{n - 1} \tag{8.1}$$

The inconsistency index of random matrix (I.I.R) is calculated as follows.

$$I.I.R = 1.98\frac{n - 2}{n} \tag{8.2}$$

Therefore, for each matrix, the inconsistency rate is obtained by the ratio of the two aforementioned indices:$I.R = \frac{I.I}{I.I.R}$.

This serves as a suitable criterion for judgment and it is determined based on the following Table 8.1.

If the inconsistency rate exceeds the predetermined threshold set by the decision-maker, it is necessary to revise the pairwise comparison matrices and perform the evaluation process again. Otherwise, we proceed to step 5.

Step 5. Calculating Final Weights

The final weight of each alternative in a hierarchical process is obtained by sum of the product of each criterion weight and the score assigned to the respective alternative. The weight for each alternative is calculated using the following equation:

$$\gamma_i = \sum_{j=1}^{n} \alpha_{ij} w_j; i = 1, ..., m. \tag{8.3}$$

Table 8.1 The inconsistency index of random matrix (I.I.R.)

15	14	13	12	11	10	9	8	7	6	5	4	3	2	1	n
1.59	1.57	1.56	1.52	1.51	1.45	1.45	1.41	1.32	1.24	1.12	0.90	0.58	0	0	I.I.R

Note that calculating the inconsistency rate for each matrix based on pairwise comparisons is not sufficient. It is also necessary to calculate the hierarchical inconsistency rate. This is obtained by comparing the weighted average ratio of the pairwise comparison inconsistency indices to the weighted average ratio of the random pairwise comparison inconsistency indices for each matrix

Where, w_j represents the importance or weight of criterion j, and α_{ij} indicates the relative importance of alternative "i" for criterion "j".

It is worth mentioning that the weighted geometric mean is used to aggregate expert opinions in this context. One of the challenges in determining weights and inconsistency rates is the limited access to experts. Obtaining multiple expert opinions in the form of pairwise comparison matrices is quite difficult in a hierarchical analysis process where access to experts is restricted. On the other hand, when the number of comparisons becomes large, experts may struggle to appropriately adjust the elements of the pairwise comparison matrix, resulting in high inconsistency rates in many cases. One method that can effectively improve the AHP structure is preference voting. In this approach, experts apply their opinions once and rank the options based on a simple ranking method. [5, 13, 16, 19–21].

8.3 Analytical Network Process (ANP)

The ANP method is more comprehensive than the AHP method. ANP allows the decision-maker to create a network instead of a hierarchy, which enables the examination of internal relationships between elements. ANP is considered as one of the most complete and effective methods for multi-criteria decision-making. As mentioned, its structure can differ from the AHP structure. The major difference lies in the definition of the model and the relationship between elements. In the AHP method, this relationship is only independent, whereas in the ANP method, it can be both independent and dependent. In other words, AHP can be considered a special case of ANP.

Another point to consider when comparing these two methods is that, during pairwise comparisons in the AHP method, the question arises as to which alternative is more important between two alternatives. In contrast, the objective of pairwise comparisons in the ANP method is to examine the degree of influence between parameters (Fig. 8.2).

The steps of the ANP method are as follows:

Step 1. Model Construction and Problem Configuration

In this step, the problem should be divided into criteria levels and, if applicable, subcriteria and alternatives. The relationships between them should be determined. An important point in this step is the existence of relationships between criteria, which can be determined through various means, such as questioning the experts.

Step 2. Formation of Pairwise Comparison Matrices and Calculation of Weight Vectors

In this stage, the elements of each level are pairwise compared to their relevant elements in the higher level, and pairwise comparison matrices are formed. Additionally, pairwise comparisons of internal relationships should also be conducted at the end.

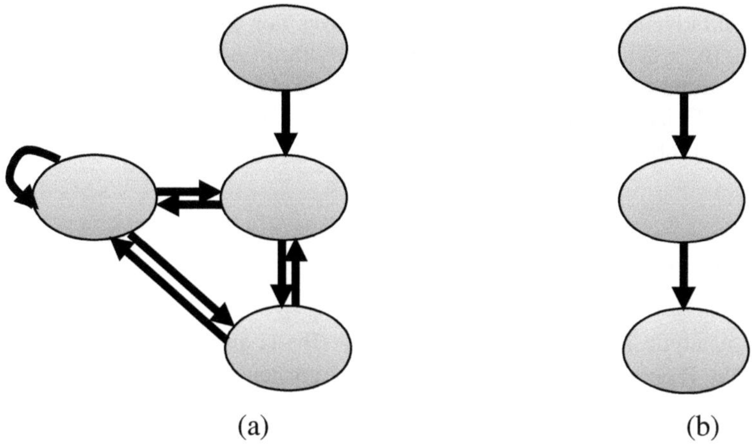

Fig. 8.2 The structure of ANP (a) compared to the structure of AHP (b)

Step 3. Inconsistency Rate Calculation

In this step, we calculate the inconsistency rate of ANP. If this rate is less than 0.1, it indicates compatibility. Otherwise, further judgments need to be obtained from the experts.

Step 4. Formation of the Supermatrix

In general, the supermatrix is formed as follows.

$$W = \begin{matrix} \text{goal} \\ \text{criteria} \\ \text{sub criteria} \\ \text{alternatives} \end{matrix} \overset{\begin{matrix} \text{goal} & \text{criteria} & \text{sub criteria} & \text{alternative} \end{matrix}}{\begin{vmatrix} W_{11} & W_{12} & W_{13} & W_{14} \\ W_{21} & W_{22} & W_{23} & W_{24} \\ W_{31} & W_{32} & W_{33} & W_{34} \\ W_{41} & W_{42} & W_{43} & W_{44} \end{vmatrix}} \tag{8.4}$$

Step 5. Calculation of the weighted supermatrix

This process involves exponentiating the supermatrix to obtain the power limit $\left(W^k\right)$. If the supermatrix converges to a limit for a large value of k, it serves as a measurement criterion for the options. However, if the convergence of the matrices is oscillatory, an arithmetic averaging will be performed.

Advantages of the ANP approach include its comprehensive nature, allowing for the consideration of all relevant criteria, including both tangible and intangible factors impacting the decision-making process. Unlike AHP, which assumes a one-way hierarchical relationship between decision levels, ANP enables more intricate relationships between decision levels and criteria without strict hierarchical constraints. This flexibility makes it suitable for exploring the interdependencies among criteria, which is vital in real-life decision-making scenarios. ANP accommodates both quantitative and qualitative characteristics, and it recognizes non-linear relationships between criteria. By providing composite scores, ANP offers a relative ranking of available options to aid decision-makers. In summary, the ANP method excels in handling complex decisions, capturing interdependencies, and providing a relative ranking based on combined scores.

8.4 Fuzzy AHP

Fuzzy Analytic Hierarchy Process (Fuzzy AHP) is an extension of the Analytic Hierarchy Process (AHP) that incorporates the concept of fuzziness and uncertainty into the decision-making process. While AHP provides a structured framework for decision-making by assigning crisp values to criteria and alternatives, Fuzzy AHP acknowledges the inherent vagueness and imprecision in decision-making and allows for the representation of subjective judgments using fuzzy numbers. By using fuzzy numbers and fuzzy arithmetic operations, Fuzzy AHP facilitates a more realistic representation of decision-makers' judgments and provides a more robust decision-making framework. In Fuzzy AHP, the process of pairwise comparisons and priority derivation is modified to accommodate fuzzy linguistic assessments. Instead of crisp numerical values, decision-makers use linguistic terms such as "very high," "high," "moderate," "low," and "very low" to express the degree of preference or importance. These linguistic terms are associated with fuzzy membership functions, which capture the degree of membership of an alternative to a particular linguistic term. The aggregation of fuzzy judgments and the calculation of priorities in Fuzzy AHP involve the use of fuzzy arithmetic operations, such as fuzzy multiplication and fuzzy normalization. These operations consider the fuzzy nature of the assessments and provide a comprehensive and coherent way of synthesizing the fuzzy judgments into overall priorities. Overall, Fuzzy AHP extends the AHP methodology by incorporating fuzzy logic and fuzzy set theory, allowing decision-makers to handle uncertainty and imprecision in a more explicit and rigorous manner. By embracing the fuzziness of decision-making, Fuzzy AHP provides a valuable tool for complex decision problems where subjective judgments and imprecise information play a significant role [3, 4, 6, 18, 23].

The method offers ease of comprehension and proficiently manages both qualitative and quantitative data in multi-attribute decision-making problems. Triangular fuzzy numbers are employed to express the preferences of one criterion over another. The extent analysis method is utilized to calculate the synthetic extent value of pairwise comparisons. Consequently, the weight vectors are determined, normalized, and thus the normalized weight vectors are obtained. After wards, the final priority weights are determined based on the varying weights of criteria and attributes [2, 9, 11].

8.4.1 Fuzzy AHP: First Approach

The steps of Fuzzy AHP is as follows:

Step 1. Constructing a Hierarchy Chart

Determine the objective, criteria, sub-criteria and alternatives and construct the hierarchy chart.

Step 2. Formation of Pairwise Comparison Matrix using fuzzy numbers

The fuzzy pairwise comparison matrix is depicted bellow.

$$\tilde{A} = \begin{bmatrix} 1 & \tilde{a}_{12} \ldots & \tilde{a}_{1n} \\ \tilde{a}_{21} & 1 \quad \ldots & \tilde{a}_{2n} \\ \vdots & \vdots \quad \ddots & \vdots \\ \tilde{a}_{n1} & \tilde{a}_{n2} \ldots & 1 \end{bmatrix}$$

Where, $\tilde{a}_{ij}$ denotes a fuzzy number, representing the preference of element "i" over element "j".

Note that if $\tilde{a}_{ij} = \left(a_{ij}^l, a_{ij}^m, a_{ij}^u\right), (a_{ij}^l > 0)$ is a triangle fuzzy number then $\tilde{a}_{ji} \approx \left(\frac{1}{a_{ij}^u}, \frac{1}{a_{ij}^m}, \frac{1}{a_{ij}^l}\right)$ and if $\tilde{a}_{ij} = \left(a_{ij}^l, a_{ij}^{m1}, a_{ij}^{m2}, a_{ij}^u\right), (a_{ij}^l > 0)$ is a trapezoidal fuzzy number then $\tilde{a}_{ji} \approx \left(\frac{1}{a_{ij}^u}, \frac{1}{a_{ij}^{m2}}, \frac{1}{a_{ij}^{m1}}, \frac{1}{a_{ij}^l}\right)$.

Note that, if there are more than one matrix which means there exist more than one decision maker, that is, $\tilde{A}^k = \left[\tilde{a}_{ijk}\right]$; $k = 1, \ldots, K$, aggregate them by arithmetic mean as follows:

$$\tilde{a}_{ij} = \left(a_{ij}^l, a_{ij}^m, a_{ij}^u\right) = \left(\frac{1}{K}\sum_{k=1}^{K} a_{ijk}^l, \frac{1}{K}\sum_{k=1}^{K} a_{ijk}^m, \frac{1}{K}\sum_{k=1}^{K} a_{ijk}^u\right).$$

Step 3. Calculating Inconsistency Ratio

To calculate inconsistency ratio, first, the decision matrix can be converted from fuzzy number to crisp by any defuzzification function.

To convert the fuzzy elements of matrix $\tilde{A}$ to crisp numbers, here, it is suggested to consider the center point of triangle fuzzy number that is a_{ij}^m and for trapezoidal fuzzy number consider $\frac{a_{ij}^{m1} + a_{ij}^{m2}}{2}$ for defuzzification.

Attaining a decision matrix that all elements are crisp, then, the inconsistency ratio can be calculated by Eq. (8.1) and Eq. (8.2). If the inconsistency rate is less than 0.1 continue to step 4, otherwise ask the decision maker(s) to revise their judgment. Next, calculate the inconsistency ratio of the pairwise comparison matrix $\tilde{A}$ again. Continue this process until the inconsistency ratio doesn't exceed 0.1.

Step 4. Calculation of the Weights

As mentioned in Chap. 3 of this book, there are various methods for determining weights from pairwise comparison matrices. Therefore, the method will be described further, assuming that the weights of elements are fuzzy numbers.

Step 5. Calculation of Final Weights

Considering that the weight of each alternative for each sub-criterion and the weight of each sub-criterion in relation to the main criteria, as well as the weight of each criterion with respect to the objectives, are all fuzzy number. The final weight of each alternative will be as follows, according to Eq. (8.5):

$$\tilde{\gamma}_i = \sum_{j=1}^{n} \tilde{a}_{ij} \otimes \tilde{w}_j; i = 1, \ldots, m. \tag{8.5}$$

Where, $\tilde{w}_j$ represents the importance or weight of jth criterion and $\tilde{\alpha}_{ij}$ indicates the relative importance of ith alternative with respect to jth criterion.

The result of the above equation is a fuzzy number. Applying any Defuzzy function, make it possible to compare the alternatives and establish their priority or ranking.

First Approach of Fuzzy AHP is summarized in Fig. (8.3).

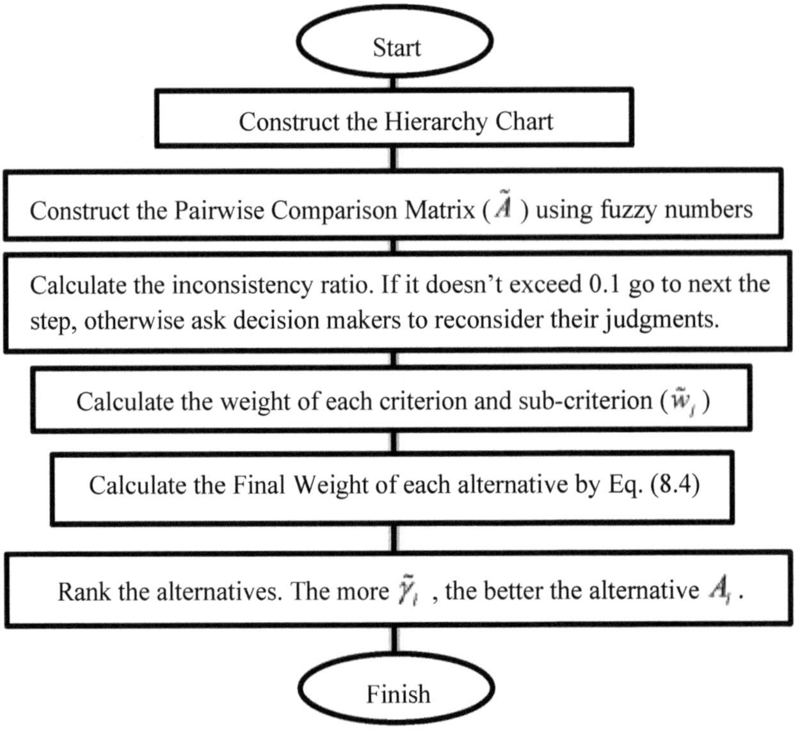

Fig. 8.3 First Approach of Fuzzy AHP

8.4.2 Fuzzy AHP: Second Approach

Based on the hierarchical structure of the process, different methods mainly differ in terms of weight determination. Below, we introduce one of the conventional methods in FAHP (Fuzzy Analytical Hierarchy Process) where the magnitude is used to determine the final weights and it is presented by Chang [2].

Step 1. Constructing a Hierarchy Tree

Determine the objective, criteria, sub-criteria and alternatives and construct the hierarchy Tree.

Step 2. Formation of Fuzzy Pairwise Comparison Matrix

The fuzzy pairwise comparison matrix is as follows.

$$\tilde{A} = \begin{bmatrix} 1 & \tilde{a}_{12} \dots & \tilde{a}_{1n} \\ \tilde{a}_{21} & 1 \quad \dots & \tilde{a}_{2n} \\ \vdots & \vdots \quad \ddots & \vdots \\ \tilde{a}_{n1} & \tilde{a}_{n2} \dots & 1 \end{bmatrix}_{n\times n}$$

Where, $\tilde{a}_{ij}$ denotes a fuzzy number, representing the preference of element "i" over element "j".

Note that if $\tilde{a}_{ij} = \left(a_{ij}^{l}, a_{ij}^{m}, a_{ij}^{u}\right)$ is a triangle fuzzy number then $\tilde{a}_{ji} = \left(\frac{1}{a_{ij}^{u}}, \frac{1}{a_{ij}^{m}}, \frac{1}{a_{ij}^{l}}\right)$ and if $\tilde{a}_{ij} = \left(a_{ij}^{l}, a_{ij}^{m1}, a_{ij}^{m2}, a_{ij}^{u}\right)$ is a trapezoidal fuzzy number then $\tilde{a}_{ji} = \left(\frac{1}{a_{ij}^{u}}, \frac{1}{a_{ij}^{m2}}, \frac{1}{a_{ij}^{m1}}, \frac{1}{a_{ij}^{l}}\right)$.

Step 3. Calculate the $\tilde{S}_i$ for each row of the Fuzzy Pairwise Comparison Matrices

The $\tilde{S}_i$ index for each row obtain as follows:

$$\tilde{S}_i = \sum_{j=1}^{n} \tilde{a}_{ij} \otimes [\sum_{i=1}^{n}\sum_{j=1}^{n} \tilde{a}_{ij}]^{-1}$$

$$= \left(\sum_{j=1}^{n} a_{ij}^{L}, \sum_{j=1}^{n} a_{ij}^{M}, \sum_{j=1}^{n} a_{ij}^{U}\right) \otimes \left(\frac{1}{\sum_{i=1}^{n}\sum_{j=1}^{n} a_{ij}^{U}}, \frac{1}{\sum_{i=1}^{n}\sum_{j=1}^{n} a_{ij}^{M}}, \frac{1}{\sum_{i=1}^{n}\sum_{j=1}^{n} a_{ij}^{L}}\right)$$

$$= \left(\frac{\sum_{j=1}^{n} a_{ij}^{L}}{\sum_{i=1}^{n}\sum_{j=1}^{n} a_{ij}^{U}}, \frac{\sum_{j=1}^{n} a_{ij}^{M}}{\sum_{i=1}^{n}\sum_{j=1}^{n} a_{ij}^{M}}, \frac{\sum_{j=1}^{n} a_{ij}^{U}}{\sum_{i=1}^{n}\sum_{j=1}^{n} a_{ij}^{L}}\right); j = 1, \dots, n. \tag{8.6}$$

Step 2. Defining the degree of possibility

The degree of possibility of $\tilde{M}_2 = (l_2, m_2, u_2) \geq \tilde{M}_1 = (l_1, m_1, u_1)$ is defined as bellow.

$$V(\tilde{M}_2 \geq \tilde{M}_1) = \sup\lfloor \min(\mu_{M_1}(d), \mu_{M_2}(d)) \rfloor$$
$$= \begin{cases} 1 & , m_2 \geq m_1, \\ 0 & , l_1 \geq u_2, \\ \frac{l_1 - u_2}{(m_2 - u_2) - (m_1 - l_1)} & , otherwise \end{cases} \quad (8.7)$$

Where, d represents the ordinate of the highest intersection between μ_{M_2} and μ_{M_1}.To calculate $\tilde{M}_1$ and $\tilde{M}_2$, it is necessary to compute the values of $V(\tilde{M}_2 \geq \tilde{M}_1)$ and $V(\tilde{M}_1 \geq \tilde{M}_2)$.

On the other hand, the magnitude of a fuzzy triangular number in relation to the rest of the fuzzy triangular numbers can be obtained from the following equation (Fig. 8.4).

$$V(M \geq M_1, M_2, ..., M_t) = V[(M \geq M_1)\&(M \geq M_2)\&...\&(M \geq M_t)] = min\{V(M \geq M_i)|i = 1, ..., t\}. \quad (8.8)$$

Step 3. Determine the final weights

$$w_i = min\{V(S_i \geq S_k)|k = 1, ..., n\&k \neq i\}, i = 1, ..., n. \quad (8.9)$$

The alternatives are ranked according to the value of w_i ; $i = 1, ..., m$. The higher the value of w_i, the better the alternative.

The second Approach of Fuzzy AHP is illustrated in Fig. 8.5.

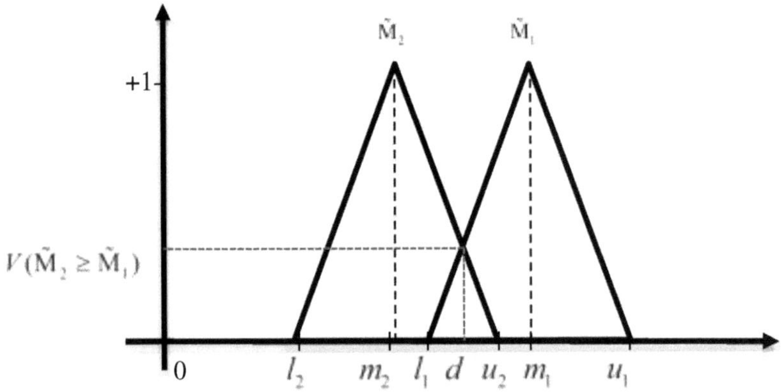

Fig. 8.4 Degree of possibility of $V(\tilde{M}_2 \geq \tilde{M}_1)$

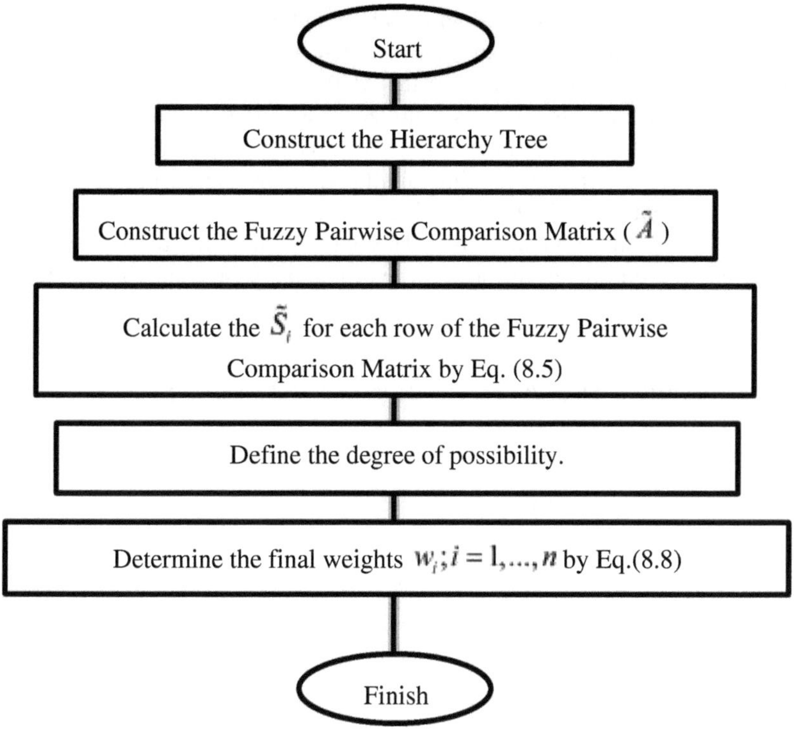

Fig. 8.5 Second Approach of Fuzzy AHP at a glance

8.4.3 Fuzzy AHP: Third Approach

In this approach, instead of using pairwise comparison matrices and dealing with inconsistency rates, experts are asked to provide qualitative rankings. This ranking is done by experts using linguistic variables to express the competency of each element in each rank using qualitative terms, as indicated in Table 8.2.

Table 8.2 Converting qualitative competence in each rank to fuzzy numbers

Importance	$\tilde{A} = (L, M, U)$
Extremely low	(0.01,0.01,0.1)
Very low	(0.01,0.1,0.3)
Low	(0.1,0.3,0.5)
Medium	(0.3,0.5,0.7)
High	(0.5,0.7,0.9)
Very high	(0.7,0.9,1)
Extremely high	(0.9,0.9,1)

This approach allows for freedom of action and eliminates the need for inconsistency rates, moving away from the traditional AHP method. In other words, to determine the weights at each level of the AHP process structure, the aggregation of rankings provided by the experts is utilized. Sharafi et al. [16] have focused on ranking candidates in preferential voting systems by proposing a method where voters use linguistic terms for ranking. Similarly, in this approach, to determine the weights of elements at each hierarchical level, weight determination through voting method with linguistic variables is used. This not only simplifies the data aggregation structure but also eliminates the need for returning pairwise comparison matrices to the experts [1, 8].

$$
\begin{aligned}
Z_p = Max \quad & \tfrac{1}{4}\left(\sum_{i=1}^{n} w_i^L y_{ip}^L + 2\sum_{i=1}^{n} w_i^M y_{ip}^M + \sum_{i=1}^{n} w_i^U y_{ip}^U\right) \\
s.t. \quad & \sum_{i=1}^{n} w_i^L y_{ij}^L \le 1, \qquad j = 1, \dots, n; \\
& \sum_{i=1}^{n} w_i^M y_{ij}^M \le 1, \qquad j = 1, \dots, n; \\
& \sum_{i=1}^{n} w_i^U y_{ij}^U \le 1, \qquad j = 1, \dots, n; \\
& w_i^L \le w_i^M \le w_i^U, \qquad i = 1, \dots, n; \\
& w_i^L - w_{i+1}^L \ge d^L(i, \varepsilon), \qquad i = 1, \dots, n-1; \\
& w_i^M - w_{i+1}^M \ge d^M(i, \varepsilon), \qquad i = 1, \dots, n-1; \\
& w_i^U - w_{i+1}^U \ge d^U(i, \varepsilon), \qquad i = 1, \dots, n-1; \\
& w_i^U - w_{i+1}^U \ge d^U(i, \varepsilon), \qquad i = 1, \dots, n-1; \\
& w_n^L \ge d^L(n, \varepsilon);\ w_n^M \ge d^M(n, \varepsilon);\ w_n^U \ge d^U(n, \varepsilon);
\end{aligned}
\tag{8.10}
$$

In the above model, the $(d^L(j, \varepsilon), d^M(j, \varepsilon), d^U(j, \varepsilon))$ is called the discrimination intensity function, which determines the minimum weighted distance between two priorities.

8.5 Fuzzy Analytic Network Process

Fuzzy ANP (Analytic Network Process) is an extension of the ANP method that incorporates fuzziness into the decision-making process. Building upon the foundations of ANP, fuzzy ANP takes into account the inherent uncertainties and vagueness that exist in real-world decision-making scenarios. By incorporating fuzzy logic, fuzzy ANP allows decision-makers to deal with imprecise and ambiguous information, providing a more robust and flexible approach to multi-criteria decision-making. The hierarchical structure and steps of the fuzzy ANP method are similar to ANP, with the additional consideration of fuzzy sets and linguistic variables to represent the uncertainties and preferences of decision-makers. Through the integration of fuzzy logic and the ANP framework, fuzzy ANP offers a powerful tool for tackling complex decision problems where uncertainty and vagueness play a significant role [10].

As mentioned earlier, the structures of AHP and ANP are similar, although they have some differences. However, their approach to determining weights is the same. Therefore, for the implementation of fuzzy ANP, it is sufficient to define the model structure, determine the levels of the network, and then establish the weights based on expert opinions. Subsequently, the supermatrix is formed, following the same process as in crisp ANP.

Another notable point is that weights can be determined using a fuzzy pairwise comparison matrix or fuzzy voting. Moreover, if the voting mechanism is utilized, there is no need to determine inconsistency rates. Determining inconsistency rates and, if necessary, conducting an evaluation process by experts can incur costs and waste time. However, if the preferential voting process is employed within the fuzzy ANP method, there is no need for inconsistency rates [17, 18].

8.6 Applications of Fuzzy AHP

When evaluating options for an ERP (Enterprise Resource Planning) system using the Analytic Hierarchy Process (AHP), several indicators are important to consider. These indicators help assess the suitability of each option based on various criteria. Here are some commonly considered indicators at different levels:

Strategic Level:

Alignment with business objectives: How well does the ERP system align with the organization's strategic goals and long-term vision?

Scalability: Can the system accommodate future growth and expansion of the organization?

Flexibility: How easily can the ERP system be customized or adapted to meet specific business requirements?

Functional Level:

Core modules and features: Does the ERP system provide essential functionalities required for day-to-day operations, such as finance, human resources, supply chain management, and manufacturing?

Integration capabilities: How well does the system integrate with existing software and systems, enabling seamless data exchange?

User-friendliness: Is the ERP system intuitive and easy to use, reducing the learning curve for employees?

Technical Level:

Technology stack: What technologies and architecture does the ERP system utilize, ensuring stability, scalability, and security?

System performance: How well does the system handle data processing, handle concurrent users, and maintain response times?

Data management: How efficiently does the ERP system handle data storage, backup, recovery, and data integrity?

The hierarchical structure of these indicators can be represented as follows (Fig. 8.6).

By considering these indicators and their hierarchical structure, decision-makers can systematically evaluate the options for an ERP system and determine the most suitable choice based on their priorities and preferences.

For this evaluation, three experts have been consulted regarding the pairwise comparison matrix. The opinions of all three experts carry equal importance for the decision-maker, and the pairwise comparison matrix for each level is provided in the specified order.

A questionnaire has been designed for experts, asking them to rank the elements using qualitative terms instead of pairwise comparisons. They are requested to express the competency of each element in each rank using qualitative terms (Tables 8.3, 8.4, 8.5, 8.6, 8.7, 8.8).

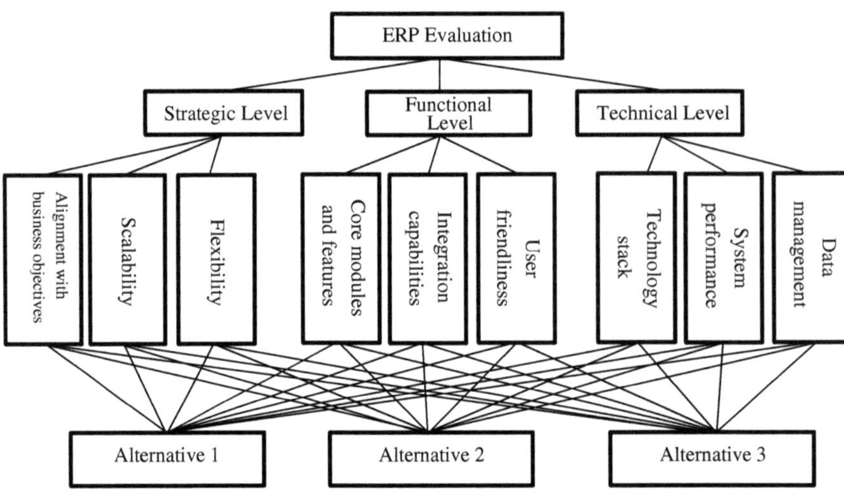

Fig. 8.6 Hierarchical structure for selecting an ERP system

Table 8.3 The results of the preference voting by experts at Level 1

	1th	2th	3th	The weights obtained by model (8.9)
Strategic Level	L,H,M	M,L,L	VH,M,VH	(0.3922,0.6245,0.8331)
Functional Level	H,EH,EH	M,H,H	M,M,H	(0.7722,0.9181,1.0000)
Technical Level	H,VH,H	EH,M,M	L,VL,M	(0.5828,0.7869,0.9559)

Table 8.4 The results of the preference voting by experts at strategic Level

	1th	2th	3th	The weights obtained by model (8.9)
Alignment with business objectives	M,VH,EL	H,L,L	EH,M,VH	(0.4871,0.6928,0.8730)
Scalability	VH,VL,EH	H,VL,H	VH,VH,H	(0.6318,0.8307,1.0000)
Flexibility	EH,H,H	L,EH,M	VH,VL,L	(0.6705,0.8530,1.0000)

Table 8.5 The results of the preference voting by experts at Functional Level

	1th	2th	3th	The weights obtained by model (8.9)
Core modules and features	VL,L,L	L,L,M	EH,VH,VL	(0.3366,0.5662,0.7978)
Integration capabilities	M,VL,H	VH,VL,H	L,EH,VH	(0.5550,0.7836,1.0000)
User-friendliness	EH,H,EH	H,M,VH	EL,VL,L	(0.7757,0.9202,1.0000)

Table 8.6 The results of the preference voting by experts at Technical Level

	1th	2th	3th	The weights obtained by model (8.9)
Technology stack	EH,H,H	VH,L,EH	H,EH,VH	(0.6938,0.8769,1.0000)
System performance	M,EH,H	H,VH,M	EH,M,H	(0.6034,0.8091,0.9772)
Data management	H,VL,VH	EH,VL,H	L,H,VH	(0.4876,0.6587,0.8289)

Table 8.7 The results of preference voting by experts regarding each sub-criterion, where the alternatives are candidates

			1th	2th	3th	The weights obtained by model (8.9)
Strategic Level	Alignment with business objectives	Alternative 1	VH,VH,EH	H,L,H	M,M,H	(0.7468,0.9199,1.0000)
		Alternative 2	M,M,VH	L,VH,M	EH,H,M	(0.5317,0.7548,0.9387)
		Alternative 3	VH,L,M	EH,VL,L	L,VL,L	(0.3882,0.5779,0.7497)
	Scalability	Alternative 1	L,M,L	L,L,H	VH,M,H	(0.3591,0.6052,0.8386)
		Alternative 2	H,EH,M	VH,VH,VL	VH,H,L	(0.6491,0.8439,1.0000)
		Alternative 3	EH,H,VL	EH,VH,EL	VH,L,M	(0.6066,0.7523,0.8777)
	Flexibility	Alternative 1	EL,H,M	VL,H,H	H,VH,L	(0.4361,0.6475,0.8752)
		Alternative 2	VH,EH,L	M,VH,M	M,H,L	(0.6681,0.8723,1.0000)
		Alternative 3	L,EL,L	M,VL,M	EH,L,H	(0.3143,0.5125,0.7269)
Functional Level	Core modules and features	Alternative 1	M,EL,VH	H,L,H	VH,H,M	(0.5526,0.7768,1.0000)
		Alternative 2	VH,VL,VL	VH,L,L	EH,M,L	(0.4303,0.6550,0.8666)
		Alternative 3	VL,EH,EL	M,VH,M	H,M,M	(0.5330,0.7275,0.9230)
	Integration capabilities	Alternative 1	L,L,M	M,M,L	H,H,VL	(0.3394,0.0.6066,0.8884)
		Alternative 2	M,M,VH	L,H,L	L,EH,L	(0.5353,0.7878,1.0000)
		Alternative 3	VL,VH,M	VL,VH,H	L,M,VH	(0.5177,0.7642,1.0000)
	User-friendliness	Alternative 1	M,L,L	H,VL,L	VH,M,H	(0.3001,0.5029,0.7067)
		Alternative 2	EH,H,H	VH,H,M	VH,EH,VH	(0.6759,0.8668,1.0000)
		Alternative 3	M,VH,M	H,M,M	H,M,H	(0.4612,0.6759,0.8579)
Technical Level	Technology stack	Alternative 1	L,M,M	L,H,M,	H,VH,H	(0.3817,0.5953,0.7981)
		Alternative 2	VH,EH,VH	VH,VH,M	EH,M,VH	(0.7576,0.9277,0.7981)
		Alternative 3	EH,L,L	VH,L,L	M,M,EH	(0.4114,0.6013,0.7556)
	System performance	Alternative 1	L,M,M	L,H,M	H,VH,H	(0.3817,0.5953,0.7981)

(continued)

Table 8.7 (continued)

			1th	2th	3th	The weights obtained by model (8.9)
		Alternative 2	VH,EH,VH	VH,VH,M	EH,M,VH	(0.7576,0.9277,1.0000)
		Alternative 3	EH,L,L	VH,L,L	M,M,EH	(0.4114,0.6013,0.7556)
	Data management	Alternative 1	VH,VH,L	VH,H,L	EH,M,M	(0.5510,0.7720,0.9220)
		Alternative 2	VH,M,M	H,H,H	M,M,VH	(0.5160,0.7380,0.9405)
		Alternative 3	M,EH,VH	M,VH,EH	EH,H,H	(0.7030,0.8880,1.0000)

Table 8.8 final ranking of alternatives

	Final fuzzy index	Defuzzy value	Rank
Alternative 1	(1.1600,3.2997,6.4596)	14.2191	3
Alternative 2	(1.9134,4.6270,7.8850)	19.0525	1
Alternative 3	(1.6153,3.7951,6.6608)	15.8663	2

8.7 Conclusion

In conclusion, this chapter has explored the use of AHP-fuzzy and ANP-fuzzy methodologies in hierarchical decision-making processes. The attitudes towards these approaches have been discussed, and a practical example has been presented to illustrate their application. Additionally, the chapter has delved into the topic of preferential voting, specifically in relation to expert opinions. Both the normal state and the use of qualitative words have been examined in this context. Overall, these discussions provide valuable insights into enhancing decision-making processes through the incorporation of fuzzy methodologies and expert preferences.

Acknowledgement A special thanks to the Iranian DEA society for their unwavering spiritual support and consensus in the writing of this book. Your invaluable support has been truly remarkable, and we are deeply grateful for the opportunity to collaborate with such esteemed professionals.

References

1. Allahviranloo, T., Lotfi, F. H., Kiasari, M. K., & Khezerloo, M. (2013). On the fuzzy solution of LR fuzzy linear systems. *Applied Mathematical Modelling, 37*(3), 1170–1176. https://doi.org/10.1016/j.apm.2012.03.037
2. Chang, D. Y. (1996). Applications of the extent analysis method on fuzzy AHP. *European journal of operational research, 95*(3), 649–655.
3. Junior, F. R., Lima, L. O., & Carpinetti, L. C. R. (2014). A comparison between Fuzzy AHP and Fuzzy TOPSIS methods to supplier selection. *Applied soft computing, 21*, 194–209.
4. Kahraman, C., Cebeci, U., & Ulukan, Z. (2003). Multi criteria supplier selection using fuzzy AHP. *Logistics information management, 16*(6), 382–394.
5. Liu, F. H. F., & Hai, H. L. (2005). The voting analytic hierarchy process method for selecting supplier. *International journal of production economics, 97*(3), 308–317.
6. Liu, Y., Eckert, C. M., & Earl, C. (2020). A review of fuzzy AHP methods for decision-making with subjective judgments. *Expert Systems with Applications, 161*, 113738.
7. Liu, F., Peng, Y., Zhang, W., & Pedrycz, W. (2017). On consistency in AHP and fuzzy AHP. *Journal of Systems Science and Information, 5*(2), 128–147.
8. Lotfi, F. H., Allahviranloo, T., Jondabeh, M. A., & Alizadeh, L. (2009). Solving a full fuzzy linear programming using lexicography method and fuzzy approximate solution. *Applied mathematical modelling, 33*(7), 3151–3156. https://doi.org/10.1016/j.apm.2008.10.020
9. Moayeri, M., Shahvarani, A., Behzadi, M. H., & Hosseinzadeh-Lotfi, F. (2015). Comparison of fuzzy AHP and fuzzy TOPSIS methods for math teachers selection. *Indian Journal of Science and Technology, 8*(13), 1.

10. Mohanty, R. P., Agarwal, R., Choudhury, A. K., & Tiwari, M. K. (2005). A fuzzy ANP-based approach to R&D project selection: A case study. *International Journal of Production Research, 43*(24), 5199–5216.
11. Pedrycz, W., & Song, M. (2014). A granulation of linguistic information in AHP decision-making problems. *Information Fusion, 17*, 93–101.
12. Pedrycz, W., & Song, M. (2011). Analytic hierarchy process (AHP) in group decision making and its optimization with an allocation of information granularity. *IEEE Transactions on Fuzzy Systems, 19*(3), 527–539.
13. Pishchulov, G., Trautrims, A., Chesney, T., Gold, S., & Schwab, L. (2019). The voting analytic hierarchy process revisited: A revised method with application to sustainable supplier selection. *International Journal of Production Economics, 211*, 166–179.
14. Saaty, T. L. (1980). The Analytic Hierarchy Process, RWS Publications
15. Saaty, T. L. (1996). Multicriteria decision making: The analytic hierarchy process. RWS Publ
16. Sharafi, H., Soltanifar, M., & Lotfi, F. H. (2022). Selecting a green supplier utilizing the new fuzzy voting model and the fuzzy combinative distance-based assessment method. *EURO Journal on Decision Processes, 10*, 100010.
17. Shariatmadari Serkani, E., Hosseinzadeh Lotfi, F., Najafi, E., & Ahadzadeh Namin, M. (2022). Efficiency measurement for hierarchical network systems using network DEA and intuitionistic fuzzy ANP. *Scientia Iranica, 29*(4), 2252–2269.
18. Shirouyehzad, H., Lotfi, F. H., Arabzad, S. M., & Dabestani, R. (2013). An AHP/DEA ranking method based on service quality approach: A case study in hotel industry. *International Journal of Productivity and Quality Management, 11*(4), 434–445.
19. Soltanifar, M. (2021). The voting linear assignment method for determining priority and weights in solving MADM problems. Journal of applied research on industrial engineering, 8(Special Issue), 1–17
20. Soltanifar, M., & Lotfi, F. H. (2011). The voting analytic hierarchy process method for discriminating among efficient decision making units in data envelopment analysis. *Computers & Industrial Engineering, 60*(4), 585–592.
21. Soltanifar, M., Sharafi, H., Hosseinzadeh Lotfi, F., Pedrycz, W., & Allahviranloo, T. (2023). Hybrid Multi-attribute Decision-Making Methods Based on Preferential Voting. *Preferential voting and applications: approaches based on data envelopment analysis* (pp. 133–164). Springer International Publishing.
22. Tavana, M., Soltanifar, M., & Santos-Arteaga, F. J. (2021). Analytical hierarchy process: revolution and evolution. Annals of Operations Research, 1–29
23. Wang, Y. M., Luo, Y., & Hua, Z. (2008). On the extent analysis method for fuzzy AHP and its applications. *European journal of operational research, 186*(2), 735–747.
24. Zhang, B., Pedrycz, W., Fayek, A. R., & Dong, Y. (2022). A differential evolution-based consistency improvement method in AHP with an optimal allocation of information granularity. *IEEE Transactions on Cybernetics, 52*(7), 6733–6744. https://doi.org/10.1109/TCYB.2020.3035909

Chapter 9
VIKOR Method in Uncertainty Environment

9.1 Introduction

The VIKOR method is a powerful decision-making technique that has gained significant popularity in the field of multi-criteria analysis. VIKOR provides a systematic framework for addressing complex decision problems where multiple criteria need to be considered simultaneously. It offers a comprehensive approach to evaluating alternatives and selecting a compromise solution that balances conflicting objectives.

At its core, the VIKOR method aims to find a compromise solution that achieves the best trade-off between the ideal and anti-ideal solutions. It combines elements of both the maximum and minimum principles to derive an optimal compromise solution that satisfies the decision-maker's preferences. The VIKOR method is particularly useful in situations where decision criteria are interrelated and there is a need to consider the overall performance of alternatives across multiple dimensions.

The VIKOR method follows a step-by-step process. First, decision criteria are identified and defined based on the problem at hand. These criteria can be quantitative or qualitative, objective or subjective. Then, performance scores or weights are assigned to the alternatives based on their performance with respect to each criterion. The VIKOR method allows decision-makers to incorporate their preferences and priorities through these scores.

Next, the VIKOR method calculates several parameters for each alternative, including the "maximum" and "minimum" values of each criterion and the "closeness coefficient" that represents the relative compromise level. The closeness coefficient is used to rank the alternatives and identify the compromise solution.

The VIKOR method also offers a sensitivity analysis to assess the stability of the chosen solution. Sensitivity analysis helps decision-makers understand the impact of changes in criteria weights or performance scores on the final ranking of alternatives. This allows for a more robust decision-making process and helps identify critical factors that influence the solution [16].

F. Hosseinzadeh Lotfi et al., *Fuzzy Decision Analysis: Multi Attribute Decision Making Approach*, Studies in Computational Intelligence 1121,
https://doi.org/10.1007/978-3-031-44742-6_9

Overall, the VIKOR method provides decision-makers with a structured and systematic approach to tackle complex decision problems involving multiple criteria. It helps in exploring trade-offs, considering conflicting objectives, and arriving at a compromise solution that aligns with the decision-maker's preferences. The VIKOR method has found applications in various fields, including engineering, management, finance, and environmental planning, where decision-makers are faced with multi-dimensional decision challenges.

In conclusion, the VIKOR method is a valuable tool for multi-criteria decision-making. Its step-by-step approach, consideration of trade-offs, and ability to handle conflicting objectives make it a practical and effective method for addressing real-world decision problems [3, 17].

9.2 VIKOR

The VIKOR method follows a systematic and structured process to facilitate multi-criteria decision-making. It involves several key steps that enable decision-makers to evaluate alternatives and identify a compromise solution. By incorporating trade-offs and balancing conflicting objectives, the VIKOR method provides a practical and effective approach to real-world decision problems. Below are the steps of this method:

Step 1. Construct the decision matrix

According to the number of criteria, the number of alternatives and the evaluation of all alternative for different criteria, the decision matrix is formed as follows.

$$X = \begin{array}{c} \begin{matrix} C_1 & \dots & C_n \end{matrix} \\ \begin{bmatrix} x_{11} & \dots & x_{1n} \\ \vdots & \ddots & \vdots \\ x_{m1} & \dots & x_{mn} \end{bmatrix} \end{array} \tag{9.1}$$

where x_{ij} is the score of the alternative $i(i = 1, ..., m)$ related to the criterion $j(j = 1, ..., n)$.

Step 2. Construct the Normalize decision matrix

At this step, based on the existing methods, the decision matrix is normalized. Naturally, all cost criterion are converted into benefit criterion and they will be treated the same.

$$\begin{cases} n_{ij} = \frac{x_{ij}}{\sum_{i=1}^{m} x_{ij}}, & i = 1, ..., m;\ j \in B, \mathrm{B} = \{\mathrm{j}|\mathrm{c_j} \text{ is benefit ceriteria}\} \\ n_{ij} = \frac{\frac{1}{x_{ij}}}{\sum_{i=1}^{m} \frac{1}{x_{ij}}}, & i = 1, ..., m;\ j \in \mathrm{C}, \mathrm{C} = \{\mathrm{j}|\mathrm{c_j} \text{ is cost ceriteria}\} \end{cases} \tag{9.2}$$

Using Eq. (9.2), the normalized matrix will be as follows.

$$N = \begin{bmatrix} n_{11} & \dots & n_{1n} \\ \vdots & \ddots & \vdots \\ n_{m1} & \cdots & n_{mn} \end{bmatrix} \tag{9.3}$$

Step 3. Determining the criteria's weights

At this step, according to the importance of different criteria in decision-making, the weight vector of the criteria is defined as $w_j(j = 1, \dots, n)$.

$$W = [\mathrm{w}_1, \mathrm{w}_2, \dots, \mathrm{w}_n]$$

Step 4. Determine the best and worst values for each criterion

Considering that in the previous step, all criteria were normalized as profit-oriented, in the following, the best values are only determined for profit indicators. If a different method is used for normalization, in this step, these values should also be determined for cost indicators.

$$\mathrm{f}_j^* = \underset{i=1,\dots,m}{\mathrm{Max}} \{\mathrm{n}_{ij}\}, \quad j = 1, \dots, n. \tag{9.4}$$

$$\mathrm{f}_j^- = \underset{i=1,\dots,m}{\mathrm{Min}} \{\mathrm{n}_{ij}\}, \quad j = 1, \dots, n. \tag{9.5}$$

Step 5. Determining the distance between the best and worst situations

The distance to the best situation (utility measure) is indicated as $\mathrm{S_i}(i = 1, \dots, m)$, and the distance to the worst situation (reject measure) is shown as $\mathrm{R_i}(i = 1, \dots, m)$.

$$\mathrm{S_i} = \sum_{\mathrm{j=1}}^{\mathrm{n}} \mathrm{w_j} \left(\frac{\left|\mathrm{f_j^*} - \mathrm{n_{ij}}\right|}{\left|\mathrm{f_j^*} - \mathrm{f_j^-}\right|} \right) \quad , i = 1, \dots, m. \tag{9.6}$$

$$\mathrm{R_i} = \underset{\mathrm{j=1,\dots,n}}{\mathrm{Max}} \left\{ \mathrm{w_j} \left(\frac{\left|\mathrm{f_j^*} - \mathrm{n_{ij}}\right|}{\left|\mathrm{f_j^*} - \mathrm{f_j^-}\right|} \right) \right\} \quad , i = 1, \dots, m. \tag{9.7}$$

In other words, the above measures represent the distance with respect to the L-1 and L-∞ norms, respectively.

Step 6. Calculation of the Q index

The Q index is calculated by combining the convex combination of the two distances in S and R, which were computed in the previous step, using the group consensus rate represented by ν, a value between 0 and 1. In other words, it can be described as a weighted average between these two values.

$$\mathrm{Q_i} = \nu\left(\frac{\mathrm{S_i} - \mathrm{S}^*}{\mathrm{S}^- - \mathrm{S}^*}\right) + (1 - \nu)\left(\frac{\mathrm{R_i} - \mathrm{R}^*}{\mathrm{R}^- - \mathrm{R}^*}\right), \quad i = 1, \ldots, m.$$
$$\mathrm{S}^* = \underset{\mathrm{i=1,\ldots,m}}{\mathrm{Min}}\ \{\mathrm{S}_i\}, \mathrm{S}^- = \underset{\mathrm{i=1,\ldots,m}}{\mathrm{Max}}\ \{\mathrm{S}_i\}, \mathrm{R}^* = \underset{\mathrm{i=1,\ldots,m}}{\mathrm{Min}}\ \{\mathrm{R}_i\}, \mathrm{R}^- = \underset{\mathrm{i=1,\ldots,m}}{\mathrm{Max}}\ \{\mathrm{R}_i\}. \tag{9.8}$$

Step 7. Rank the alternatives

Alternatives are ranked using the R, S, and Q ranking system. As a result, three ranking lists were obtained, with the lowest values indicating the best alternatives.

It should be noted that in the ranking based on the Q index, the top-ranked alternative is chosen if it satisfies the following conditions. For example, considering the first alternative as A_{r1} and the second alternative as A_{r2}, it should:

Condition 1:

$$\mathrm{Q}(\mathrm{A}_{r2}) - \mathrm{Q}(\mathrm{A}_{r1}) \geq \frac{1}{m - 1}. \tag{9.9}$$

Condition 2:

Alternative A1 must be recognized as superior in at least one of the R and S groups.

When Condition 1 is not met, a set of Alternatives are recognized as the top choice, which is determined by the following relationship with the highest count:

$$\begin{aligned} &\text{Top alternatives} = A_{r1}, A_{r2}, A_{r3}, \ldots, A_{rt} \\ &Q(A_{rt}) - Q(A_{r1}) < \tfrac{1}{m-1} \end{aligned} \tag{9.10}$$

When Condition 2 is not met, both alternatives A_{r1} and A_{r2} are selected as the top alternatives.

It is evident that the existence of three ranking lists can result in various final ranking scenarios. This can be considered as a voting situation with three voters and m candidates, and preference voting methods can be utilized for ranking [5, 22].

9.3 The Evaluation of Insurance Companies

The evaluation of insurance companies plays a crucial role in assessing their performance and determining their reliability. In this regard, the VIKOR (VlseKriterijumska Optimizacija I Kompromisno Resenje) method provides a comprehensive approach to rating insurance companies. By considering four key areas and their respective sub-criteria, this evaluation aims to provide a holistic analysis of an insurance company's strengths and weaknesses.

I. Financial Field:
The financial field focuses on the financial stability and prosperity of insurance institutions. It encompasses various criteria, including the damage ratio, which measures the relationship between incurred damages and earned insurance premiums. The ratio of financial prosperity evaluates the financial health of insurance institutions, while the debt ratio assesses the relationship between total liabilities and total assets. Furthermore, the ratio of administrative and general expenses to the number of employees, along with the ratio of profit before tax deduction to the number of employees, shed light on the efficiency and productivity of the company's financial operations.

II. Customer Field:
The customer field evaluates the company's performance in terms of customer satisfaction and outreach. It considers factors such as the representation of the company in different cities, aiming for comprehensive coverage nationwide. The geographic distribution of branches is also examined to ensure equal accessibility to insurance services across all cities. Moreover, the number of complaints received by the Central Insurance Company per thousand claims provides insight into the company's commitment to resolving customer issues promptly. The evaluation index of customer satisfaction measures the overall satisfaction levels of policyholders and reflects the quality of the company's customer service.

III. Internal Processes:
The field of internal processes assesses the company's operational efficiency and resource allocation. The ratio of the education budget to the total current budget highlights the importance placed on continuous learning and professional development within the organization. Additionally, the expenses incurred in the field of information technology (IT) are examined in relation to the company's overall expenses, reflecting its commitment to technological advancements and digitalization. The ratio of issued insurance premiums to the number of employees indicates the efficiency of the company's underwriting processes.

IV. Growth and Learning:
The area of growth and learning focuses on the company's human resources and their potential for advancement. The number of employees with a bachelor's degree and above, in proportion to the total number of employees, demonstrates the company's investment in a highly educated workforce. The average education level of employees provides insights into the overall educational qualifications within the organization. Furthermore, the number of employees with work experience exceeding ten years, relative to the total number of employees, reflects the company's ability to retain experienced professionals. The average years of experience of employees further showcases the company's commitment to nurturing and retaining skilled talent.

By employing the VIKOR method and considering these four areas and their respective sub-criteria, this evaluation aims to provide valuable insights into the

Table 9.1 VIKOR method results for insurance companies

Insurance companies	S	R	Q
	Rank		
A1	1	3	1
A2	3	1	2
A3	4	4	3
A4	5	3	4
A5	2	5	5

Table 9.2 Preference voting table

Insurance companies	First place	Second place	Third place	Fourth place	Fifth place
A1	2	0	1	0	0
A2	1	1	1	0	0
A3	0	0	1	2	0
A4	0	0	1	1	1
A5	0	1	0	0	2

performance and competitive positioning of insurance companies, aiding readers in making informed decisions regarding their insurance needs.

Based on the criteria and the simple steps of the VIKOR method, the results in terms of S, R, and Q are presented. The main task in implementing the VIKOR method is to determine the final ranking based on these three indicators. As mentioned before, the Q indicator is essentially a convex combination of the other two indicators, and preference voting methods can be used for the final ranking, which will be discussed in this matter.

Based on the above information, another type of ranking aggregation can be considered based on preference voting. For this purpose, each alternative is ranked in each of the three ranking criteria, and then a preference voting table is formed. In other words, the number of votes in each position is determined based on the Table 9.1 [15].

As an example, the third alternative has been ranked third once and fourth twice. Therefore, it is considered as 1 in the third place and 2 in the fourth place for that alternative see Table 9.2. The preference voting table is completed based on the places of each insurance companies attained by ranks mentioned in Table 9.1.

To aggregate these data and reach a final ranking, various methods have been introduced. One of them is the Cook et al. (1990) method, which is mentioned below:

Table 9.3 Final ranking based on preference voting

Insurance companies	Z_p	Rank
A1	1.00000	1
A2	0.99000	2
A3	0.96333	3
A4	0.96083	5
A5	0.96333	3

$$\begin{aligned} Z_p &= Max \sum_{r=1}^{k} w_r v_{rp} \\ s.t. \quad & \sum_{r=1}^{k} w_r v_{rj} \leq 1, \qquad j = 1, \ldots, n \\ & w_r - w_{r+1} \geq d(r, \varepsilon), \quad r = 1, \ldots, k-1 \\ & w_k \geq d(k, \varepsilon) \end{aligned} \tag{9.11}$$

Equation (9.11) is expanded for the third alternative as bellow:

$$\begin{aligned} Z_3 = Max \quad & w_3 + 2w_4 \\ s.t. \quad & 2w_1 + w_3 \leq 1, \\ & w_1 + w_2 + w_3 \leq 1, \\ & w_3 + 2w_4 \leq 1, \\ & w_3 + w_4 + w_5 \leq 1, \\ & w_2 + 2w_5 \leq 1, \\ & w_r - w_{r+1} \geq d(r, \varepsilon), \quad r = 1, \ldots, 4 \\ & w_5 \geq d(5, \varepsilon) \end{aligned} \tag{9.12}$$

After solving model (9.11) for all alternatives, the following results will be obtained, and the final ranking will be presented based on them (Table 9.3).

9.4 Fuzzy VOKIR

In the ever-evolving field of decision-making and multi-criteria analysis, the VIKOR (VlseKriterijumska Optimizacija I Kompromisno Resenje) method has gained significant attention and recognition for its effectiveness in handling complex decision problems. However, traditional VIKOR has its limitations when dealing with situations that involve uncertainty, ambiguity, and imprecision in decision criteria and data. This led to the emergence of a variant known as Fuzzy VIKOR, which incorporates fuzzy logic to handle such complexities and provide more robust decision-making outcomes [2, 4, 5, 7].

Fuzzy VIKOR extends the original VIKOR method by incorporating fuzzy set theory, which allows for the representation of imprecise and uncertain information in a decision-making process. By considering linguistic variables, membership functions, and fuzzy weights, Fuzzy VIKOR provides a more flexible and realistic framework for decision-making under vagueness and ambiguity. The key advantage of Fuzzy VIKOR lies in its ability to handle subjective and uncertain information effectively. It takes into account the imprecision and ambiguity inherent in decision criteria and data, enabling decision-makers to make informed and rational choices even in situations where precise measurements or crisp boundaries are not available. This makes it particularly useful in real-world applications where decision-making involves human judgment, expert opinions, and qualitative assessments [12,19].

In this section, we discuss about the principles and methodologies of Fuzzy VIKOR, exploring its theoretical foundations and computational procedures. We also examine its applications in various domains, highlighting its advantages and showcasing how it overcomes the limitations of traditional VIKOR. Furthermore, we discuss the challenges and future directions of Fuzzy VIKOR, paving the way for further research and advancements in the field of fuzzy multi-criteria decision analysis. This section serves as a comprehensive guide to Fuzzy VIKOR, providing insights and practical knowledge that can be applied in a wide range of decision-making scenarios [13, 20, 25].

The decision matrix is derived by assessing the rating of each criterion for every index. Occasionally, the rating system employed may be fuzzy, while in other cases, it involves converting linguistic variables. In both instances, the decision matrix is established as follows. It is important to note that when utilizing this approach, any specific values must be transformed into a fuzzy format. Moreover, we will consider $z = (a, b, c)$ as $z = (a, b, b, c)$, meaning that all values will be converted into trapezoidal fuzzy representation.

Let's consider a scenario in which the decision maker (DM) has defined a weight vector and a decision matrix for a multi-criteria decision problem, involving m alternatives and m criteria. In this particular scenario, certain values are expressed qualitatively using linguistic variables. The weights and ratings ($h = 1,\ldots,k$) of DMh are combined as follows, the weight vectors and ratings denoted by $\tilde{V}_{ij}^{h} = (v_{ij}^{h,L}, v_{ij}^{h,M_1}, v_{ij}^{h,M_2}, v_{ij}^{h,U}), (v_{ij}^{h,L} > 0)$ and $\tilde{Y}_j^h = (y_j^{h,L}, y_j^{h,M_1}, y_j^{h,M_2}, y_j^{h,U}), (y_j^{h,L})$, respectively.

$$\widetilde{w}_j = (w_j^L, w_j^{M_1}, w_j^{M_2}, w_j^U); \; j = 1, \ldots, n.$$

$$w_j^L = \underset{h=1,\ldots,k}{\text{Min}} \{y_j^{h,L}\}, \quad w_j^{M_1} = \frac{1}{k}\sum_{h=1}^{k} y_j^{h,M_1}, \quad w_j^{M_2} = \frac{1}{k}\sum_{h=1}^{k} y_j^{h,M_2}, \quad w_j^U = \underset{h=1,\ldots,k}{\text{Max}} \{y_j^{h,U}\}$$

$$\widetilde{W} = (\tilde{w}_1, \tilde{w}_2, \ldots, \tilde{w}_n) \tag{9.13}$$

Furthermore, the fuzzy decision matrix $\widetilde{X}$ is aggregated and constructed as follows.

$$\tilde{x}_{ij} = (x_{ij}^L, x_{ij}^{M_1}, x_{ij}^{M_2}, x_{ij}^U); \; j = 1, \ldots, n.$$

$$\mathrm{x}_{\mathrm{ij}}^{\mathrm{L}} = \underset{\mathrm{h}=1,\ldots,\mathrm{k}}{\mathrm{Min}}\{\mathrm{v}_{\mathrm{ij}}^{\mathrm{h,L}}\},\ \mathrm{x}_{\mathrm{ij}}^{\mathrm{M_1}} = \frac{1}{\mathrm{k}}\sum_{\mathrm{h}=1}^{\mathrm{k}}\mathrm{v}_{\mathrm{ij}}^{\mathrm{h,M_1}},\ \mathrm{x}_{\mathrm{ij}}^{\mathrm{M_2}} = \frac{1}{\mathrm{k}}\sum_{\mathrm{h}=1}^{\mathrm{k}}\mathrm{v}_{\mathrm{ij}}^{\mathrm{h,M_2}},\ \mathrm{x}_{\mathrm{ij}}^{\mathrm{U}} = \underset{\mathrm{h}=1,\ldots,\mathrm{k}}{\mathrm{Max}}\{\mathrm{v}_{\mathrm{ij}}^{\mathrm{h,U}}\}$$

$$\widetilde{\mathrm{X}} = \begin{bmatrix} \tilde{\mathrm{x}}_{11} & \ldots & \tilde{\mathrm{x}}_{1\mathrm{n}} \\ \vdots & \ddots & \ldots \\ \tilde{\mathrm{x}}_{\mathrm{m}1} & \ldots & \tilde{\mathrm{x}}_{\mathrm{mn}} \end{bmatrix} \tag{9.14}$$

Normalizing the fuzzy decision matrix would result in the following representation, which is displayed using $\widetilde{\mathrm{N}}$:

$$\tilde{n}_{ij} = \left(n_{ij}^{L}, n_{ij}^{M_1}, n_{ij}^{M_2}, n_{ij}^{U}\right) = \left(\frac{x_{ij}^{L}}{x_j^{Max}}, \frac{x_{ij}^{M_1}}{x_j^{Max}}, \frac{x_{ij}^{M_2}}{x_j^{Max}}, \frac{x_{ij}^{U}}{x_j^{Max}}\right), \quad i = 1, \ldots, m \ \ ; j \in B(\text{Benefit criteria})$$

$$\tilde{n}_{ij} = \left(n_{ij}^{L}, n_{ij}^{M_1}, n_{ij}^{M_2}, n_{ij}^{U}\right) = \left(\frac{x_j^{Min}}{x_{ij}^{L}}, \frac{x_j^{Min}}{x_{ij}^{M_1}}, \frac{x_j^{Min}}{x_{ij}^{M_2}}, \frac{x_j^{Min}}{x_{ij}^{U}}\right), \quad i = 1, \ldots, m; \ j \in C(\text{Cost criteria})$$

$$\widetilde{\mathrm{N}} = \begin{bmatrix} \tilde{n}_{11} & \ldots & \tilde{n}_{1\mathrm{n}} \\ \vdots & \ddots & \ldots \\ \tilde{n}_{\mathrm{m}1} & \ldots & \tilde{n}_{\mathrm{mn}} \end{bmatrix} \tag{9.15}$$

In which $x_j^{Max} = \underset{i}{Max}\{x_{ij}^{U}\}, (j = 1, \ldots, n)$ and $x_j^{Min} = \underset{i}{Min}\{x_{ij}^{L}\}, (j = 1, \ldots, n)$.

Step 1: Formation of the decision matrix from group decision-making.

Using Eqs. (9.9–9.13), the decision matrix is determined by aggregating the opinions of experts.

Step 2: Normalization of the decision matrix

Using Eqs. (9.9–9.14), the normalized decision matrix is constructed.

Step 3: Determining the criteria weights

Using Eqs. (9.9–9.12) or using weight determination methods discussed in Chapter 3 of the book, the weight vector is determined.

Step 4. Determine the best and worst values for each criterion

Since in the previous step all elements of the decision matrix have been normalized and are between zero and one.

$$\tilde{\mathrm{f}}_j^{*} = (1, 1, 1, 1), \quad \mathrm{j} = 1, \ldots, n. \tag{9.16}$$

$$\tilde{\mathrm{f}}_j^{-} = (0, 0, 0, 0), \quad \mathrm{j} = 1, \ldots, n. \tag{9.17}$$

Step 5. Determining the distance between the best and worst situations

The distance to the best situation (utility measure) is indicated as $S_i(i = 1, ..., m)$, and the distance to the worst situation (reject measure) is shown as $R_i(i = 1, ..., m)$.

$$\widetilde{S}_i = \sum_{j=1}^{n} \tilde{w}_j \left(\frac{(\tilde{f}_j^* - \tilde{n}_{ij})}{(\tilde{f}_j^* - \tilde{f}_j^-)} \right), i = 1, ..., m. \tag{9.18}$$

$$\widetilde{R}_i = \underset{j=1,...,n}{\text{Max}} \left\{ \tilde{w}_j \left(\frac{(\tilde{f}_j^* - \tilde{n}_{ij})}{(\tilde{f}_j^* - \tilde{f}_j^-)} \right) \right\}, i = 1, ..., m. \tag{9.19}$$

Step 6. Calculation of the Q index

The Q index is calculated by combining the convex combination of the two distances in S and R, which were computed in the previous step, using the group consensus rate represented by ν, a value between 0 and 1. In other words, it can be described as a weighted average between these two value [15].

$$\begin{aligned} \widetilde{Q}_i &= \nu\left(\frac{\tilde{S}_i - \tilde{S}^*}{\tilde{S}^- - \tilde{S}^*}\right) + (1 - \nu)\left(\frac{\tilde{R}_i - \tilde{R}^*}{\tilde{R}^- - \tilde{R}^*}\right), \quad i = 1, ..., m. \\ \widetilde{S}^-, \widetilde{R}^- &= (1, 1, 1, 1); \quad \widetilde{S}^*, \widetilde{R}^* = (0, 0, 0, 0). \end{aligned} \tag{9.20}$$

Step 7. Rank the alternatives

In this step, the final ranking of each alternative is determined using preferential voting. If Eq. (9.15) is used for normalization in step 3, the Cook et al. [6] model can be utilized. Alternatively, if there is a cost-oriented criterion after normalization, the Soltanifar et al. [22] method can be employed.

9.5 Choosing a Suitable Tourism Location with Fuzzy VIKOR Method

When planning a trip or vacation, selecting the right tourism location is crucial to ensure an enjoyable and fulfilling experience. However, with countless destinations to choose from, travelers often face the challenge of making an informed decision. In such cases, employing decision-making methods can assist in evaluating various factors and arriving at a suitable choice [1].

One such method is the Fuzzy VIKOR method, which incorporates qualitative opinions from experts and facilitates group decision making. By considering multiple criteria and categorizing them into economic, social, environmental, and political factors, the Fuzzy VIKOR method offers a comprehensive framework for selecting the ideal tourism location.

Economic Factors: (C1)

The economic category encompasses a range of factors that influence the financial aspect of tourism. These factors include the fairness of prices, the efficiency of public transportation, the quality and availability of food, the services offered to tourists, the quality of accommodation facilities, and the prevailing currency exchange rates. Assessing these economic factors allows travelers to consider their budget constraints, access to amenities, and overall cost-effectiveness of the chosen location.

Social Factors: (C2)

The social category involves evaluating the sociocultural aspects of a destination. Factors within this category include the attitudes of locals towards tourists, the knowledge level of tour leaders, the lifestyle and customs of the area, the presence of impoverished regions or beggars, and the architectural design of the surroundings. Understanding these social factors helps travelers gauge the level of hospitality, cultural richness, and authenticity they can expect from the chosen location.

Environmental Factors: (C3)

The environmental category focuses on the natural and environmental aspects of a destination. Elements such as cleanliness in public spaces, the scenic beauty of parks and mountains, weather conditions, and the level of urban traffic fall under this category. By considering these environmental factors, travelers can prioritize destinations that offer clean and well-maintained surroundings, stunning landscapes, and favorable climatic conditions.

Political Factors: (C4)

The political category pertains to the governance and security aspects of a tourism location. Factors within this category include the entry process for tourists, the overall safety and security measures in place, and the level of reassurance offered to travelers. Evaluating these political factors allows tourists to make informed decisions based on the ease of access, stability, and safety of the destination.

By employing the Fuzzy VIKOR method, which combines qualitative expert opinions and group decision-making techniques, travelers can objectively assess these four categories of criteria when choosing a suitable tourism location. The method enables a holistic evaluation process, taking into account diverse perspectives and facilitating a well-rounded decision that aligns with the preferences and requirements of the travelers.

Assume there are five Locations and three decision-makers. Each decision-maker has evaluated the managers using linguistic variables and assigned weights to them. The results of their evaluations are as follows. Before proceeding, let's define the criteria for converting linguistic variables to fuzzy numbers, as shown below (Tables 9.4, 9.5).

Decision makers have expressed their opinions for each criterion, which are listed in Table 9.6.

Table 9.4 Transformation of linguistic variables in determining the importance of criteria weight

Linguistic terms	Fuzzy number
Very low importance (VL)	(0.1,0.1,0.1,0.2)
Low importance (L)	(0.1,0.2,0.2,0.3)
Medium low importance (ML)	(0.2,0.3,0.4,0.5)
Medium importance (M)	(0.4,0.5,0.5,0.6)
Medium high importance (MH)	(0.5,0.6,0.7,0.8)
High importance (H)	(0.7,0.8,0.8,0.9)
Very High importance (VH)	(0.9,0.9,1,1)

Table 9.5 Transformation of linguistic variables in determining the performance of alternatives

Linguistic terms	Fuzzy number
Very Poor (VP)	(0.1,0,1,2)
Poor (P)	(1,2,2,3)
Medium Poor (MP)	(2,3,4,5)
Fair (F)	(4,5,5,6)
Medium Good (MG)	(5,6,7,8)
Good (G)	(7,8,8,9)
Very Good (VG)	(8,9,10,10)

Table 9.6 Opinions of decision makers on the importance of each criterion

Criteria	Opinions of decision maker
C1	M,MH,L
C2	MH,ML,H
C3	MH,H,MH
C4	VH,H,H

Table 9.7 Decision makers' opinions on the performance of each alternative on each criterion

Alternative	C1	C2	C3	C4
A1	F,MP,MG	MP,F,MP	F,MG,F	P,F,P
A2	G,F,MG	MG,MG,G	G,MG,G	VP,MP,F
A3	MG,P,F	F,G,MG	G,MG,F	F,MG,P
A4	G,G,MG	F,MP,MG	F,G,F	MG,F,MP
A5	VG,,MG,G	MP,MG,G	MP,F,P	VG,G,F

Table 9.8 The normalized decision matrix

Alternative	C1	C2	C3	C4
A1	(0.407,0.519,0.593,0.704)	(0.32,0.44,0.52,0.64)	(0.5,0.61,0.65,0.77)	(0.25,0.375,0.375,0.5)
A2	(0.593,0.704,0.741,0.852)	(0.68,0.8,0.88,1)	(0.65,0.77,0.85,0.96)	(0.25,0.333,0.417,0.542)
A3	(0.37,0.481,0.519,0.63)	(0.64,0.76,0.8,0.92)	(0.73,0.84,0.88,1)	(0.417,0.542,0.583,0.708)
A4	(0.704,0.815,0.852,0.963)	(0.44,0.56,0.64,0.76)	(0.57,0.69,0.69,0.80)	(0.458,0.583,0.667,0.792)
A5	(0.741,0.852,0.926,1)	(0.56,0.68,0.76,0.88)	(0.26,0.38,0.42,0.53)	(0.708,0.833,0.917,1)

Table 9.9 Ranking indices ($\nu = 0.5$)

Alternative	S	Rank S	R	Rank R	Q	Rank Q
A1	(1.091,1.290,1.303,1.373)	5	(0.467,0.542,0.5420.575)	5	(0.779,0.912,0.916,0.974)	5
A2	(0.544,0.815,0.9751,1.056)	4	(0.428,0.5060.556,0.575)	4	(0.486,0.660,0.753,0.816)	4
A3	(0.541,0.797,0.845,0.978)	2	(0.272,0.361,0.382,0.447)	2	(0.406,579,0.614,0.712)	3
A4	(0.552,0.812,0.882,1.015)	3	(0.194,0.289,0.347,0.415)	1	(0.373,0.550,0.615,0.715)	2
A5	(0.473,0.682,0.795,0.929)	1	(0.385,0.410,0.414,0.423)	3	(0.429,0.552,0.602,0.672)	1

Table 9.10 Final rank by preference voting

Alternative	Z_p	Rank
A1	0.9525	5
A2	0.9600	4
A3	0.9800	3
A4	0.9900	2
A5	1.0000	1

The opinions of decision makers regarding the performance of each alternative on each criterion are as Table 9.7.

The normalized decision matrix and the criteria weight vector are as (Table 9.8).

The indices S, Q, and R are calculated using Eqs. (9.18), (9.19), and (9.20) respectively, and they are presented below. It should be noted that they have been ranked using a ranking function, and the rank of each alternative is determined based on each index (Table 9.9).

After calculating the above ranking indices and determining the rank of each alternative in them, a preferential voting table is formed. Then, model (9.11) is solved to aggregate the rankings (Table 9.10).

To solve model (9.11) in this case, intensity discrimination function measure $d(\mathrm{r}, \varepsilon) = \frac{\varepsilon}{r}$ and discernment factor ($\varepsilon = 0.1$) have been used.

9.6 Data Envelopment Analysis and VIKOR

Data Envelopment Analysis (DEA) is a widely used mathematical modeling technique in the field of operations research and management science. It provides a systematic approach for evaluating the relative efficiency of multiple decision-making units (DMUs) or entities that convert inputs into outputs, such as organizations, companies, or departments [9, 24].

The fundamental concept behind DEA is to measure the efficiency of each DMU by comparing its performance to that of others in the same industry or sector. Unlike traditional efficiency measures that rely on a single input–output ratio, DEA considers multiple inputs and outputs simultaneously, making it suitable for evaluating complex systems with diverse inputs and outputs. DEA calculates efficiency scores by constructing an enveloping frontier, known as the efficient frontier or production possibility set, which represents the best achievable performance for a given set of inputs and outputs. DMUs that lie on the efficient frontier are considered fully efficient, while those falling below the frontier are deemed inefficient and can be potential targets for improvement [8, 10, 18].

One key concept in DEA is cross efficiency, which extends the analysis beyond measuring the efficiency of individual DMUs. Cross efficiency evaluates the performance of each DMU relative to a composite benchmark unit created from the efficient frontier. It provides a measure of how efficiently a DMU performs in comparison to the best-performing units in the industry, enabling identification of areas for improvement and best practices [11,14].

The cross efficiency structure is essentially a matrix in which the efficiency of each decision-making unit is constructed with optimal weights from other decision-making units. In some production technologies, such as variable returns to scale technology, negative numbers are encountered in the cross efficiency matrix, indicating inefficient efficiency. One possible method to address this issue is to use the absolute maximum in each column, which is added to all elements in that column. The closer the resulting number, the higher the eligibility for a better rank. Soltanifar et al. [23] employed this technique and then transformed these eligibilities into fuzzy numbers and used fuzzy VIKOR to rank the decision-making units. Certainly, many discussions can be raised here, such as considering the input and output matrices as criteria from the beginning, similar to the decision matrix, and ranking from the start using a method like VIKOR. However, it should be noted that the numbers in the cross efficiency matrix are the result of evaluating each decision-making unit based on the accessible pattern. That is, the weakness that exists in many multi-criteria decision-making methods is the lack of an accessible (attainable) pattern in the ranking process. On the other hand, some existing methods in data envelopment analysis are not capable of ranking decision-making units. Therefore, combining data envelopment analysis with multi-criteria decision-making methods can cover the weaknesses and create a less flawed evaluation and ranking process.

9.7 Conclusion

In this section, the VIKOR method was explained and a methodology for utilizing the fuzzy VIKOR method was presented. An application of the method for selecting a tourist destination was described. Furthermore, in aggregating the rankings, a preferential voting approach was employed. The VIKOR method has been widely used in decision-making processes to effectively handle multiple conflicting criteria. Fuzzy VIKOR approach has proven to be valuable in situations where uncertainties and vagueness are present, providing more flexibility in decision-making. Applications of VIKOR and fuzzy VIKOR can be found in various fields such as finance, engineering, and environmental management, showcasing their versatility and practicality. In conclusion, in this section of the book, the VIKOR method was introduced and its application in the context of tourist destination selection was demonstrated. The fuzzy VIKOR method was utilized, and the rankings were aggregated using a preferential voting perspective.

Acknowledgement A special thanks to the Iranian DEA society for their unwavering spiritual support and consensus in the writing of this book. Your invaluable support has been truly remarkable, and we are deeply grateful for the opportunity to collaborate with such esteemed professionals.

References

1. Adabi, G., Hajiha, A., & Hosseinzadeh Lotfi, F. (2022). Designing a tourism industry model in Iran with the approach of the spatial correlation structure. *Journal of Industrial Engineering and Management Studies*, *8*(2), 261-276
2. Banaeian, N., Mobli, H., Fahimnia, B., Nielsen, I. E., & Omid, M. (2018). Green supplier selection using fuzzy group decision making methods: A case study from the agri-food industry. *Computers & Operations Research, 89*, 337–347.
3. Bazzazi, A. A., Osanloo, M., & Karimi, B. (2011). Deriving preference order of open pit mines equipment through MADM methods: Application of modified VIKOR method. *Expert Systems with Applications, 38*(3), 2550–2556.
4. Chen, L. Y., & Wang, T. C. (2009). Optimizing partners' choice in IS/IT outsourcing projects: The strategic decision of fuzzy VIKOR. *International Journal of Production Economics, 120*(1), 233–242.
5. Chu, M. T., Shyu, J., Tzeng, G. H., & Khosla, R. (2007). Comparison among three analytical methods for knowledge communities group-decision analysis. *Expert systems with applications, 33*(4), 1011–1024.
6. Cook, W. D., & Kress, M. (1990). A data envelopment model for aggregating preference rankings. *Management Science, 36*(11), 1302–1310.
7. Deveci, M., Gokasar, I., Pamucar, D., Zaidan, A. A., Wen, X., & Gupta, B. B. (2023). Evaluation of cooperative intelligent transportation system scenarios for resilience in transportation using type-2 neutrosophic fuzzy VIKOR. *Transportation Research Part A: Policy and Practice, 172*, 103666.
8. Farhad, Hosseinzadeh Lotfi Ali, Ebrahimnejad Mohsen, Vaez-Ghasemi Zohreh, Moghaddas Farhad, Hosseinzadeh Lotfi Ali, Ebrahimnejad Mohsen, Vaez-Ghasemi Zohreh, Moghaddas (2020) Data Envelopment Analysis with R Fuzzy Data Envelopment Analysis Models with R Codes Springer International Publishing Cham 163-236
9. Hosseinzadeh Lotfi, F., Jahanshahloo, G. R., Khodabakhshi, M., Rostamy-Malkhlifeh, M., Moghaddas, Z., & Vaez-Ghasemi, M.(2013). A review of ranking models in data envelopment analysis. *Journal of Applied Mathematics*, 20131–20 https://doi.org/10.1155/2013/492421
10. Hosseinzadeh Lotfi, Farhad, Masoud Sanei, Ali Asghar Hosseinzadeh, Sadegh Niroomand, Ali Mahmoodirad, Chapter Two - Introduction to data envelopment analysis,Uncertainty in Data Envelopment Analysis, Academic Press, 2023, Pages 39–83,ISBN 9780323994446.
11. Jahanshahloo, G. R., Hosseinzadeh Lotfi, F., Alimardani Jondabeh, M., Banihashemi, S., & Lakzaie, L. (2008). Cost efficiency measurement with certain price on fuzzy data and application in insurance organization. *Applied Mathematical Sciences*, *2*(1), 1-18
12. Kaya, T., & Kahraman, C. (2010). Multicriteria renewable energy planning using an integrated fuzzy VIKOR & AHP methodology: The case of Istanbul. *Energy, 35*(6), 2517–2527.
13. Liu, G., & Wang, X. (2023). A trapezoidal fuzzy Number-Based VIKOR method with completely unknown weight information. *Symmetry, 15*(2), 559.
14. Mehdi, Soltanifar Hamid, Sharafi (2022) A modified DEA cross efficiency method with negative data and its application in supplier selection. *Journal of Combinatorial Optimization, 43*(1), 265–296. https://doi.org/10.1007/s10878-021-00765-7
15. Mehdi, Soltanifar Hamid, Sharafi Farhad, Hosseinzadeh Lotfi Witold, Pedrycz Tofigh, Allahviranloo (2023) Preferential Voting and Applications: Approaches Based on Data Envelopment Analysis Springer International Publishing Cham

16. Opricovic, S. (1998). Multicriteria optimization of civil engineering systems. *Faculty of Civil Engineering, Belgrade, 2*(1), 5–21.
17. Opricovic, S., & Tzeng, G. H. (2007). Extended VIKOR method in comparison with outranking methods. *European Journal of Operational Research, 178*(2), 514–529.
18. Pejman, Peykani Farhad, Hosseinzadeh Lotfi Seyed Jafar, Sadjadi Ali, Ebrahimnejad Emran, Mohammadi (2022) Fuzzy chance-constrained data envelopment analysis: a structured literature review current trends and future directions. *Fuzzy Optimization and Decision Making, 21*(2), 197–261. https://doi.org/10.1007/s10700-021-09364-x
19. Qin, J., Liu, X., & Pedrycz, W. (2015). An extended VIKOR method based on prospect theory for multiple attribute decision making under interval type-2 fuzzy environment. *Knowledge-Based Systems, 86*, 116–130.
20. Roostaee, R., Izadikhah, M., Lotfi, F. H., & Rostamy-Malkhalifeh, M. (2012). A multi-criteria intuitionistic fuzzy group decision making method for supplier selection with VIKOR method. *International Journal of Fuzzy System Applications (IJFSA), 2*(1), 1–17.
21. Shariari, M., & Pilevari, N. (2016). Agile supplier selection in sanitation supply chain using fuzzy VIKOR method. *Journal of Optimization in Industrial Engineering, 10*(21), 19–28.
22. Soltanifar, M., & Sharafi, H. (2022). Preferential voting in the presence of undesirable voters. *International Journal of Applied Decision Sciences, 15*(6), 753–766.
23. Soltanifar, M., & Sharafi, H. (2022). A modified DEA cross efficiency method with negative data and its application in supplier selection. *Journal of Combinatorial Optimization, 43*(1), 265–296. https://doi.org/10.1007/s10878-021-00765-7
24. Somayeh, Razipour-GhalehJough Farhad, Hosseinzadeh Lotfi Gholamreza, Jahanshahloo Mohsen, Rostamy-malkhalifeh Hamid, Sharafi (2020) Finding closest target for bank branches in the presence of weight restrictions using data envelopment analysis. *Annals of Operations Research, 288*(2), 755–787. https://doi.org/10.1007/s10479-019-03166-6
25. Wan, S. P., Wang, Q. Y., & Dong, J. Y. (2013). The extended VIKOR method for multi-attribute group decision making with triangular intuitionistic fuzzy numbers. *Knowledge-Based Systems, 52*, 65–77.

Chapter 10
The Measuring Attractiveness by a Categorical Based Evaluation Technique (MACBETH) in Uncertainty Environment

10.1 Introduction of the MACBETH method

MACBETH (Measuring Attractiveness by a Categorical Based Evaluation Technique) is a multi-criteria decision making approach which firstly presented by Bana e Costa, Vansnick and De Corte in 1990s.

MACBETH is an interactive approach that helps a decision maker or a group of decision makers to evaluate alternatives by qualitatively comparing their differences in attractiveness in a multi-criteria analysis and what distinguishes MACBETH from other multi-criteria models is that it only requires qualitative judgments about the difference in attractiveness between two elements at a time to generate numerical scores for the options of each criterion and weight the criteria.

Bana e Costa et al. [1] presented M-MACBETH software that automatically verifies the consistency of the judgments expressed by the decision maker and suggests resolving inconsistencies if it occurs. It also provides the sensitivity analysis and robustness of the model's results.

MACBETH method is applied to variety of fields. For instance, Bana e Costa et al. [2] solved complex strategic problems of Santa Catarina textile industry by integrating several decision support systems. Bana e Costa and Chagas [4] used MACBETH method and software to solve career choice problem. Aslo, Bana Costa, et al. [3] presented an up-to-date comprehensive overview of the MACBETH approach to multi-criteria decision making. Meanwhile, Cliville et al. [6] used MACBETH method for determining the industrial performance expressions. Moreover, Karande and Chakraborty [11] applied this method in supplier selection problems. Rodrigues [18] used MACBETH method to build a multidimensional value-based population health indices. Karande and Chakraborty [12] solved facility layout selection problems with MACBETH method. In addition, ranking air compressor for a textile company, Kundakcı and Işık [13]. Applied MACBETH method to determine the weight of criteria and ranked the alternatives by COPRAS method. Ferreira et al. [9] studied papers published from 1994 to 2016 about MACBETH technique.

F. Hosseinzadeh Lotfi et al., *Fuzzy Decision Analysis: Multi Attribute Decision Making Approach*, Studies in Computational Intelligence 1121,
https://doi.org/10.1007/978-3-031-44742-6_10

10.2 Description of the MACBETH Method

Comparing to other MADM methods, MACBETH method uses a semantic scale to express the assessment of the decision makers, to compare pairs of alternatives, using the categories indifferent, very weak, weak, moderate, strong, very strong, and extreme. Decision makers require to assign an absolute verbal judgment about the difference in attractiveness between two alternatives or actions.

Ordinal and cardinal value information:

Ordinal measurement: Let X be a finite set of elements, ordinal measurement of the attractiveness of the element x of X is a real number denoted by $\nu(x)$ that satisfies the following ordinal measurement conditions:

$$\forall x \forall y (x, y \in X; [x \text{ is more attractive than } y(xPy) \Leftrightarrow \upsilon(x) > \upsilon(y)]) \tag{10.1}$$

$$\forall x \forall y (x, y \in X; [x \text{ is as attractive as } y(xIy) \Leftrightarrow \nu(x) = \upsilon(y)]) \tag{10.2}$$

This numerical scale can be obtained by asking decision makers to rank the elements of X in order of decreasing attractiveness.

Cardinal measurement: Cardinal measurement of attractiveness of element x of X is a real number denoted by $\nu(x)$ that satisfies condition (10.3) in addition to conditions (10.1) and (10.2).

$for\ all\, x, y, w, z\ of\ X$ if x is more attractive than y and w is more attractive than z, then the ratio (10.3) measures the difference in attractiveness between x and y when the difference in attractiveness between w and z is taken as the measurement unit.

$$(\nu(x) - \nu(y)) / (\nu(w) - \nu(z)) \tag{10.3}$$

To obtain the ordinal information, decision maker can check the attractiveness of two elements at a time or simply ranking the elements of X by decreasing attractiveness. If x is more attractive than y, then it is denoted by (xPy). To obtain pre-cardinal information, seven categories are defined: Null (h_0), very weak (h_1), weak (h_2), moderate (h_3), strong (h_4), very strong (h_5) and extreme (h_6). For qualitative judgments of decision maker it is necessary to check the consistency of it with cardinal information. That is, whether it is possible to associate with each element x of X a number $\nu(x)$ satisfying the following conditions:

$$\forall x \forall y\, (x, y \in X; \quad xIy \;\Rightarrow\; \nu(x) = \nu(y)) \tag{10.4}$$

$$\forall x \forall y\, (x, y \in X; \quad xPy \;\Rightarrow\; \nu(x) > \nu(y)) \tag{10.5}$$

$$\begin{aligned}&\forall x\forall y((x, y) \in h_i \cup ... \cup h_s \; and \; \forall(w, z) \in h_{i'} \cup ... \cup h_{s'}\\&\quad with \; i, s, i', s' \in \{1, 2, 3, 4, 5, 6\},\\&i \le s, \; i' \le s' \; : \; i > s' \;\Rightarrow\; \nu(x) - \nu(y) \ge \nu(w) - \nu(z)).\end{aligned} \tag{10.6}$$

Condition (10.4) means if x and y are equally attractive then x and y are at the same level, and (10.5) means if decision maker judged x more attractive than y then x is placed above y. Condition (10.6) express that for each x, y, z, w with x more attractive than y and w more attractive than z, if the decision maker judged that the difference of attractiveness between x and y is greater than the difference of attractiveness between w and z then the distance between x and y is greater than the distance between w and z.

Pre-cardinal scale, called basic MACBETH scale can be obtained from the following linear programming problem called LP-MACBETH, where $\nu(x)$ is the score assigned to element x of X and x^+ and x^- are respectively the most and the least attractive elements of X. Model 10.7 is for determining the score of each alternative according to their categories.

10.2.1 Lp-Macbeth

$$\begin{aligned}&Min\ (v(x^+) - v(x^-))\\&s.t.\\&v(x^-) = 0 \qquad (arbitrary\ score)\\&v(x) - v(y) = 0 \qquad ;\ \forall(x, y) \in h_0\\&v(x) - v(y) \ge i \qquad ;\ \forall(x, y) \in h_i \cup \ldots \cup h_s,\\&\qquad with\ i,\ s \in \{1,\ 2,\ 3,\ 4,\ 5,\ 6\}\ and\ i \le s.\\&v(x) - v(y) \ge v(w) - v(z) + i - s'\ ;\ \forall(x, y) \in h_i \cup \ldots \cup h_s\ and\ \forall(w,\ z) \in h_i' \cup \ldots \cup h_s'\\&\qquad with\ i,\ s,\ i',\ s' \in \{1,\ 2,\ 3,\ 4,\ 5,\ 6\}\ and\ i \le s,\ i' \le s'\ and\ i > s'\end{aligned} \tag{10.7}$$

Bana e Costa et al. [3] summarized the steps of MACBETH method as follows:

Step 1. Introducing the decision criteria $C = \{c_1, ..., c_j, ..., c_n\}$. The criteria can be displayed in the form of a value tree.

Step 2. Defining the alternatives $A = \{A_1, ..., A_i, ..., A_m\}$, and then ordering the alternatives in a $m \times m$ matrix form left to right based on their importance with respect to each criterion. Also, the same procedure is applied for the criteria.

Step 3. Forming pairwise comparisons for the criteria and the alternatives based on difference of attractiveness. MACBETH method uses a semantic scale set with seven categories to illustrate the difference of attractiveness. The MACBETH semantic scale is shown in Table 10.1.

Step 4. Checking the consistency of the decision makers' judgments. If the judgments are inconsistent, M-MACBETH software (http://www.m-macbeth.com) discover the

Table 10.1 Semantic Scale in MACBETH method

Semantic scale	Significance	Set h_j	Numerical scale
Null	Indifference between alternatives	h_0	0
Very weak	An alternative is very weakly attractive over another	h_1	1
Weak	An alternative is weakly attractive over another	h_2	2
Moderate	An alternative is moderately attractive over another	h_3	3
Strong	An alternative is strongly attractive over another	h_4	4
Very strong	An alternative is very strongly attractive over another	h_5	5
Extreme	An alternative is extremely attractive over another	h_6	6

source of inconsistency graphically and suggests possible changes to make the judgments consistent.

Step 5. Solving the LP-MACBETH problem (10.7) to obtain the MACBETH scale. The optimal solution of LP-MACBETH is not necessarily unique and different values may obtain from the problem, hence the basic MACBETH scale is anchored by values $\nu(x^-) = 0$ and $\nu(x^+) = 100$.

Step 6. Finally, the weighted global score that represent the overall attractiveness of the alternatives is calculated by the following additive model.

$$
\begin{aligned}
&V(A_i) = \sum_{j=1}^{n} w_j.\nu_j(A_i) \\
&where \ \sum_{j=1}^{n} w_j = 1, \ \ w_j > 0 \ , \ \ \nu_j(A_i^+) = 100 \ , \nu_j(A_i^-) = 0
\end{aligned}
\tag{10.8}
$$

That w_j denotes the weight of the jth criterion and $\nu_j(A_i)$ indicate the value obtained from (10.7) for i^{th} alternative with respect to j^{th} criterion. The additive value $V(A_i)$ shows the overall attractiveness of ith alternative, which applied to rank the alternatives.

10.3 Example of the MACBETH Method

Selecting among 3 houses, considering price, age, brightness and the floor of the apartment as 4 criteria, the MACBETH value tree is illustrated in Fig. 10.1.

Figure 10.2 shows the judgement of criteria. The weights obtained by M-Macbeth software are illustrated in Fig. 10.3. Figure 10.4 presents the performance of each alternative with respect to each criterion and Figs. 10.5, 10.6, 10.7 and 10.8 show the judgment matrix of each criterion.

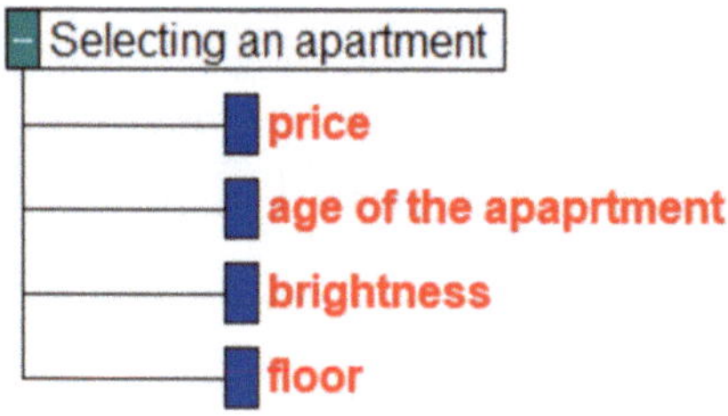

Fig. 10.1 MACBETH value tree for selecting an apartment

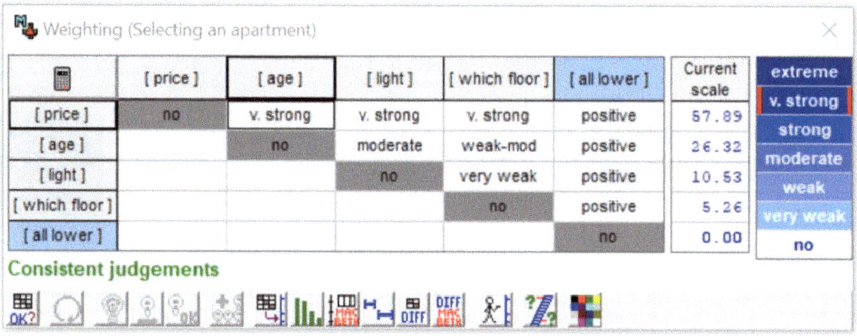

	[price]	[age]	[light]	[which floor]	[all lower]	Current scale
[price]	no	v. strong	v. strong	v. strong	positive	57.89
[age]		no	moderate	weak-mod	positive	26.32
[light]			no	very weak	positive	10.53
[which floor]				no	positive	5.26
[all lower]					no	0.00

Fig. 10.2 Judgment matrix of criteria

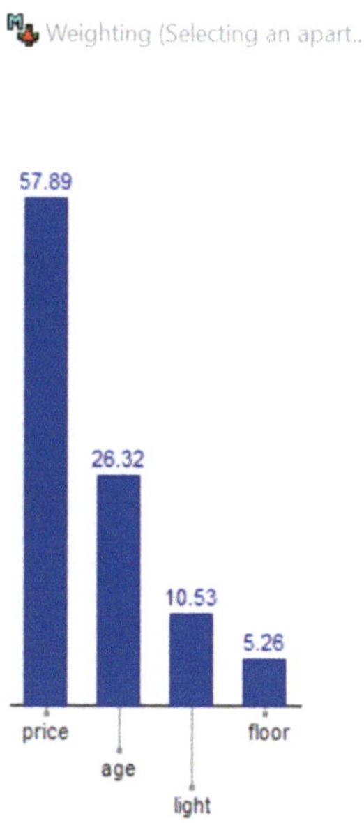

Fig. 10.3 Weights of criteria obtained by M-Macbeth software

Table of performances

Options	price	age	light	which floor
A1	Very Expensive	New	Bright	F2
A2	Very Low	>10	Dim	F1
A3	Expensive	5< <10	Bright	F3

Fig. 10.4 Qualitative performance of alternatives

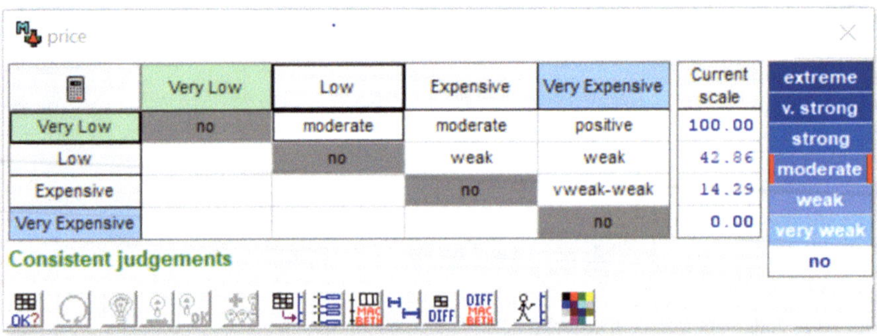

Fig. 10.5 Judgment matrix of price

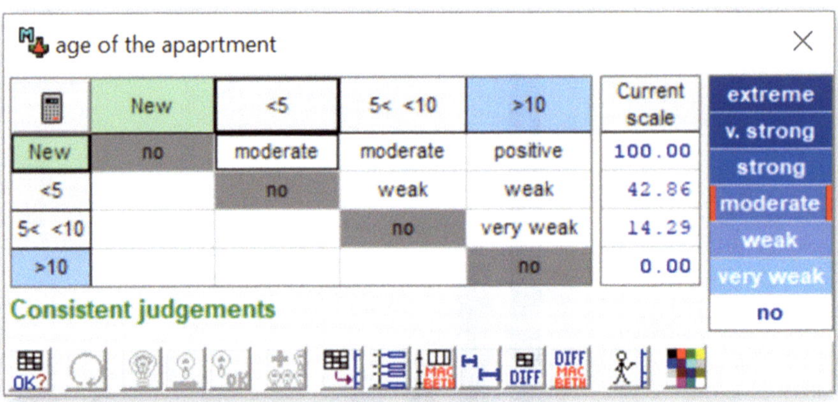

Fig. 10.6 Judgment matrix of age

Figure 10.9 represent the ranking of each alternative with respect to each criterion. The overall score of each alternative are mentioned in Fig. 10.10 $V(A_1) = 34.66$, $V(A_2) = 61.40$, $V(A_3) = 22.21$ and result in the following ranking.

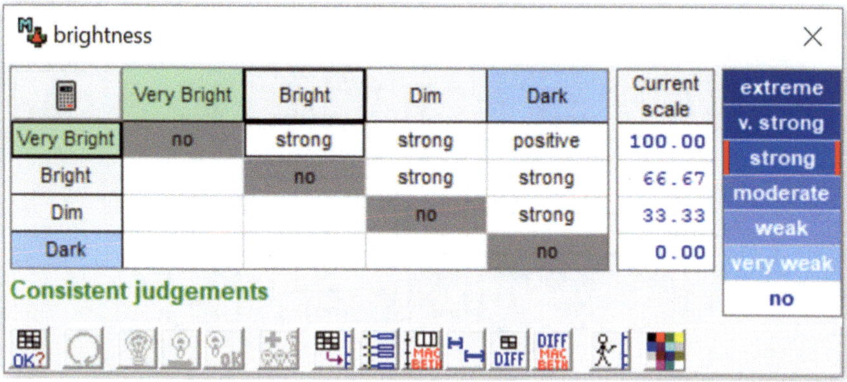

Fig. 10.7 Judgment matrix of brightness

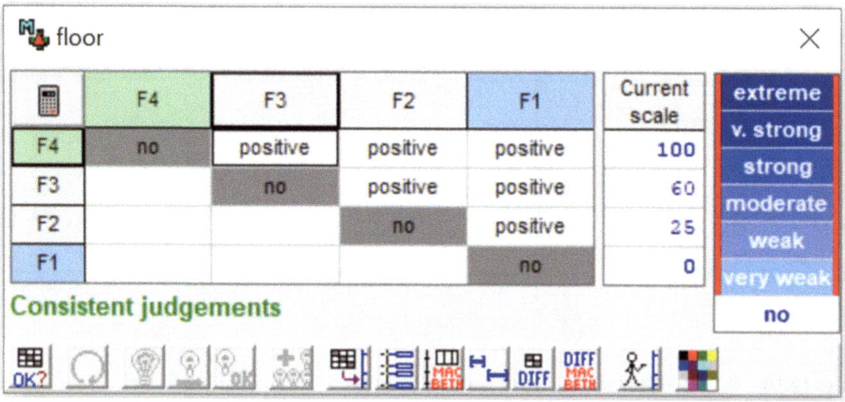

Fig. 10.8 Judgment matrix of floor

$A_2 \succ A_1 \succ A_3$.Therefore, apartment A_2 is the picked up first then A_1 and apartment A_3 is the last choice.

10.4 The Fuzzy MACBETH Method: Introduction

Lastly, Dhouib [7] applied fuzzy 2-tuple MACBETH method to evaluate reverse logistics alternatives for the automobile tire wastes. Ertay et al. [8] evaluated the renewable energy alternatives for resolving the Turkey's energy-related challenges, under fuzziness. They applied MACBETH and AHP-Fuzzy methods based on linguistic evaluations to evaluate the alternatives of wind, solar, biomass, geothermal and hydroelectric renewable energies. Pishdar et al. [17] applied Fuzzy Type-2 method to attain the criteria weights and evaluated the Iranian international airports

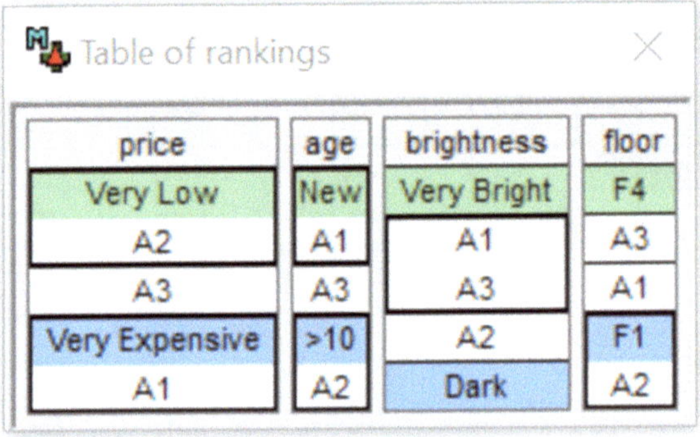
Table of rankings

price	age	brightness	floor
Very Low	New	Very Bright	F4
A2	A1	A1	A3
A3	A3	A3	A1
Very Expensive	>10	A2	F1
A1	A2	Dark	A2

Fig. 10.9 Ranking the alternatives with respect to each criterion

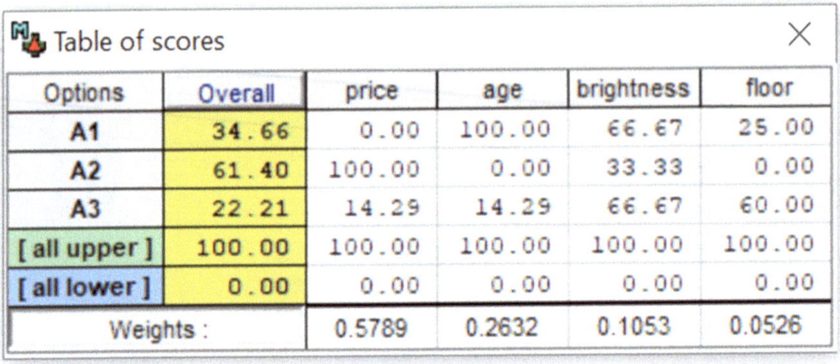
Table of scores

Options	Overall	price	age	brightness	floor
A1	34.66	0.00	100.00	66.67	25.00
A2	61.40	100.00	0.00	33.33	0.00
A3	22.21	14.29	14.29	66.67	60.00
[all upper]	100.00	100.00	100.00	100.00	100.00
[all lower]	0.00	0.00	0.00	0.00	0.00
Weights :		0.5789	0.2632	0.1053	0.0526

Fig. 10.10 MACBETH overall score of each alternative

by the MACBETH method. Yurtyapan et al. [20] utilized a grey number scale to MACBETH semantic scale to generate a new numeric scale for the method. Pamucar et al. [16] developed a decision making approach using MACBETH to address a supplier selection problem during the COVID-19 pandemic. Their approach is implemented under fuzzy rough number due to high uncertainty. Bastos et al. [5] developed a triangular fuzzy number scale based on the semantic scale of MACBETH [14].

10.5 Description of Fuzzy MACBETH Method

Step 1: First, the pairwise comparison matrix is determined based on the relative superiority of alternatives to each other.

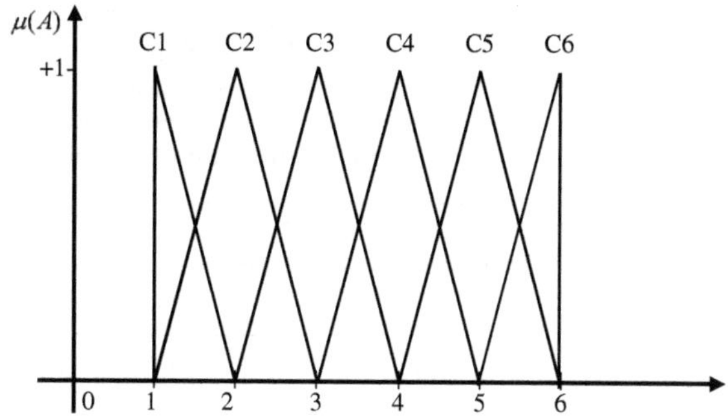

Fig. 10.11 Fuzzy values corresponding to Fuzzy MACBETH category

Step 2: For each pair of arbitrary alternatives concerning a criterion (e.g., j^th^ criterion), the preferences of the alternatives over each other are obtained based on the following relationships. (Let's assume that $\tilde{v}(A_i)$ represents the fuzzy score of ith alternative relative to the jth criterion.)

Indifference between alternatives $(A_i, A_t) \in C_0 \Leftrightarrow \tilde{v}(A_i) = \tilde{v}(A_t)$.

Difference concerning very weak attractiveness $(A_i, A_t) \in C_1 \Leftrightarrow \tilde{v}(A_i) \geq \tilde{v}(A_t) + (1, 1, 2)$.

Difference concerning weak attractiveness $(A_i, A_t) \in C_2 \Leftrightarrow \tilde{v}(A_i) \geq \tilde{v}(A_t) + (1, 2, 3)$.

Difference concerning moderate attractiveness $(A_i, A_t) \in C_3 \Leftrightarrow \tilde{v}(A_i) \geq \tilde{v}(A_t) + (2, 3, 4)$.

Difference concerning strong attractiveness $(A_i, A_t) \in C_4 \Leftrightarrow \tilde{v}(A_i) \geq \tilde{v}(A_t) + (3, 4, 5)$.

Difference concerning very strong attractiveness $(A_i, A_t) \in C_5 \Leftrightarrow \tilde{v}(A_i) \geq \tilde{v}(A_t) + (4, 5, 6)$.

Difference concerning extreme attractiveness $(A_i, A_t) \in C_6 \Leftrightarrow \tilde{v}(A_i) \geq \tilde{v}(A_t) + (5, 6, 6)$.

This can be graphically represented as shown in Fig. 10.11 (Table 10.2 show the semantic fuzzy Scale in MACBETH method).

Step 3: The higher the levels, the more important they are. Therefore, based on this principle, create the following relationships.

$$\begin{gathered}\tilde{v}(A_i) - \tilde{v}(A_t) \geq \tilde{v}(A_k) - \tilde{v}(A_l), (A_i, A_t) \in C_b \cdot \& \\ (A_k, A_l) \in C_{b''} \,\&\, b', b'' \in \{1, 2, \ldots, 6\} \,\&\, b' > b''\end{gathered} \tag{10.9}$$

Step 4: Solve the following optimization problem to determine the score of each alternative in this criterion

$$\begin{aligned}&\text{Min}\tilde{v}\left(A^+\right) - \tilde{v}\left(A^-\right)\\&\text{s.t. } \tilde{v}\left(A^-\right) = (0, 0, 0)\\&\quad \tilde{v}(A_i) = \tilde{v}(A_t) \quad ; (A_i, A_t) \in C_0\end{aligned}$$

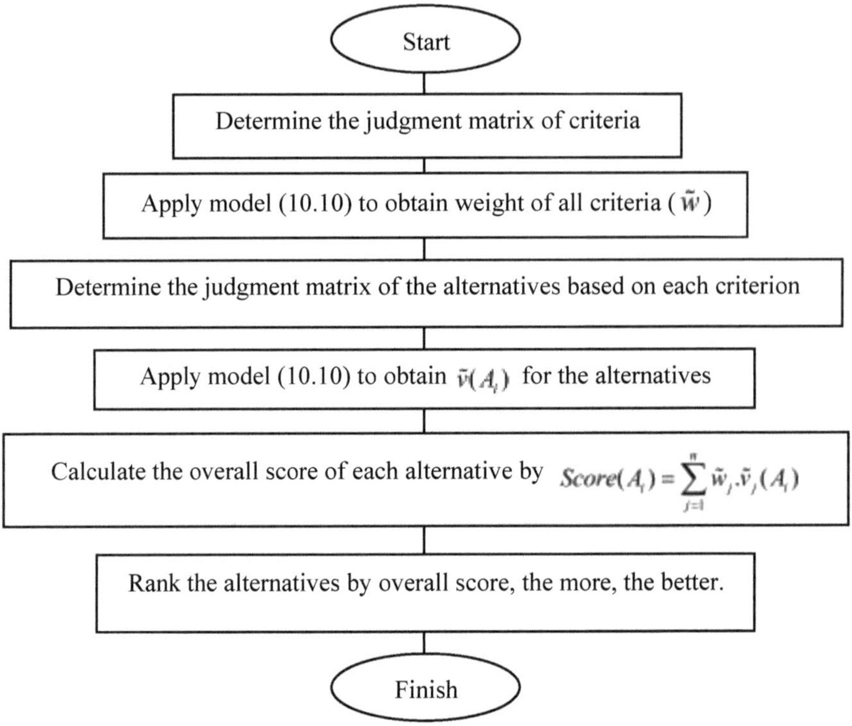

Fig. 10.12 Fuzzy MACBETH method at a glance

Table 10.2 Semantic fuzzy Scale in MACBETH method

Semantic scale	Significance	Set h_j	Numerical scale
Null	Indifference between alternatives	h_0	(0,0,0)
Very weak	An alternative is very weakly attractive over another	h_1	(1,1,2)
Weak	An alternative is weakly attractive over another	h_2	(1,2,3)
Moderate	An alternative is moderately attractive over another	h_3	(2,3,4)
Strong	An alternative is strongly attractive over another	h_4	(3,4,5)
Very strong	An alternative is very strongly attractive over another	h_5	(4,5,6)
Extreme	An alternative is extremely attractive over another	h_6	(5,6,6)

$$\begin{aligned}
\tilde{v}(A_i) &\geq \tilde{v}(A_t) + (1,1,2) \quad ; (A_i, A_t) \in C_1 \\
\tilde{v}(A_i) &\geq \tilde{v}(A_t) + (1,2,3) \quad ; (A_i, A_t) \in C_2 \\
\tilde{v}(A_i) &\geq \tilde{v}(A_t) + (2,3,4) \quad ; (A_i, A_t) \in C_3 \\
\tilde{v}(A_i) &\geq \tilde{v}(A_t) + (3,4,5) \quad ; (A_i, A_t) \in C_4 \\
\tilde{v}(A_i) &\geq \tilde{v}(A_t) + (4,5,6) \quad ; (A_i, A_t) \in C_5
\end{aligned}$$

$$\begin{aligned}
&\tilde{v}(A_i) \geq \tilde{v}(A_t) + (5, 6, 6) \quad ; (A_i, A_t) \in C_6 \\
&\tilde{v}(A_i) - \tilde{v}(A_t) \geq \tilde{v}(A_k) - \tilde{v}(A_l) \quad ; (A_i, A_t) \in C_{b'} \& \\
&(A_k, A_l) \in C_{b''} \& b', b'' \in \{1, 2, \ldots, 6\} \& b' > b'' \\
&0 \leq v^L(A_i) \leq v^M(A_i) \leq v^U(A_i); i = 1, \ldots, m
\end{aligned} \tag{10.10}$$

Step 5: By solving the problems in Step 4 for each criterion, the complete decision matrix table will be obtained. Note that the elements of the decision matrix are triangular fuzzy numbers, and the overall score of each alternative is obtained from the following linear combination.

$$Score(A_i) = \sum_{j=1}^{n} \tilde{w}_j.\tilde{v}_j(A_i) \simeq (w_j^l.v_j^l(A_i), w_j^m.v_j^m(A_i), w_j^u.v_j^u(A_i)) \tag{10.11}$$

where $\tilde{v}_j(A_i)$ is the optimal solution of the optimization problem in Step 3 for ith alternative and jth criterion.

Step 6: Since $score(A_i)$; $i = 1, \ldots, m$ are triangular fuzzy numbers, they can be ranked using a fuzzy ranking function, sorted in descending order, and then ranked.

The algorithm of fuzzy MACBETH is illustrated in Fig. 10.12.

10.6 Applications and Example of the Fuzzy MACBETH Method

When selecting a manager for the research and development (R&D) department of a company using the MACBETH (Measuring Attractiveness by a Categorical Based Evaluation Technique) method [16], qualitative criteria play a crucial role. These criteria focus on the skills, qualities, and experience that are essential for effectively leading and managing the R&D team. Here are some qualitative criteria that can be considered using the MACBETH method:

I. Technical expertise: The manager's technical knowledge and understanding of the industry and specific R&D field are critical for informed decision-making, providing guidance, and contributing to the department's strategic direction. (C1)
II. Leadership abilities: Effective leadership qualities, such as the ability to inspire and empower team members, foster collaboration, and communicate the department's vision and goals, are important for guiding and motivating the R&D team. (C2)

III. Innovation and creativity: The manager's ability to promote and nurture creativity within the team, encourage experimentation, and foster an environment that supports innovation is essential for driving new ideas, innovative products or solutions, and solving complex problems. (C3)
IV. Project management skills: Strong project management skills, including resource allocation, planning, and timely project delivery, are vital for handling the multiple tasks, deadlines, and stakeholders involved in R&D projects. (C4)
V. Strategic thinking: The manager should demonstrate a strategic mindset, aligning the department's activities with the overall business objectives. This involves understanding market trends, identifying opportunities, and making strategic decisions to drive long-term growth and competitive advantage. (C5)
VI. Communication and interpersonal skills: Effective communication skills are crucial for collaborating with cross-functional teams, presenting ideas and findings to stakeholders, and building relationships within and outside the department. Interpersonal skills, such as empathy, active listening, and negotiation, are also important for resolving conflicts and maintaining a positive work environment. (C6)
VII. Adaptability and resilience: The manager should exhibit adaptability to new technologies, evolving market conditions, and changing priorities, while maintaining resilience and problem-solving abilities in the face of challenges inherent in R&D processes. (C7)

Below, the matrices resulting from the expert judgments regarding each alternative in each criterion are displayed (Tables 10.3, 10.4, 10.5, 10.6, 10.7, 10.8 and 10.9).

In setting up the above tables, care has been taken to ensure that the compared alternatives are sorted in descending order. This will prevent any inconsistencies.

Table10.3 Preferences based on the first criterion

C1	A1	A2	A3	A4	A5
A1	No	W	M	S	VS
A2		No	VW	W	S
A3			No	VW	M
A4				No	W
A5					No

Table 10.4 Preferences based on the second criterion

C2	A3	A5	A1	A4	A2
A3	No	VW	W	M	E
A5		No	VW	W	VS
A1			No	VW	S
A4				No	W
A2					No

Table 10.5 Preferences based on the third criterion

C3	A2	A1	A3	A5	A4
A2	No	W	M	S	E
A1		No	VW	W	VS
A3			No	VW	S
A5				No	M
A4					No

Table 10.6 Preferences based on the fourth criterion

C4	A1	A4	A3	A2	A5
A1	No	M	S	VS	E
A4		No	W	S	VS
A3			No	M	S
A2				No	W
A5					No

Table 10.7 Preferences based on the fifth criterion

C5	A4	A1	A2	A5	A3
A4	No	VW	W	S	E
A1		No	VW	M	VS
A2			No	VW	M
A5				No	VW
A3					No

Table 10.8 Preferences based on the sixth criterion

C6	A1	A3	A2	A5	A4
A1	No	VW	M	S	E
A3		No	W	M	E
A2			No	W	S
A5				No	VW
A4					No

Table 10.9 Preferences based on the seventh criterion

C7	A4	A2	A1	A5	A3
A4	No	W	M	VS	E
A2		No	VW	M	VS
A1			No	VW	M
A5				No	VW
A3					No

After solving model for each of the above tables and determining the score of each alternatives in each criterion, Table 10.10 is obtained.

The weights obtained based on the decision maker's opinion are presented in Table 10.11. However, the Macbeth method can also be used for weight determination, which requires establishing preferences and their intensities among the criteria.

In this step, it is sufficient to multiply the weights obtained from Table 10.11 by the corresponding values in Table 10.10, and using Eq. (10.11) to calculate the final score for each alternative. These results are mentioned in Table 10.12. Additionally, fuzzy ranking functions has been utilized to determine the ranking of each alternative, and the final rankings of the alternatives are provided in the last column of Table 10.12, where A1 represents the best choice based on this ranking. Note that if group decision making is used, the matrix of pairwise comparisons of experts can be aggregated using arithmetic mean

Table 10.10 Score of each alternatives based on qualitative preferences

	A1	A2	A3	A4	A5
C1	(8.00, 8.00, 10.00)	(5.00, 5.00, 7.00)	(4.00, 4.00, 5.00)	(2.00, 2.00, 3.00)	(0.00, 0.00, 0.00)
C2	(6.00, 6.00, 6.00)	(0.00, 0.00, 0.00)	(8.00, 8.00, 10.0)	(4.00, 4.00, 4.00)	(7.00, 7.00, 8.00)
C3	(7.00, 7.00, 8.00)	(9.00, 9.00, 11.00)	(5.00, 5.00, 6.00)	(0.00, 0.00, 0.00)	(4.00, 4.00, 4.00)
C4	(11.00, 11.00, 14.00)	(4.00, 5.00, 7.00)	(1.00, 2.00, 3.00)	(7.00, 7.00, 10.00)	(0.00, 0.00, 0.00)
C5	(5.00, 5.00, 6.00)	(4.00, 4.00, 4.00)	(0.00, 0.00, 0.00)	(6.00, 6.00, 8.00)	(2.00, 2.00, 2.00)
C6	(10.00, 10.00, 11.00)	(6.00, 6.00, 6.00)	(9.00, 9.00, 9.00)	(0.00, 0.00, 0.00)	(3.00, 3.00, 3.00)
C7	(3.00, 3.00, 4.00)	(5.00, 5.00, 6.00)	(0.00, 0.00, 0.00)	(7.00, 7.00, 9.00)	(1.00, 1.00, 2.00)

Table 10.11 Score of each criterion based on qualitative preferences

	$\tilde{w} = (w^l, w^m, w^u)$
C1	(0.0203,0.0407,0.1017)
C2	(0.0814,0.1017,0.1221)
C3	(0.1424,0.1628,0.2035)
C4	(0.0610,0.1017,0.1628)
C5	(0.1424,0.1729,0.1831)
C6	(0.0712,0.0977,0.1455)
C7	(0.0203,0.0610,0.0814)

Table 10.12 Sum weighed and final ranking

	$Score(A_i)$			Rank
	L	M	U	
A1	3.8047	5.2187	8.6806	1
A2	2.7263	3.7599	6.1831	2
A3	2.1465	2.8728	4.7477	3
A4	1.7904	2.6653	4.6185	4
A5	1.6582	2.0631	2.7558	5

10.7 Macbeth and DEA

In this subsection, we discuss several points. First and foremost, it should be noted that the weight determination structure in the Macbeth method is very similar to the weight determination structure in preference voting methods based on data envelopment analysis (DEA). In preference voting methods, the minimum difference between two weights in different priorities is determined based on the discernment intensity function, whereas in Macbeth, fixed values are always used.

Additionally, Tavana et al. [19] utilized the combination of cross-efficiency and data envelopment analysis for ranking decision-making units in DEA. It is evident that this method can be extended to a fuzzy environment applying model (10.10). Furthermore, Gkouvitsos et al. [10] proposed a multi-stage ranking method based on the Macbeth approach for ranking decision-making units.

10.8 Conclusion

This chapter highlights the significance of MACBETH as an MADM technique and its extension to the fuzzy MACBETH method. The applications presented in selecting an apartment and choosing an R&D manager demonstrate the efficacy and versatility of these methods in complex decision-making scenarios. The integration of MACBETH and fuzzy logic provides decision-makers with a powerful tool to tackle ambiguity and imprecise information, enabling them to make well-informed choices based on multiple criteria. Overall, this chapter serves as a valuable resource for researchers, practitioners, and decision-makers interested in understanding and applying the MACBETH and fuzzy MACBETH methods in diverse decision-making contexts.

Acknowledgement A special thanks to the Iranian DEA society for their unwavering spiritual support and consensus in the writing of this book. Your invaluable support has been truly remarkable, and we are deeply grateful for the opportunity to collaborate with such esteemed professionals.

References

1. Bana e Costa, C. A., Vansnick, J. C., De Corte, J. M., MACBETH. (2003). Working paper LSEOR 03.56, London School of Economics, London. http://www.m-macbeth.com/en/downloads.html.
2. Bana e Costa, C. A., & Vansnick, J. C. (1999). The MACBETH approach: Basic ideas, software, and an application. *Advances in Decision Analysis,* 131–157.
3. Bana e Costa, C. A., De Corte, J. M. & Vansnick, J. C. (2012). MACBETH. *International Journal of Information Technology & Decision Making, 11*(2), 359À387.
4. Bana e Costa, C. A., & Chagas, M. P. (2004). A career choice problem: An example of how to use MACBETH to build a quantitative value model based on qualitative value judgments. *European journal of operational research, 153*(2), 323-331.
5. Bastos, T. R., Longaray, A. A., dos Santos Machado, C. M., Ensslin, L., Ensslin, S. R., & Dutra, A. (2023). Fuzzy-MACBETH hybrid method: Mathematical treatment of a qualitative scale using the fuzzy theory. *International Journal of Computational Intelligence Systems, 16*(1), 21.
6. Clivillé, V., Berrah, L., & Mauris, G. (2007). Quantitative expression and aggregation of performance measurements based on the MACBETH multi-criteria method. *International Journal of Production Economics, 105*(1), 171–189.
7. Dhouib, D. (2014). An extension of MACBETH method for a fuzzy environment to analyze alternatives in reverse logistics for automobile tire wastes. *Omega, 42*(1), 25–32.
8. Ertay, T., Kahraman, C., & Kaya, İ. (2013). Evaluation of renewable energy alternatives using MACBETH and fuzzy AHP multicriteria methods: The case of Turkey. *Technological and Economic Development of Economy, 19*(1), 38–62.
9. Ferreira, F. A., & Santos, S. P. (2021). Two decades on the MACBETH approach: A bibliometric analysis. *Annals of Operations Research, 296*, 901–925.
10. Gkouvitsos, I., & Giannikos, I. (2022). Using a MACBETH based multicriteria approach for virtual weight restrictions in each stage of a DEA multi-stage ranking process. *Operational Research, 22*(3), 1787–1811.
11. Karande, P., & Chakraborty, S. (2013). Using MACBETH method for supplier selection in manufacturing environment. *International Journal of Industrial Engineering Computations, 4*(2), 259–279.
12. Karande, P., & Chakraborty, S. (2014). A facility layout selection model using MACBETH method. In *Bali, Indonesia: Proceedings of the 2014 International Conference on Industrial Engineering and Operations Management (January 7–9).*
13. Kundakcı, N., & Işık, A. (2016). Integration of MACBETH and COPRAS methods to select air compressor for a textile company. *Decision Science Letters, 5*(3), 381–394.
14. Lotfi, F. H., Allahviranloo, T., Jondabeh, M. A., & Alizadeh, L. (2009). Solving a full fuzzy linear programming using lexicography method and fuzzy approximate solution. *Applied mathematical modelling, 33*(7), 3151–3156. https://doi.org/10.1016/j.apm.2008.10.020.
15. Mohagheghi, V., Mousavi, S. M., Vahdani, B., & Shahriari, M. R. (2017). R&D project evaluation and project portfolio selection by a new interval type-2 fuzzy optimization approach. *Neural Computing and Applications, 28*, 3869–3888.
16. Pamucar, D., Torkayesh, A. E., & Biswas, S. (2022). Supplier selection in healthcare supply chain management during the COVID-19 pandemic: a novel fuzzy rough decision-making approach. *Annals of Operations Research,* 1–43.
17. Pishdar, M., Ghasemzadeh, F., & Antuchevičienė, J. (2019). A mixed interval type-2 fuzzy best-worst MACBETH approach to choose hub airport in developing countries: Case of Iranian passenger airports. *Transport, 34*(6), 639–651.
18. Rodrigues, T. C. (2014). The MACBETH approach to health value measurement: Building a population health index in group processes. *Procedia Technology, 16*, 1361–1366.

19. Tavana, M., Soltanifar, M., Santos-Arteaga, F. J., & Sharafi, H. (2023). Analytic hierarchy process and data envelopment analysis: A match made in heaven. *Expert Systems with Applications, 223*, 119902.
20. Yurtyapan, M. S., & Aydemir, E. (2022). ERP software selection using intuitionistic fuzzy and interval grey number-based MACBETH method. *Grey Systems: Theory and Application, 12*(1), 78–100.

Chapter 11
Multi Attributive Border Approximation Area Comparison (MABAC) in Uncertainty Environment

11.1 Introduction: The Power of MABAC

In the realm of decision-making, where complexities abound and choices are aplenty, having a robust methodology to evaluate and compare options becomes imperative. This chapter delves into the intricacies of the Multi Attributive Border Approximation Area Comparison (MABAC) method, a powerful approach that enables systematic analysis and prioritization across multiple attributes.

Within this chapter, we will examine the step-by-step process of implementing the MABAC method, from defining criteria to assigning weights and calculating final rankings. Additionally, we will explore real-world examples and case studies that highlight the effectiveness and versatility of MABAC in diverse decision-making scenarios. By understanding the nuances of MABAC, readers will gain a valuable tool to navigate complex decision landscapes, effectively compare alternatives, and prioritize options based on their specific needs and objectives. The chapter aims to equip readers with the knowledge and skills necessary to harness the power of MABAC and make well-informed decisions that drive success in their personal and professional lives.

MABAC is one of the new method of multi-attribute decision methods in the last decade. It is propounded by the research center at the University of Defence in Belgrade by Pamučar and Ćirović [7] with an approach that has low computational complexity and reasonable logic that confirm the rationale of human decision making. Xue et al. [16] introduced an approach based on interval-valued intuitionistic fuzzy set and MABAC methods to handle material selection problems with incomplete weight information. Peng et al. [10] applied Pythagorean fuzzy Choquet integral based MABAC method for multiple attribute group decision making. Pamučar et al. [8] modified the Best–Worst and MABAC methods based on interval-valued fuzzy-rough numbers. To raise the performance of Failure mode and effect analysis Liu [5] presented an integrated risk prioritization approach using interval-valued intuitionistic fuzzy sets (IVIFSs) and the MABAC method. Gupta et al. [2] applied

F. Hosseinzadeh Lotfi et al., *Fuzzy Decision Analysis: Multi Attribute Decision Making Approach*, Studies in Computational Intelligence 1121,
https://doi.org/10.1007/978-3-031-44742-6_11

different multi attribute decision making methods such as MABAC to evaluate green supplier selection. Liang et al. [4] studied about assessing the risk of rockburst in complicated decision-making situations and applied MABAC method corresponding to triangular fuzzy numbers to obtain the ranking results under fuzzy environment. Pamučar et al. [9] presented MABAC method based on D-numbers to select an infectious waste treatment facility to provide a solution related to the health care waste (HCW) management problem. Torkayesh et al. [14] presented a literature review of 117 articles on recent developments and applications of MABAC method. The advantages of using the MABAC method are as follows: First, it has a simple mathematical framework and yields stable results. Second, it allows for obtaining comprehensive results easily as it considers the probabilistic values of profit and loss. Third, it can be combined with other approaches, making it versatile. Therefore, the MABAC method possesses the capability to fulfill the requirements of a reliable prioritization tool [3, 6].

11.2 Description of the MABAC Method

Consisting $A_1, A_2, ..., A_m$ as "m" alternatives and $C_1, C_2, ..., C_n$ as "n" attributes, the MABAC method is processed as the following six steps.

Step 1. The Decision Matrix

The decision matrix $X = \left[x_{ij}\right]_{m\times n}$ determined based on the information from decision maker, where x_{ij} denotes the value of ith alternative according to jth criterion for $i = 1, ..., m$ and $j = 1, ..., n$.

Step 2. The Normalized Decision Matrix

The Normalized decision matrix is constructed as bellow.

$$N = \begin{matrix} & \begin{matrix} C_1 & C_2 & \dots & C_n \end{matrix} \\ \begin{matrix} A_1 \\ A_2 \\ \vdots \\ A_m \end{matrix} & \begin{bmatrix} n_{11} & n_{12} & \dots & n_{1n} \\ n_{21} & n_{22} & \dots & n_{2n} \\ \vdots & \vdots & \ddots & \vdots \\ n_{m1} & n_{m2} & \dots & n_{mn} \end{bmatrix} \end{matrix} \tag{11.1}$$

where each element n_{ij} obtained by Eqs. (11.2 and 11.3).

For benefit type criteria (a higher value of the criterion is preferable):

$$n_{ij} = \frac{x_{ij} - x_j^-}{x_j^+ - x_j^-} \tag{11.2}$$

For cost type criteria (a lower value of the criterion is preferable):

$$n_{ij} = \frac{x_{ij} - x_j^+}{x_j^- - x_j^+} \tag{11.3}$$

That $x_j^+ = \max\{x_{ij} | i = 1, ..., m\}$ and $x_j^- = \min\{x_{ij} | i = 1, ..., m\}$ for $j = 1, ..., n$.

Step 3. The Weighted Normalized Decision Matrix

The elements of the weighted normalized decision matrix obtained according to Eq. (11.4)

$$v_{ij} = w_j.(n_{ij} + 1) \quad ; i = 1, ..., m, \;\; j = 1, ..., n. \tag{11.4}$$

where $(w_1, ..., w_j, ..., w_n)$ is the weight vector of criteria that obtained by one of the weight determination methods mentioned in Chap. 3.

Using Eq. (11.4) the weighted normalized decision matrix V is as bellow.

$$V = \begin{bmatrix} v_{11} & v_{12} & \dots & v_{1n} \\ v_{21} & v_{22} & \dots & v_{2n} \\ \vdots & \vdots & \ddots & \vdots \\ v_{m1} & v_{m2} & \dots & v_{mn} \end{bmatrix} = \begin{bmatrix} w_1.(n_{11}+1) & w_2.(n_{12}+1) & \dots & w_n.(n_{1n}+1) \\ w_1.(n_{21}+1) & w_2.(n_{22}+1) & \dots & w_n.(n_{2n}+1) \\ \vdots & \vdots & \ddots & \vdots \\ w_1.(n_{m1}+1) & w_2.(n_{m2}+1) & \dots & w_n.(n_{mn}+1) \end{bmatrix} \tag{11.5}$$

Step 4. The Border Approximation Area Matrix

The border approximation area for each criterion is equal to geometric average of the components of matrix V for each criterion, which determined by Eq. (11.6)

$$g_j = \sqrt[m]{\prod_{i=1}^{m} v_{ij}} \quad ; j = 1, ..., n. \tag{11.6}$$

The border approximation area matrix is as bellow.

$$G = \begin{bmatrix} g_1, & g_2, & ..., & g_j, & ..., & g_n \\ \vdots & \vdots & & \vdots & & \vdots \\ g_1, & g_2, & ..., & g_j, & ..., & g_n \end{bmatrix}_{m \times n} \tag{11.7}$$

Step 5. Distance from the Border Approximation Area

The distance of the alternatives from the border approximation area (q_{ij}) is determined as bellow:

$$Q = V - G \tag{11.8}$$

$$Q = \begin{bmatrix} q_{11} \; q_{12} \cdots q_{1n} \\ q_{21} \; q_{22} \cdots q_{2n} \\ \vdots \;\; \vdots \;\; \ddots \;\; \vdots \\ q_{m1} \; q_{m2} \cdots q_{mn} \end{bmatrix} = \begin{bmatrix} v_{11} - g_1 \; v_{12} - g_2 \cdots v_{1n} - g_n \\ v_{21} - g_1 \; v_{22} - g_2 \cdots v_{2n} - g_n \\ \vdots \;\; \vdots \;\; \ddots \;\; \vdots \\ v_{m1} - g_1 \; v_{m2} - g_2 \cdots v_{mn} - g_n \end{bmatrix} \tag{11.9}$$

The approximation area is divided to three parts, upper approximation area denote by G^+, lower approximation area denote by G^- and the border approximation area denote by G. The Alternative A_i ; $i = 1, \ldots, m$ could belongs to either of them as Eq. (11.10)

$$A_i \in \begin{cases} G^+ \; if \; q_{ij} > 0 \\ G \; if \; q_{ij} = 0 \\ G^- \; if \; q_{ij} < 0 \end{cases} \; ; i = 1, \ldots, m \tag{11.10}$$

Hence, the ideal alternative (A^+) belongs to G^+ and the anti-ideal alternative (A^-) belongs to G^- which is illustrated in Fig. 11.1.

Step 6. Ranking the Alternatives

Rank the alternatives according to Eq. (11.11)

$$S_i = \sum_{j=1}^{n} q_{ij} \; ; i = 1, \ldots, m. \tag{11.11}$$

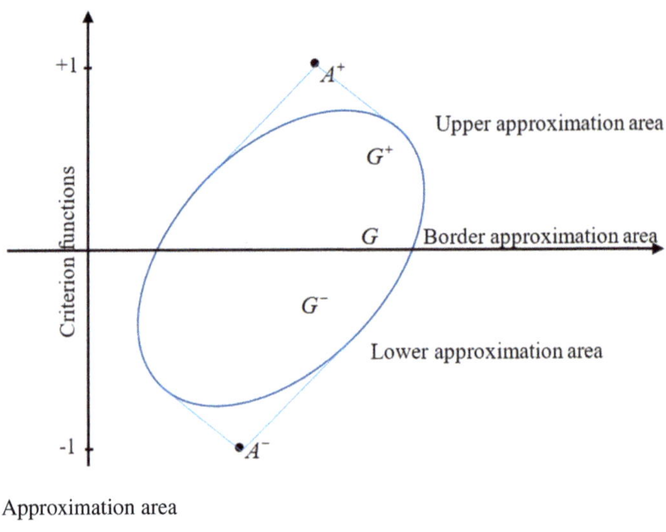

Fig. 11.1 Approximation area

The higher the resulting value (S_i), the closer alternative A_i is to the ideal alternative.

11.3 Numerical Example of the MABAC Method

In the competitive world of business, making informed decisions is crucial for success. Whether it's choosing the right vendors, suppliers, or partners, selecting the most suitable option can significantly impact an organization's performance. One effective approach to aid in this decision-making process is the Multi-Attribute Border Approximation and Comparison (MABAC) method.

In the following analysis, we will evaluate and prioritize four vendors, namely A1, A2, A3, and A4, using the MABAC method. The MABAC method provides a structured approach to decision-making by considering multiple criteria and assigning weights to each criterion based on their relative importance.

Each vendor will be assessed based on the following criteria:

Number of items sold in a day (C1): The volume of items sold is a key indicator of a vendor's productivity and market demand. It directly reflects the vendor's ability to meet customer needs efficiently.

Daily sales' profit (C2): Profitability is a fundamental aspect of any business. By assessing the daily sales' profit, we can gauge the financial performance of each vendor and identify those who bring the most value to the organization.

Missed customers (C3): Understanding the number of customers who did not make a purchase is crucial in identifying potential areas of improvement. This criterion helps us evaluate the vendors' effectiveness in attracting and retaining customers.

Number of rejected items due to customer dissatisfaction (C4): Customer satisfaction plays a vital role in maintaining a positive brand image. By examining the number of rejected items due to dissatisfaction, we can assess the vendors' commitment to quality and customer-centricity.

Number of daily online deliveries (C5): In today's digital age, online sales and deliveries have become increasingly important. Evaluating the vendors' ability to meet the demands of online customers can provide valuable insights into their operational efficiency and adaptability.

By employing the MABAC method, we can effectively analyze and compare the vendors based on these criteria. The MABAC method offers a systematic and quantitative approach to decision-making, allowing us to prioritize vendors based on their performance across multiple attributes. This enables organizations to make informed choices that align with their goals, maximize efficiency, and ultimately enhance overall business success.

Step-by-step process of applying the MABAC method to evaluate and prioritize the four vendors based on the aforementioned criteria is discussed in as follows. Table 11.1 denotes the value of alternative(vendors) according to the five criteria.

Applying Eq. (11.2) for C1, C2, C5 and Eq. (11.3) for C3 and C4 the normalized decision matrix is obtained Table 11.2.

Table 11.1 Decision matrix of evaluating four venders

	C1 (benefit)	C2 (benefit)	C3 (cost)	C4 (cost)	C5 (benefit)
A1	33	10,000	12	9	15
A2	23	5,100	17	7	11
A3	25	6,800	13	4	13
A4	17	4,900	19	4	9

Table 11.2 Normalized decision matrix of evaluating four vendors

	C1 (benefit)	C2 (benefit)	C3 (cost)	C4 (cost)	C5 (benefit)
A1	1.000	1.000	1.000	0.000	1.000
A2	0.375	0.039	0.286	0.400	0.333
A3	0.500	0.373	0.857	1.000	0.667
A4	0.000	0.000	0.000	1.000	0.000

The weight vector that shows the importance of criterion with respect to decision makers' opinion, is equal to $w = (w_1, w_2, w_3, w_4, w_5) = (0.10, 0.45, 0.17, 0.13, 0.15)$. Using Eq. (11.4) the weighted normalized decision matrix obtains as illustrated in Table 11.3.

Applying Eq. (11.6) the border approximation area equals to $g = (g_1, g_2, g_3, g_4, g_5) = (0.143, 0.585, 0.251, 2.000, 0.218)$. Applying Eq.(11.9) the distance from the border approximation area is determined which is shown in Table 11.4.

According to Table 11.4 the value of Eq. (11.11) for each alternatives are calculated as follows:

Table 11.3 Weighted normalized decision matrix of evaluating four vendors

	C1 (benefit)	C2 (benefit)	C3 (cost)	C4 (cost)	C5 (benefit)
A1	0.200	0.900	0.340	0.130	0.300
A2	0.138	0.468	0.219	0.182	0.200
A3	0.150	0.618	0.316	0.260	0.250
A4	0.100	0.450	0.170	0.260	0.150

Table 11.4 Distance from the border approximation area

	C1 (benefit)	C2 (benefit)	C3 (cost)	C4 (cost)	C5 (benefit)
A1	0.057	0.315	0.089	−0.070	0.082
A2	−0.005	−0.117	−0.033	−0.018	−0.018
A3	0.007	0.033	0.064	0.060	0.032
A4	−0.043	−0.135	−0.081	0.060	−0.068

$$S_1 = \sum_{j=1}^{5} q_{1j} = 0.057 + 0.315 + 0.089 - 0.070 + 0.082 = 0.4735$$

$$S_2 = \sum_{j=1}^{5} q_{2j} = -0.005 - 0.117 - 0.033 - 0.018 - 0.018 = -0.1907$$

$$S_3 = \sum_{j=1}^{5} q_{3j} = 0.007 + 0.033 + 0.064 + 0.060 + 0.032 = 0.1969$$

$$S_4 = \sum_{j=1}^{5} q_{4j} = -0.043 - 0.135 - 0.081 + 0.060 - 0.068 = -0.2664$$

Vendors are ranked based on descending order of S_i values:

$$A_1 \succ A_3 \succ A_2 \succ A_4$$

11.4 The Fuzzy MABAC Method

This section of the book explores the MABAC method in the context of a fuzzy environment, where imprecise and uncertain data prevail. Fuzziness allows for a more realistic representation of real-world scenarios, capturing the inherent vagueness and subjectivity present in decision-making processes. By integrating fuzzy logic into MABAC, we gain a more nuanced understanding of the criteria and their relative importance. Throughout this section, we will delve into the steps involved in implementing the MABAC method within a fuzzy environment, ensuring a comprehensive grasp of the methodology. By utilizing a practical example from the real world, we will demonstrate the application of the MABAC method and guide readers through the process of understanding and implementing each step. Some studies conducted in the field of MABAC in uncertainty environment include: Yu et al. [15], Wang et al. [17], Büyüközkan et al. [1], Pamučar et al. [9].

The objective of this section is to equip readers with the necessary knowledge and practical skills to confidently apply the MABAC method within fuzzy environments. By merging theory and real world examples, we aim to empower readers to make informed decisions in complex and uncertain situations, driving success in their personal and professional endeavors.

The fuzzy MABAC method is summarized as following steps.

Step 1. The Decision Matrix

Considering qualitative information, the decision matrix constructed with triangular fuzzy numbers is as bellow:

$$\tilde{X} = \begin{array}{c} \\ A_1 \\ A_2 \\ \vdots \\ A_m \end{array} \begin{array}{c} \begin{array}{cccc} C_1 & C_2 & \ldots & C_n \end{array} \\ \begin{bmatrix} \tilde{x}_{11} & \tilde{x}_{12} & \ldots & \tilde{x}_{1n} \\ \tilde{x}_{21} & \tilde{x}_{22} & \ldots & \tilde{x}_{2n} \\ \vdots & \vdots & \ddots & \vdots \\ \tilde{x}_{m1} & \tilde{x}_{m2} & \ldots & \tilde{x}_{mn} \end{bmatrix} \end{array} \tag{11.12}$$

where $\tilde{x}_{ij} = \left(x_{ij}^l, x_{ij}^m, x_{ij}^u\right)$ is triangular fuzzy number.

Step 2. The Normalized Decision Matrix

The Normalized decision matrix is as bellow.

$$\tilde{N} = \begin{bmatrix} \tilde{n}_{11} & \tilde{n}_{12} & \ldots & \tilde{n}_{1n} \\ \tilde{n}_{21} & \tilde{n}_{22} & \ldots & \tilde{n}_{2n} \\ \vdots & \vdots & \ddots & \vdots \\ \tilde{n}_{m1} & \tilde{n}_{m2} & \ldots & \tilde{n}_{mn} \end{bmatrix} \tag{11.13}$$

where each fuzzy element $\tilde{n}_{ij}$ attained by Eqs. (11.14 and 11.15).

For benefit type criteria:

$$\tilde{n}_{ij} = (n_{ij}^l, n_{ij}^m, n_{ij}^u) = (\frac{x_{ij}^l}{x_j^+}, \frac{x_{ij}^m}{x_j^+}, \frac{x_{ij}^u}{x_j^+}) \tag{11.14}$$

where, $x_j^+ = \max_i \left\{x_{ij}^u\right\}; \ j = 1, \ldots, n.$

For cost type criteria:

$$\tilde{n}_{ij} = (n_{ij}^l, n_{ij}^m, n_{ij}^u) = (\frac{x_j^-}{x_{ij}^u}, \frac{x_j^-}{x_{ij}^m}, \frac{x_j^-}{x_{ij}^l}) \tag{11.15}$$

where, $x_j^- = \min_i \left\{x_{ij}^l\right\}; \ j = 1, \ldots, n.$

Note that if $x_j^- = 0$, that is, if the minimum amount of x_{ij}^l for any alternative equals to zero, use another type of normalization. For example, you can use the following normalization:

For benefit type criteria:

$$\tilde{n}_{ij} = \frac{\tilde{x}_{ij} - \tilde{x}_j^{Max}}{\tilde{x}_j^{Max} - \tilde{x}_j^{Min}} = \left(\frac{x_{ij}^l - x_j^+}{x_j^+ - x_j^-}, \frac{x_{ij}^m - x_j^+}{x_j^+ - x_j^-}, \frac{x_{ij}^u - x_j^+}{x_j^+ - x_j^-}\right) \tag{11.16}$$

For cost type criteria:

$$\tilde{n}_{ij} = \frac{\tilde{x}_j^{Max} - \tilde{x}_{ij}}{\tilde{x}_j^{Max} - \tilde{x}_j^{Min}} = \left(\frac{x_j^+ - x_{ij}^l}{x_j^+ - x_j^-}, \frac{x_j^+ - x_{ij}^m}{x_j^+ - x_j^-}, \frac{x_j^+ - x_{ij}^u}{x_j^+ - x_j^-}\right) \tag{11.17}$$

where, $\tilde{x}_j^{Max} = \left(x_j^+, x_j^+, x_j^+\right)$, $\tilde{x}_j^{Min} = \left(x_j^-, x_j^-, x_j^-\right)$ and $x_j^+ = \max_i\left\{x_{ij}^u\right\}$,$x_j^- = \min_i\left\{x_{ij}^l\right\}$.

Step 3. The Weighted Normalized Decision Matrix

Considering $\tilde{w} = \left(\tilde{w}_1, ..., \tilde{w}_j, ..., \tilde{w}_n\right)$ as the weight vector of criteria that obtained by one of the weight determination methods and linguistic terms. The elements of the weighted normalized decision matrix obtains by Eq. (11.18)

$$\tilde{v}_{ij} \simeq \tilde{w}_j.(\tilde{n}_{ij} + \tilde{1}) \quad ; i = 1, ..., m, \ \ j = 1, ..., n. \tag{11.18}$$

where $\tilde{1} = (1, 1, 1)$ and $\tilde{w}_j = (w_j^l, w_j^m, w_j^u)$ for $j = 1, ..., n$ are triangular fuzzy numbers.

Applying Eq. (11.18) the weighted normalized decision matrix $\tilde{V}$ is as bellow.

$$\tilde{V} = \begin{bmatrix} \tilde{v}_{11} \ \tilde{v}_{12} \ \ldots \ \tilde{v}_{1n} \\ \tilde{v}_{21} \ \tilde{v}_{22} \ \ldots \ \tilde{v}_{2n} \\ \vdots \ \ \vdots \ \ddots \ \vdots \\ \tilde{v}_{m1} \ \tilde{v}_{m2} \ \ldots \ \tilde{v}_{mn} \end{bmatrix} = \begin{bmatrix} \tilde{w}_1.(\tilde{n}_{11} + \tilde{1}) & \tilde{w}_2.(\tilde{n}_{12} + \tilde{1}) & \ldots & \tilde{w}_n.(\tilde{n}_{1n} + \tilde{1}) \\ \tilde{w}_1.(\tilde{n}_{21} + \tilde{1}) & \tilde{w}_2.(\tilde{n}_{22} + \tilde{1}) & \ldots & \tilde{w}_n.(\tilde{n}_{2n} + \tilde{1}) \\ \vdots & \vdots & \ddots & \vdots \\ \tilde{w}_1.(\tilde{n}_{m1} + \tilde{1}) & \tilde{w}_2.(\tilde{n}_{m2} + \tilde{1}) & \ldots & \tilde{w}_n.(\tilde{n}_{mn} + \tilde{1}) \end{bmatrix} \tag{11.19}$$

Step 4. The Border Approximation Area Matrix

The border approximation area for each criterion is determined by Eq. (11.20)

$$\tilde{g}_j = \sqrt[m]{\prod_{i=1}^{m} \tilde{v}_{ij}} = \left(\sqrt[m]{\prod_{i=1}^{m} v_{ij}^L}, \sqrt[m]{\prod_{i=1}^{m} v_{ij}^M}, \sqrt[m]{\prod_{i=1}^{m} v_{ij}^U}\right) \quad ; j = 1, ..., n. \tag{11.20}$$

The border approximation area matrix is as bellow.

$$\begin{array}{c} C_1, \ldots, C_j, \ldots, C_n \\ \tilde{G} = \begin{bmatrix} \tilde{g}_1, \ldots, \tilde{g}_j, \ldots, \tilde{g}_n \\ \vdots \qquad \vdots \qquad \vdots \\ \tilde{g}_1, \ldots, \tilde{g}_j, \ldots, \tilde{g}_n \end{bmatrix}_{m \times n} \end{array} \tag{11.21}$$

Step 5. Distance from the Border Approximation Area

The distance of the alternatives from the border approximation area $(\tilde{q}_{ij})$ is determined as bellow:

$$\tilde{Q} = \tilde{V} - \tilde{G} = \begin{bmatrix} \tilde{q}_{11} & \tilde{q}_{12} & \cdots & \tilde{q}_{1n} \\ \tilde{q}_{21} & \tilde{q}_{22} & \cdots & \tilde{q}_{2n} \\ \vdots & \vdots & \ddots & \vdots \\ \tilde{q}_{m1} & \tilde{q}_{m2} & \cdots & \tilde{q}_{mn} \end{bmatrix} \tag{11.22}$$

where,

$$\tilde{q}_{ij} = \begin{cases} d(\tilde{v}_{ij}, \tilde{g}_j) & for \ \ \tilde{v}_{ij} > \tilde{g}_j \\ 0 & for \ \ \tilde{v}_{ij} = \tilde{g}_j \\ -d(\tilde{v}_{ij}, \tilde{g}_j) & for \ \ \tilde{v}_{ij} < \tilde{g}_j \end{cases} \tag{11.23}$$

And

$$d(\tilde{v}_{ij}, \tilde{g}_j) = \sqrt[2]{\frac{1}{3}[(v_{ij}^l - g_j^l)^2 + (v_{ij}^m - g_j^m)^2 + (v_{ij}^u - g_j^u)^2]} \tag{11.24}$$

Note that for $\tilde{v}_{ij} = \left(v_{ij}^l, v_{ij}^m, v_{ij}^u\right)$ and $\tilde{g}_j = \left(g_j^l, g_j^m, g_j^u\right)$ if $\frac{v_{ij}^l+4v_{ij}^m+v_{ij}^u}{6} > \frac{g_j^l+4g_j^m+g_j^u}{6}$ then $\tilde{v}_{ij} > \tilde{g}_j$.

The approximation area is divided into three parts, the upper approximation area denote by $\tilde{G}^+$, lower approximation area denote by $\tilde{G}^-$ and the border approximation area denote by $\tilde{G}$. The Alternative A_i ; $i = 1, ..., m$ could belongs to either of them as Eq. (11.25)

$$A_i \in \begin{cases} \tilde{G}^+ \ if \ \ \tilde{q}_{ij} > 0 \\ \tilde{G} \ \ if \ \ \tilde{q}_{ij} = 0 \\ \tilde{G}^- \ if \ \ \tilde{q}_{ij} < 0 \end{cases} \quad ; i = 1, ..., m \tag{11.25}$$

Hence, the ideal alternative (A^+) belongs to $\tilde{G}^+$ and the anti-ideal alternative (A^-) belongs to $\tilde{G}^-$.

Step 6. Ranking the Alternatives

If the total distance of all criteria for an alternative is larger, it shows that this alternative is more ideal. Hence, the rank of each alternative can be obtained by the Eq. (11.26).

$$S_i = \sum_{j=1}^{n} q_{ij} \ \ ; i = 1, ..., m. \tag{11.26}$$

The higher the resulting value (S_i), the closer alternative A_i is to the ideal alternative. Therefore, the larger the value of S_i, the better the alternative A_i.

11.5 Applications and Example of the Fuzzy MABAC Method

In the realm of banking, the selection of a productive and reliable bank clerk is of utmost importance. To navigate this decision-making process effectively, the application of a robust methodology becomes essential [11, 12]. This part of the book focuses on utilizing the Fuzzy Multi Attributive Border Approximation Area Comparison (Fuzzy MABAC) method to identify the most suitable bank clerk based on a set of critical criteria [13].

The criteria used for evaluation are as follows.

Quality of work (C1): This criterion encompasses the accuracy and responsibility exhibited by the bank clerk in their daily tasks. The ability to perform duties without mistakes or errors is crucial in maintaining the integrity and efficiency of banking operations.

Quantity of work (C2): The quantity of work refers to the bank clerk's ability to stay up-to-date with tasks and avoid backlogs. This criterion highlights the importance of timely completion of responsibilities to ensure smooth workflow and customer satisfaction.

Independency (C3): Independence denotes the bank clerk's capability to carry out their responsibilities without constant supervision from superiors. It showcases their autonomy and self-reliance in performing tasks, which contributes to overall operational efficiency.

Reliability (C4): Reliability encompasses the bank clerk's sense of responsibility towards superiors and colleagues, as well as their commitment to maintaining the confidentiality of sensitive information. This criterion reflects the trustworthiness and professionalism required in the banking industry.

By applying the Fuzzy MABAC method, this part of the book provides a structured framework to assess and prioritize the bank clerks based on the aforementioned criteria. The Fuzzy MABAC method allows for the consideration of uncertain and imprecise data, capturing the inherent fuzziness present in real-world decision-making scenarios.

Using a practical example from the banking sector, we illustrate the step-by-step implementation of the Fuzzy MABAC method. By showcasing the process and showcasing the outcomes, readers will gain a deeper understanding of how to effectively utilize this methodology in similar decision-making contexts.

The decision matrix, depicted in Table 11.5, incorporates linguistic variables, while the scale from Table 11.6 is employed to fuzzify these linguistic variables.

Table 11.5 Decision matrix using linguistic variable to rank bank clerks

	C1 (benefit)	C2 (benefit)	C3 (cost)	C4 (benefit)
A1	H	VH	M	H
A2	VH	M	H	M
A3	M	H	M	H
A4	L	VL	VH	L

Table 11.6 Linguistic rating for evaluating the alternatives

Linguistic terms	Fuzzy number
Very low (VL)	(0,0,0.25)
Low (L)	(0,0.25,0.5)
Medium (M)	(0.25,0.5,0.75)
High (H)	(0.5,0.75,1)
Very High (VH)	(0.75,1,1)

Table 11.7 shows the decision matrix by triangular fuzzy number. Table 11.8 is the normalized fuzzy decision matrix obtained by Eq. (11.14) for C1, C2, C2 as beneficial criteria type and Eq. (11.15) for C3 as a cost criteria.

Table 11.9 shows the weighted normalized fuzzy decision matrix, regarding the fuzzy weight vector equal to.

$$\tilde{w} = (\tilde{w}_1, \tilde{w}_2, \tilde{w}_3, \tilde{w}_4) = ((0.2, 0.27, 0.27), (0.13, 0.2, 0.27), (0.07, 0.13, 0.2), (0.13, 0.2, 0.27))$$

Table 11.7 Fuzzy decision matrix for ranking bank clerks

	C1 (benefit)	C2 (benefit)	C3 (cost)	C4 (benefit)
A1	(0.5,0.75,1)	(0.75,1,1)	(0.25,0.5,0.75)	(0.5,0.75,1)
A2	(0.75,1,1)	(0.25,0.5,0.75)	(0.5,0.75,1)	(0.25,0.5,0.75)
A3	(0.25,0.5,0.75)	(0.5,0.75,1)	(0.25,0.5,0.75)	(0.5,0.75,1)
A4	(0,0.25,0.5)	(0,0,0.25)	(0.75,1,1)	(0,0.25,0.5)

Table 11.8 Normalized fuzzy decision matrix for ranking bank clerks

	C1 (benefit)	C2 (benefit)	C3 (cost)	C4 (benefit)
A1	(0.5,0.75,1)	(0.75,1,1)	(0.33,0.5,1)	(0.5,0.75,1)
A2	(0.75,1,1)	(0.25,0.5,0.75)	(0.25,0.33,0.5)	(0.25,0.5,0.75)
A3	(0.25,0.5,0.75)	(0.5,0.75,1)	(0.33,0.5,1)	(0.5,0.75,1)
A4	(0,0.25,0.5)	(0,0,0.25)	(0.25,0.25,0.33)	(0,0.25,0.5)

Applying Eq. (11.20) the border approximation area is equal to.

$$\tilde{g} = (\tilde{g}_1, \tilde{g}_2, \tilde{g}_3, \tilde{g}_4) = ((0.27, 0.43, 0.48), (0.18, 0.30, 0.46), (0.09, 0.19, 0.34), (0.17, 0.31, 0.48))$$

.

Using Eqs. (11.23 and 11.24) the distance of the alternatives from the border approximation area is as Table 11.10. Note that the elements of Table 11.10 are crisp and the value of (S_i) for each alternative obtains by Eq. (11.26) as bellow:

$$S_1 = \sum_{j=1}^{4} q_{1j} \quad = 0.04 + 0.08 + 0.04 + 0.04 = 0.20$$

$$S_2 = \sum_{j=1}^{4} q_{2j} \quad = 0.08 - 0.01 - 0.02 - 0.01 = 0.04$$

$$S_3 = \sum_{j=1}^{4} q_{3j} \quad = -0.02 + 0.05 + 0.04 + 0.04 = 0.11$$

$$S_4 = \sum_{j=1}^{4} q_{4j} \quad = -0.08 - 0.10 - 0.04 - 0.06 = -0.28$$

The larger the value of S_i, the better the alternative A_i. Hence, bank clerks are ranked as $A_1 \succ A_3 \succ A_2 \succ A_4$.

Table 11.9 Weighted normalized fuzzy decision matrix for ranking bank clerks

	C1 (benefit)	C2 (benefit)	C3 (cost)	C4 (benefit)
A1	(0.30,0.47,0.53)	(0.23,0.40,0.53)	(0.09,0.20,0.40)	(0.20,0.35,0.53)
A2	(0.35,0.53,0.53)	(0.17,0.30,0.47)	(0.08,0.18,0.30)	(0.17,0.30,0.47)
A3	(0.25,0.40,0.47)	(0.20,0.35,0.53)	(0.09,0.20,0.40)	(0.20,0.35,0.53)
A4	(0.20,0.33,0.40)	(0.13,0.20,0.33)	(0.08,0.17,0.27)	(0.13,0.25,0.40)

Table 11.10 Distance from the border approximation area

	C1 (benefit)	C2 (benefit)	C3 (cost)	C4 (benefit)
A1	0.04	0.08	0.04	0.04
A2	0.08	−0.01	−0.02	−0.01
A3	−0.02	0.05	0.04	0.04
A4	−0.08	−0.10	−0.04	−0.06

11.6 Conclusion

The chapter journeyed through the realm of decision-making, unveiling the power and versatility of the Multi Attributive Border Approximation Area Comparison (MABAC) method. We began by introducing the method, exploring its fundamental concepts, and highlighting its significance in making well-informed decisions. The steps of the MABAC method were meticulously presented, equipping readers with a systematic framework to evaluate and compare alternatives across multiple criteria. Through a real-world example, we witnessed the method's practical application, witnessing its ability to guide decision-makers towards optimal choices. Recognizing the complexities of decision-making in uncertain and imprecise environments, we extended our exploration to the fuzzy domain. In this fuzzy environment, the MABAC method demonstrated its adaptability, enabling the analysis of subjective and ambiguous data. By solving a real-world example within this context, we showcased how the method thrives in capturing the nuances of decision-making, leading to more informed and reliable outcomes. By considering multiple criteria, incorporating linguistic variables, and embracing uncertainty, MABAC opens doors to a deeper understanding of choices and their consequences. We encourage readers to embrace the knowledge gained within this chapter and apply it to their own decision-making processes. By integrating the MABAC method into their arsenal of analytical tools, they can embark on a journey of informed decision-making, maximizing outcomes and achieving remarkable results. In the ever-evolving landscape of choices, the Multi Attributive Border Approximation Area Comparison (MABAC) method remains a steadfast companion. It offers a robust framework, blending objectivity and subjectivity, precision and fuzziness, to help navigate the complexities of decision-making.

Acknowledgement A special thanks to the Iranian DEA society for their unwavering spiritual support and consensus in the writing of this book. Your invaluable support has been truly remarkable, and we are deeply grateful for the opportunity to collaborate with such esteemed professionals.

References

1. Büyüközkan, G., Mukul, E., & Kongar, E. (2021). Health tourism strategy selection via SWOT analysis and integrated hesitant fuzzy linguistic AHP-MABAC approach. *Socio-Economic Planning Sciences, 74*, 100929.
2. Gupta, S., Soni, U., & Kumar, G. (2019). Green supplier selection using multi-criterion decision making under fuzzy environment: A case study in automotive industry. *Computers & Industrial Engineering, 136*, 663–680.
3. Huang, G., Xiao, L., Pedrycz, W., Pamucar, D., Zhang, G., & Martínez, L. (2022). Design alternative assessment and selection: A novel Z-cloud rough number-based BWM-MABAC model. *Information Sciences, 603*, 149–189.
4. Liang, W., Zhao, G., Wu, H., & Dai, B. (2019). Risk assessment of rockburst via an extended MABAC method under fuzzy environment. *Tunnelling and Underground Space Technology, 83*, 533–544.

5. Liu, H. C., You, J. X., & Duan, C. Y. (2019). An integrated approach for failure mode and effect analysis under interval-valued intuitionistic fuzzy environment. *International Journal of Production Economics, 207*, 163–172.
6. Luo, S., Pedrycz, W., & Xing, L. (2022). Selection of data products: a hybrid AFSA-MABAC approach. *International Journal of Machine Learning and Cybernetics*, 1–19.
7. Pamučar, D., & Ćirović, G. (2015). The selection of transport and handling resources in logistics centers using Multi-Attributive border approximation area comparison (MABAC). *Expert Systems with Applications, 42*(6), 3016–3028.
8. Pamučar, D., Petrović, I., & Ćirović, G. (2018). Modification of the Best-Worst and MABAC methods: A novel approach based on interval-valued fuzzy-rough numbers. *Expert Systems with Applications, 91*, 89–106.
9. Pamučar, D., Puška, A., Stević, Ž., & Ćirović, G. (2021). A new intelligent MCDM model for HCW management: The integrated BWM–MABAC model based on D numbers. *Expert Systems with Applications, 175*, 114862.
10. Peng, X., & Yang, Y. (2016). Pythagorean fuzzy Choquet integral based MABAC method for multiple attribute group decision making. *International Journal of Intelligent Systems, 31*(10), 989–1020.
11. Razipour-GhalehJough, S., Hosseinzadeh Lotfi, F., Jahanshahloo, G., Rostamy Malkhalifeh, M., & Sharafi, H. (2020). Finding closest target for bank branches in the presence of weight restrictions using data envelopment analysis. *Annals of Operations Research, 288*, 755–787.
12. Razipour-GhalehJough, S., Hosseinzadeh Lotfi, F., Rostamy-Malkhalifeh, M., & Sharafi, H. (2021). Benchmarking bank branches: A dynamic DEA approach. *Journal of Information and Optimization Sciences, 42*(6), 1203–1236.
13. Shafiee, M., Hosseinzadeh Lotfi, F., Saleh, H., & Ghaderi, M. (2016). A mixed integer bi-level DEA model for bank branch performance evaluation by Stackelberg approach. *Journal of Industrial Engineering International, 12*, 81–91.
14. Torkayesh, A. E., Tirkolaee, E. B., Bahrini, A., Pamucar, D., & Khakbaz, A. (2023). A systematic literature review of MABAC method and applications: An outlook for sustainability and circularity. *Informatica*, 1–34.
15. Wang, J., Wei, G., Wei, C., & Wei, Y. (2020). MABAC method for multiple attribute group decision making under q-rung orthopair fuzzy environment. *Defence Technology, 16*(1), 208–216.
16. Xue, Y. X., You, J. X., Lai, X. D., & Liu, H. C. (2016). An interval-valued intuitionistic fuzzy MABAC approach for material selection with incomplete weight information. *Applied Soft Computing, 38*, 703–713.
17. Yu, S. M., Wang, J., & Wang, J. Q. (2017). An interval type-2 fuzzy likelihood-based MABAC approach and its application in selecting hotels on a tourism website. *International Journal of Fuzzy Systems, 19*, 47–61.

Chapter 12
The Complex Proportional Assessment (COPRAS) in Uncertainty Environment

12.1 Introduction

In the domain of decision-making, where uncertainty and complexity often prevail, finding a reliable and effective method for evaluating alternatives can be a difficult task. This is where the Complex Proportional Assessment (COPRAS), offers a structured approach that navigates the intricate web of competing factors and delivers insightful results. In this chapter, we talk about COPRAS method, explore its various applications across diverse domains. COPRAS is more than just a decision-making tool; it is a systematic framework that empowers individuals and organizations to make informed choices, even when faced with intricate and multifaceted challenges. Throughout this chapter, we will unravel its underlying principles, analyze its unique features, and illustrate its practical implementation through real-world examples. By the end, you will have a comprehensive grasp of this powerful methodology, ready to apply it confidently in your own decision-making processes. Whether you are a researcher seeking to optimize resource allocation, a business executive evaluating investment opportunities, or a policy-maker grappling with complex societal issues, COPRAS holds the key to unlocking clarity amidst the chaos. It empowers you to navigate the intricate landscape of alternatives, evaluate their strengths and weaknesses, and ultimately make choices that align with your objectives and values.

The COmplex PRoportional ASsessment (COPRAS) method was first put forward by Zavadskas et al. in1994 [20]. It is applied to rank the alternatives on the basis of several criteria along with the weight of related criteria and it is one of the innovative methods of decision making, prioritizing, ranking and generally choosing the best alternatives. This method is widely applied when a decision-maker has to select the optimal alternative among a pool of alternatives by considering a set of evaluation criteria.

The COPRAS method is applied in wide variety of areas such as: Selecting air conditioner by Adali and Işık [1]. Garg and Jain [10] applied COPRAS method to rank e-learning websites. Valipour et al. [17] and Alinezhad et al. [2] utilized the COPRAS

F. Hosseinzadeh Lotfi et al., *Fuzzy Decision Analysis: Multi Attribute Decision Making Approach*, Studies in Computational Intelligence 1121,
https://doi.org/10.1007/978-3-031-44742-6_12

method in risk assessment area. Xia et al. [18] improved COPRAS method in material selection problem. Yazdani et al. [19] proposed an integrated COPRAS-AHP methodology to select the best renewable energy project. Ramiah and Pandian [16] studying about machine learning presented a novel hybrid multi objective optimization technique to optimize the process parameters in 3D printing applying F-AHP COPRAS method.

Alinezhad and Khalili [3] discussed about COPRAS method and its application in a chapter of comprehensive book on the MADM methods. They mentioned the following features for COPRAS method:

- It is a compensatory method;
- Attributes are independent;
- The qualitative attributes are converted into the quantitative attributes.

12.2 Description of the COPRAS Method

The steps of COPRAS method are as bellow:

Step 1. The Decision Matrix.

Based on the information received from decision maker, the decision matrix, consisting "m" alternatives and "n" attributes is formed as follows:

$$X = \begin{bmatrix} x_{11} & \dots & x_{1j} & \dots & x_{1n} \\ \vdots & \ddots & \vdots & \ddots & \vdots \\ x_{i1} & \dots & x_{ij} & \dots & x_{in} \\ \vdots & \ddots & \vdots & \ddots & \vdots \\ x_{m1} & \dots & x_{mj} & \dots & x_{mn} \end{bmatrix} \tag{12.1}$$

Also, the weight vector of attributes is denoted by $W = \left(w_1, \dots, w_j, \dots w_n\right)$.

Step 2. The Normalized Decision Matrix.

Normalization of the decision matrix can be achieved using the following equation.

$$\overline{x}_{ij} = \frac{x_{ij}}{\sum_{i=1}^{m} x_{ij}}; \ j = 1, \dots, n. \tag{12.2}$$

Step 3. The weighted normalized decision matrix.

The elements of the weighted normalized decision matrix are calculated as bellow:

$$\hat{x}_{ij} = \overline{x}_{ij}.w_j; \ i = 1, \dots, m, \ j = 1, \dots, n. \tag{12.3}$$

Step 4. The R_i and P_i values.

For benefit indices: J^+ = {j | indices of criteria which higher values are more desirable}.

For cost indices: J^- = {j | indices of criteria which lower values are more desirable}

$$P_i = \sum_{j \in J+} \hat{x}_{ij};\ i = 1, \ldots, m. \tag{12.4}$$

$$R_i = \sum_{j \in J^-} \hat{x}_{ij};\ i = 1, \ldots, m. \tag{12.5}$$

$$R_{\min} = \min_i R_i. \tag{12.6}$$

Step 5. The relative weight of each alternative (Q_i).

The relative weight of each alternative is calculated through following equations.

$$Q_i = P_i + \frac{R_{\min} \sum_{i=1}^{m} R_i}{R_i \sum_{i=1}^{m} \frac{R_{\min}}{R_i}};\ i = 1, .., m. \tag{12.7}$$

It can also be written as follow:

$$Q_i = P_i + \frac{\sum_{i=1}^{m} R_i}{R_i \sum_{i=1}^{m} \frac{1}{R_i}};\ i = 1, .., m. \tag{12.8}$$

Step 6. Ranking of the Alternatives.

Ranking the alternatives in descending order of the relative weight, the alternative with the highest score is selected as the preferred one.

Note: the utility degree of the alternatives can be calculated applying the following equation.

$$N_i = \frac{Q_i}{Q_{\max}}.100\%;\ i = 1, \ldots, m. \tag{12.9}$$

where $Q_{\max} = \max_i Q_i$. and obviously, the higher is the utility degree of alternative, the higher is the priority for that alternative.

12.3 Solving Multi-Criteria Decision Making for Smart Phone Selection by COPRAS

In the age of technology, where an array of smart phones flood the market, choosing the perfect device can be an overwhelming task. With numerous factors to consider, such as memory, battery life, camera quality, processor performance, and cost, customers face a complex decision-making process. To address this challenge, we turn to the powerful Complex Proportional Assessment (COPRAS) method, which enables us to navigate the maze of criteria and arrive at an optimal choice.

In this section, a practical example is presented that examines the application of COPRAS in solving a multi-criteria decision-making problem for smart phone selection. We introduce a customer who is on the hunt for the most suitable smart phone among three types: A1, A2, and A3. Understanding the significance of memory (C1), battery (C2), camera (C3), processor (C4), and cost (C5) as essential criteria, the customer seeks a device that excels in these areas while considering the budgetary constraints. COPRAS leverages the power of complex proportional assessment, which takes into account the relative importance of each criterion and assigns weights accordingly. By employing this comprehensive approach, COPRAS not only facilitates informed decision-making but also reveals the strengths and weaknesses of each alternative, aiding in the selection process.

Throughout this section, we explore the intricate calculations and decision matrices involved in applying COPRAS to the smart phone selection problem. We analyze and discuss the significance of each criterion and demonstrate how COPRAS seamlessly integrates these elements into a cohesive decision-making framework. By the end, by utilizing COPRAS, customers can confidently select a smart phone that aligns with their preferences, strikes the perfect balance between performance and cost, and ultimately enhances their overall user experience.

Table 12.1 shows the decision matrix and Table 12.2 represents the weight of each criteria.

Table 12.1 Decision matrix of choosing smart phone

Alternative	C1 (benefit)	C2 (benefit)	C3 (benefit)	C4 (benefit)	C5 (cost)
A1	0.810	1.150	5.410	0.530	0.340
A2	1.030	0.920	6.820	0.990	0.250
A3	1.500	0.840	4.210	0.240	0.490

Table 12.2 Criteria weight of choosing smart phone

Criteria	C1	C2	C3	C4	C5
Weight	0.190	0.180	0.220	0.210	0.200

Table 12.3 The normalized decision matrix of choosing smart phone

Alternative	C1	C2	C3	C4	C5
A1	0.242515	0.395189	0.329075	0.301136	0.314815
A2	0.308383	0.316151	0.414842	0.5625	0.231481
A3	0.449102	0.28866	0.256083	0.136364	0.453704

Table 12.4 The weighted normalized decision matrix of choosing smart phone

Alternative	C1	C2	C3	C4	C5
A1	0.046078	0.071134	0.072397	0.063239	0.062963
A2	0.058593	0.056907	0.091265	0.118125	0.046296
A3	0.085329	0.051959	0.056338	0.028636	0.090741

Table 12.5 Result of COPRAS method for choosing smart phone

Alternative	Pi	Ri	Qi	Rank
A1	0.252847	0.062963	0.318338	2
A2	0.32489	0.046296	0.413957	1
A3	0.222263	0.090741	0.267705	3

The normalized decision matrix is expressed in Table 12.3, using Eq. (12.2).

By Eq. (12.3), the weighted decision matrix is as Table 12.4. Result are summarized in Table 12.5.

Second smart phone (A_2) is the best alternative and the alternatives are ranked as $A_2 \succ A_1 \succ A_3$.

12.4 Fuzzy COPRAS Method

As decision-making processes often involve uncertain and imprecise information, the integration of COPRAS with fuzzy logic provides a robust framework for tackling real-world scenarios. In this section, we unravel the intricacies of decision-making in an environment where ambiguity prevails. We talk about the principles and techniques that enable COPRAS to adapt to fuzzy information, capturing the nuances of imprecision and vagueness inherent in many real-world situations. By embracing the power of fuzzy COPRAS, decision-makers gain a deeper understanding of the complexities at play and can make more informed choices in uncertain contexts.

In the classical COPRAS method, the weights of the criteria and the values are clear and crisp. However, under many conditions modeling real life decision problems with crisp data is not able and sometimes determining the exact weights is difficult. The fuzzy approach helps to determine the relative importance of features by using fuzzy numbers instead of exact numbers [4, 15].

Fouladgar et al. [9] applied Fuzzy COPRAS method for selecting the optimal maintenance strategy. Masoomi et al. [13] evaluated set of strategic green suppliers based on environmental criteria applying fuzzy BWM, WASPAS and COPRAS approaches. Dhiman et al. [8] applied Data Envelopment Analysis (DEA) and Fuzzy TOPSIS and DEA-Fuzzy COPRAS to rank solar energy projects. Buying tractors is a big investment for farmers, Özdağoğlu et al. [14] utilized Integrated PIPRECIA and COPRAS Method under Fuzzy Environment for truck tractor selection. Wind energy is one of the renewable energies and wind farm are designed to provide electricity, Chatterjee and Bose [6, 7] applied fuzzy COPRAS method for selecting and ranking of vendors of a wind farm. Keshavarz Ghorabaee et al. [11] applied COPRAS method with interval type-2 fuzzy sets in supplier selection. Mardani et al. [12] reviewed systematically the applications and methodologies of the fuzzy multi decision-making (FMCDM) techniques.

The steps of fuzzy COPRAS method can be described as follows:

Step 1. The Linguistic Rating.

In this step, first fuzzy numbers suitable for the assessment of alternatives and determining the importance of criteria is selected according to experts' opinion. The importance weights of evaluation criteria and the rating of alternatives are considered as linguistic terms, Table 12.6 shows the linguistic rating for criteria and Table 12.7 shows the linguistic rating for alternatives.

Hence Decision makers can assess each alternative by a given criteria, utilizing linguistic rating of Table 12.7. Linguistic values for importance weight of each criteria are presented in Fig. 12.1 and linguistic values for preference rating of each alternative are shown in Fig. 12.2.

Table 12.6 Linguistic rating of criteria

Linguistic terms	Fuzzy number
Very low importance (VL)	(0,0,0.25)
Low importance (L)	(0,0.25,0.5)
Medium importance (M)	(0.25,0.5,0.75)
High importance (H)	(0.5,0.75,1)
Very High importance (VH)	(0.75,1,1)

Table 12.7 Linguistic rating of alternatives

Linguistic terms	Fuzzy number
Very Poor (VP)	(0,0,2.5)
Poor (P)	(0,2.5,5)
Fair (F)	(2.5,5,7.5)
Good (G)	(5,7.5,10)
Very Good (VG)	(7.5,10,10)

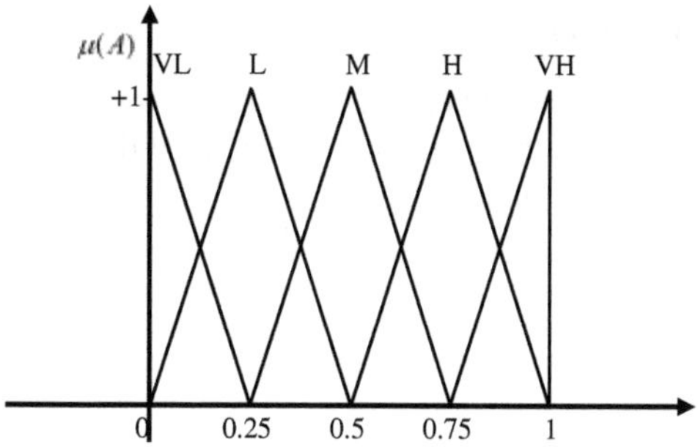

Fig. 12.1 Linguistic values for importance of weight of criteria

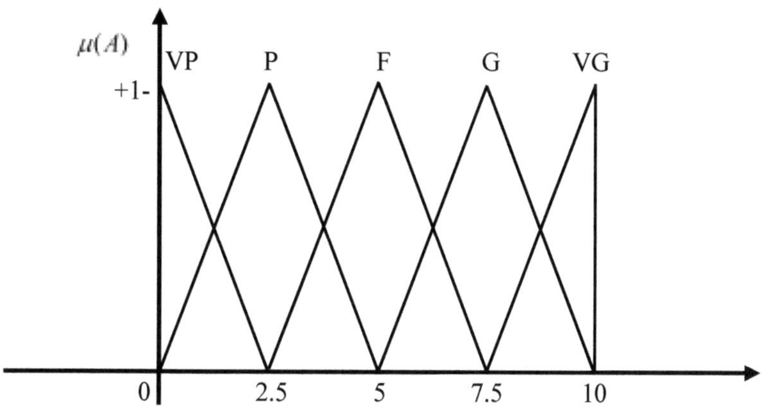

Fig. 12.2 Linguistic values for preferences of alternatives

Step 2. Fuzzy Decision Matrix.

Considering "n" criteria ($C_1, C_2, \ldots, C_n$) and "m" alternatives ($A_1, A_2, \ldots, A_m$), the fuzzy decision matrix is a as following:

$$\tilde{X} = \begin{array}{c} \\ A_1 \\ A_2 \\ \vdots \\ A_m \end{array} \begin{array}{c} C_1 \; C_2 \; \ldots \; C_n \\ \begin{bmatrix} \tilde{x}_{11} & \tilde{x}_{12} & \ldots & \tilde{x}_{1n} \\ \tilde{x}_{21} & \tilde{x}_{22} & \ldots & \tilde{x}_{2n} \\ \vdots & \vdots & \ddots & \vdots \\ \tilde{x}_{m1} & \tilde{x}_{m2} & \ldots & \tilde{x}_{mn} \end{bmatrix} \end{array} \tag{12.10}$$

where $\tilde{x}_{ij} = \left(x_{ij}^L, x_{ij}^m, x_{ij}^U\right)$ is a triangle fuzzy number (TFN).

Step 3. Defuzzify into crisp values.

In order to deffuzify fuzzy weight vector and fuzzy decision matrix into crisp values, the center of area (COA) method can be applied since the method is simple and it doesn't need the preferences of any evaluators.

For any fuzzy number $\tilde{x}_{ij} = (a_{ij}, b_{ij}, c_{ij})$ the defuzzified number is as

$$x_{ij} = \frac{(c_{ij} - a_{ij}) + (b_{ij} - a_{ij})}{3} + a_{ij} \tag{12.11}$$

At this stage, the elements of fuzzy decision matrix and the fuzzy weight vector are defuzzified into crisp numbers.

Step 4. Normalization of the defuzzified decision matrix.

In this step the defuzzified decision matrix is normalized as follows:

$$\overline{x}_{ij} = \frac{x_{ij}}{\sum_{i=1}^{m} x_{ij}}\ , \ i = 1, \ldots, m,\ \ j = 1, \ldots, n\,. \tag{12.12}$$

where $\overline{x}_{ij}$ are crisp values, and the normalized decision matrix is as below:

$$\overline{X} = \begin{bmatrix} \overline{x}_{11} & \overline{x}_{12} & \ldots & \overline{x}_{1n} \\ \overline{x}_{21} & \overline{x}_{22} & \ldots & \overline{x}_{2n} \\ \vdots & \vdots & \ddots & \vdots \\ \overline{x}_{m1} & \overline{x}_{m2} & \ldots & \overline{x}_{mn} \end{bmatrix} \tag{12.13}$$

Step 5. The weighted Normalized Decision Matrix.

The weight of the criteria can be derived using BWM, SWARA or Shannon entropy. Assume that $\tilde{W} = (\tilde{w}_1, \tilde{w}_2, \ldots, \tilde{w}_n)$ is the weight vector that denotes the importance of criteria.

The weighted normalized decision matrix is calculated as bellow:

$$\hat{x}_{ij} = \overline{x}_{ij}.w_j\ , \ i = 1, \ldots, m,\ \ j = 1, \ldots, n\,. \tag{12.14}$$

where $w_j,\ j = 1, \ldots, n$ is the defuzzified weight of the jth criterion.

$$\hat{X} = \begin{bmatrix} \hat{x}_{11} & \hat{x}_{12} & \ldots & \hat{x}_{1n} \\ \hat{x}_{21} & \hat{x}_{22} & \ldots & \hat{x}_{2n} \\ \vdots & \vdots & \ddots & \vdots \\ \hat{x}_{m1} & \hat{x}_{m2} & \ldots & \hat{x}_{mn} \end{bmatrix} \tag{12.15}$$

$\hat{X}$ is the weighted normalized decision matrix.

Step 6. Calculation of P_i and R_i values.

For benefit indices: $J^+ =$ {j | indices of criteria which higher values are more desirable}.

For cost indices: $J^- =$ {j | indices of criteria which lower values are more desirable}

$$P_i = \sum_{j \in J+} \hat{x}_{ij};\ i = 1, \ldots, m. \tag{12.16}$$

$$R_i = \sum_{j \in J^-} \hat{x}_{ij};\ i = 1, \ldots, m. \tag{12.17}$$

$$R_{\min} = \min_i R_i. \tag{12.18}$$

Step 7. Q_i Values -The relative weight of alternatives.

$$Q_i = P_i + \frac{R_{\min} \sum_{i=1}^{m} R_i}{R_i \sum_{i=1}^{m} \frac{R_{\min}}{R_i}};\ i = 1, .., m. \tag{12.19}$$

It can also be written as follow:

$$Q_i = P_i + \frac{\sum_{i=1}^{m} R_i}{R_i \sum_{i=1}^{m} \frac{1}{R_i}};\ i = 1, .., m. \tag{12.20}$$

and

$$Q_{\max} = \max_i Q_i. \tag{12.21}$$

Step 8. Calculate the utility degree and priority of the alternatives.

The N_i index which shows the utility degree of alternatives is calculated as below:

$$N_i = \frac{Q_i}{Q_{\max}}.100\%;\ i = 1, \ldots, m. \tag{12.22}$$

The higher is the utility degree of alternative, the higher is the priority for that alternative.

Figure 12.3 shows the COPRAS Method.

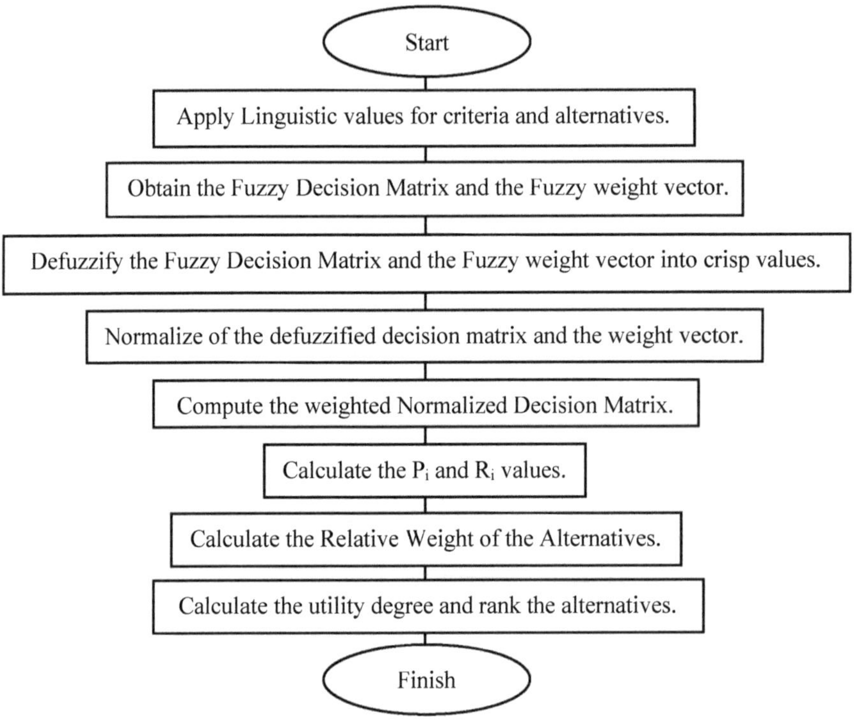

Fig. 12.3 The Fuzzy COPRAS Method at a Glance

12.5 Fuzzy COPRAS Approach Under Group Decision Making

In group decision making problems, a committee of decision makers is formed to rank the alternatives. Considering knowledge and experience of a group of experts means a more intellectual judgment [5].

In the group fuzzy COPRAS method, first, the decision matrix of all decision makers is aggregated.

Step 1. Fuzzy Aggregated Decision Matrix.

Considering “k” decision makers, there will be “k” fuzzy decision matrices as follows.

$$DM^k = \begin{array}{c} \\ A_1 \\ \vdots \\ A_i \\ \vdots \\ A_m \end{array} \overset{\begin{array}{ccccc} C_1 & \dots & C_j & \dots & C_n \end{array}}{\begin{bmatrix} \tilde{x}_{11k} & \dots & \tilde{x}_{1jk} & \dots & \tilde{x}_{1nk} \\ \vdots & \ddots & \vdots & \ddots & \vdots \\ \tilde{x}_{i1k} & \dots & \tilde{x}_{ijk} & \dots & \tilde{x}_{ink} \\ \vdots & \ddots & \vdots & \ddots & \vdots \\ \tilde{x}_{m1k} & \dots & \tilde{x}_{mjk} & \dots & \tilde{x}_{mnk} \end{bmatrix}}; k = 1, ..., K. \quad (12.23)$$

where $\tilde{x}_{ijk} = \left(x_{ijk}^L, x_{ijk}^m, x_{ijk}^U\right)$ is the rating of alternative A_i with respect to criterion C_j evaluated by kth decision maker.

The decision makers may not be at the same level, depending on their experiences and expertise. Hence, let $\tilde{q}_k$ be the fuzzy coefficient of significance for the kth decision maker. Note that if experts are homogeneous then,$\tilde{q}_k = \left(\frac{1}{k}, \frac{1}{k}, \frac{1}{k}\right)$.

Each element of the aggregated fuzzy matrix is calculated by following equation:

$$\tilde{x}_{ij} = \frac{\sum_{k=1}^{k} \tilde{q}_k \tilde{x}_{ijk}}{\sum_{k=1}^{k} \tilde{q}_k} \quad (12.24)$$

The aggregated fuzzy decision matrix is a as following:

$$\tilde{X} = \begin{array}{c} A_1 \\ A_2 \\ \vdots \\ A_m \end{array} \overset{\begin{array}{cccc} C_1 & C_2 & \dots & C_n \end{array}}{\begin{bmatrix} \tilde{x}_{11} & \tilde{x}_{12} & \dots & \tilde{x}_{1n} \\ \tilde{x}_{21} & \tilde{x}_{22} & \dots & \tilde{x}_{2n} \\ \vdots & \vdots & \ddots & \vdots \\ \tilde{x}_{m1} & \tilde{x}_{m2} & \dots & \tilde{x}_{mn} \end{bmatrix}} \quad (12.25)$$

where $\tilde{x}_{ij} = \left(x_{ij}^L, x_{ij}^M, x_{ij}^U\right)$ is a triangle fuzzy number (TFN).

Step 2. Defuzzify into crisp values.

In order to deffuzify fuzzy decision matrix into crisp values, the center of area (COA) method can be applied. For any fuzzy number $\tilde{x}_{ij} = \left(x_{ij}^L, x_{ij}^M, x_{ij}^U\right)$ the defuzzified number is as

$$x_{ij} = \frac{(x_{ij}^U - x_{ij}^L) + (x_{ij}^M - x_{ij}^L)}{3} + x_{ij}^L \quad (12.26)$$

Step 3. Normalization of the defuzzified aggregated decision matrix.

In order to normalize the defuzzified decision matrix the following equation may applied.

$$\overline{x}_{ij} = \frac{x_{ij}}{\sum_{i=1}^{m} x_{ij}} \ , i = 1, \ldots, m, \ j = 1, \ldots, n. \tag{12.27}$$

where $\overline{x}_{ij}$ are crisp values, and the normalized decision matrix is as below:

$$\overline{X} = \begin{bmatrix} \overline{x}_{11} & \overline{x}_{12} & \ldots & \overline{x}_{1n} \\ \overline{x}_{21} & \overline{x}_{22} & \ldots & \overline{x}_{2n} \\ \vdots & \vdots & \ddots & \vdots \\ \overline{x}_{m1} & \overline{x}_{m2} & \ldots & \overline{x}_{mn} \end{bmatrix} \tag{12.28}$$

Step 4. The Criteria Weight.

Assume that $\tilde{W}_k = (\tilde{w}_{1k}, \tilde{w}_{2k}, \ldots, \tilde{w}_{nk})$ is the weight vector that denotes the importance of criteria, presented by kth decision maker. The aggregated weight vector is as follows:

$$\tilde{W} = \frac{\sum_{k=1}^{k} \tilde{q}_k \tilde{W}_k}{\sum_{k=1}^{k} \tilde{q}_k} \tag{12.29}$$

Each element of the aggregated weight vector $\tilde{W} = (\tilde{w}_1, \tilde{w}_2, \ldots, \tilde{w}_n)$ is a triangular fuzzy number.

Defuzify each TFN $w_j = \left(w_j^L, w_j^m, w_j^U\right)$ to crisp value by following equation.

$$w_j = \frac{\left(w_j^U - w_j^L\right) + \left(w_j^M - w_j^L\right)}{3} + w_j^L \tag{12.30}$$

The normalized weight vector is equal to $\overline{W} = \left(\overline{w}_1, \ldots, \overline{w}_j, \ldots, \overline{w}_n\right)$ where $\overline{w}_j = \frac{w_j}{\sum_{j=1}^{n} w_j}; \ j = 1, \ldots, n.$

Step 5. The Weighted Normalized Aggregated Decision Matrix.

Each element of the weighted normalized decision matrix is calculated as bellow:

$$\hat{x}_{ij} = \overline{x}_{ij}.\overline{w}_j \ , i = 1, \ldots, m, \ j = 1, \ldots, n. \tag{12.31}$$

$$\hat{X} = \begin{bmatrix} \hat{x}_{11} & \hat{x}_{12} & \ldots & \hat{x}_{1n} \\ \hat{x}_{21} & \hat{x}_{22} & \ldots & \hat{x}_{2n} \\ \vdots & \vdots & \ddots & \vdots \\ \hat{x}_{m1} & \hat{x}_{m2} & \ldots & \hat{x}_{mn} \end{bmatrix} \tag{12.32}$$

Step 6. The P_i and R_i values.

For benefit indices: $J^+ = \{j \mid$ indices of criteria which higher values are more desirable$\}$.

For cost indices: $J^- = \{j \mid$ indices of criteria which lower values are more desirable$\}$

$$P_i = \sum_{j \in J+} \hat{x}_{ij};\ i = 1, \ldots, m. \tag{12.33}$$

$$R_i = \sum_{j \in J-} \hat{x}_{ij};\ i = 1, \ldots, m. \tag{12.34}$$

$$R_{\min} = \min_i R_i. \tag{12.35}$$

Step 7. The relative Weight of Alternatives (Q_i values).

$$Q_i = P_i + \frac{R_{\min} \sum_{i=1}^{m} R_i}{R_i \sum_{i=1}^{m} \frac{R_{\min}}{R_i}};\ i = 1, .., m. \tag{12.36}$$

It can also be written as follow:

$$Q_i = P_i + \frac{\sum_{i=1}^{m} R_i}{R_i \sum_{i=1}^{m} \frac{1}{R_i}};\ i = 1, .., m. \tag{12.37}$$

Step 8. The utility degree and ranking the alternatives.

The N_i index which shows the utility degree of alternatives is calculated as below:

$$N_i = \frac{Q_i}{Q_{\max}}.100\%;\ i = 1, \ldots, m. \tag{12.38}$$

where,

$$Q_{\max} = \max_i Q_i\,. \tag{12.39}$$

The higher is the utility degree of alternative, the higher is the priority for that alternative.

12.6 Solving Investment Selection with Fuzzy COPRAS: Navigating Complex Criteria in Decision-Making

In the field of investment, where countless opportunities present themselves, to aid investors in this decision-making process, we turn to the powerful combination of fuzzy logic and the COPRAS method. We assess and rank the investment alternatives through the fuzzy COPRAS method. We consider the varying degrees of benefit and cost across the criteria, capturing the fuzziness inherent in the decision-making process. Using fuzzy sets and linguistic variables, fuzzy COPRAS provides a comprehensive framework for evaluating and selecting the investment option that aligns with the investor's objectives and risk appetite.

In a practical application, we explore how fuzzy COPRAS guides an investor in choosing among three investment options: A1, A2, and A3. In this scenario, the investor seeks guidance from three experts, who provide advice based on seven critical criteria. Four of these criteria, C1, C2, C3, and C6, represent the benefits associated with the investments, while C4, C5, and C7 represent the associated costs.

Using fuzzy logic the experts' evaluation with respect to each criterion are presented in Table 12.8.

Respecting linguistic rating of Table 12.7, the fuzzy group decision making matrix is summarized in Table 12.9. The fuzzy numbers assigned to each criteria by decision makers are aggregated using Eq. (12.24). The Fuzzy aggregated decision Matrix is shown in Table 12.10.

The linguistic variable assigned to each criterion by decision maker are presented in Table 12.11 and the fuzzy triangular number of criteria's weight are displayed in Table 12.12. Using Eq. (12.29) the weights are aggregated and applying Eq. (12.30), fuzzy triangular numbers are deffuzified to crisp values in Table 12.13.

Moreover, applying Eq. (12.26) the Decision Matrix is deffuzified to crisp values. The normalized decision matrix is expressed in Table 12.14.

By Eq. (12.31), the weighted decision matrix is as Table 12.15. Result are summarized in Table 12.16.

Table 12.8 Evaluation of 3 investments using linguistic variables

Alternative	Decision maker	C1	C2	C3	C4	C5	C6	C7
A1	DM1	F	G	VG	P	VG	P	F
	DM2	G	G	F	VP	G	P	G
	DM3	VG	F	G	P	VG	VP	F
A2	DM1	G	G	F	VG	VP	F	P
	DM2	VG	F	G	G	P	G	VP
	DM3	G	G	F	F	P	G	VP
A3	DM1	P	G	G	P	G	F	F
	DM2	F	VG	F	P	G	G	F
	DM3	P	VG	G	P	G	F	F

Table 12.9 Fuzzy group decision matrix

Alternative	Decision maker	C1	C2	C3	C4	C5	C6	C7
A1	DM1	(2.5,5,7.5)	(5,7.5,10)	(7.5,10,10)	(0,2.5,5)	(7.5,10,10)	(0,2.5,5)	(2.5,5,7.5)
	DM2	(5,7.5,10)	(5,7.5,10)	(2.5,5,7.5)	(0,0,2.5)	(5,7.5,10)	(0,2.5,5)	(5,7.5,10)
	DM3	(7.5,10,10)	(2.5,5,7.5)	(5,7.5,10)	(0,2.5,5)	(7.5,10,10)	(0,0,2.5)	(2.5,5,7.5)
A2	DM1	(5,7.5,10)	(5,7.5,10)	(2.5,5,7.5)	(7.5,10,10)	(0,0,2.5)	(2.5,5,7.5)	(0,2.5,5)
	DM2	(7.5,10,10)	(2.5,5,7.5)	(5,7.5,10)	(5,7.5,10)	(0,2.5,5)	(5,7.5,10)	(0,0,2.5)
	DM3	(5,7.5,10)	(5,7.5,10)	(2.5,5,7.5)	(2.5,5,7.5)	(0,2.5,5)	(5,7.5,10)	(0,0,2.5)
A3	DM1	(0,2.5,5)	(5,7.5,10)	(5,7.5,10)	(0,2.5,5)	(5,7.5,10)	(2.5,5,7.5)	(2.5,5,7.5)
	DM2	(2.5,5,7.5)	(5,7.5,10)	(2.5,5,7.5)	(0,2.5,5)	(5,7.5,10)	(5,7.5,10)	(2.5,5,7.5)
	DM3	(0,2.5,5)	(5,7.5,10)	(5,7.5,10)	(0,2.5,5)	(5,7.5,10)	(2.5,5,7.5)	(2.5,5,7.5)

Table 12.10 Fuzzy aggregated decision matrix

Alternative	C1	C2	C3	C4	C5	C6	C7
A1	(5,7.5,9.2)	(4.2,6.7,9.2)	(5,7.5,9.2)	(0,1.7,4.2)	(6.7,9.2,10)	(0,1.7,4.2)	(3.3,5.8,8.3)
A2	(5.8,8.3,10)	(4.2,6.7,9.2)	(3.3,5.8,8.3)	(5,7.5,9.2)	(0,1.7,4.2)	(4.2,6.7,9.2)	(0,0.8,3.3)
A3	(0.8,3.3,5.8)	(6.7,9.2,10)	(4.2,6.7,9.2)	(0,2.5,0)	(5,7.5,10)	(3.3,5.8,8.3)	(2.5,5,7.5)

Table 12.11 Weight of criteria using linguistic variables

	C1	C2	C3	C4	C5	C6	C7
DM1	VH	VL	H	M	H	VH	L
DM2	H	L	VH	M	M	H	L
DM3	VH	L	H	H	H	H	VL

Table 12.12 Fuzzy weight of criteria

	C1	C2	C3	C4	C5	C6	C7
DM1	(0.75,1,1)	(0,0,0.25)	(0.5,0.75,1)	(0.25,0.5,0.75)	(0.5,0.75,1)	(0.75,1,1)	(0,0.25,0.5)
DM2	(0.5,0.75,1)	(0,0.25,0.5)	(0.75,1,1)	(0.25,0.5,0.75)	(0.25,0.5,0.75)	(0.5,0.75,1)	(0,0.25,0.5)
DM3	(0.75,1,1)	(0,0.25,0.5)	(0.5,0.75,1)	(0.5,0.75,1)	(0.5,0.75,1)	(0.5,0.75,1)	(0,0,0.25)

Table 12.13 Aggregated weight of criteria

	C1	C2	C3	C4	C5	C6	C7
Fuzzy	(0.67, 0.92, 1)	(0, 0.17, 0.42)	(0.58, 0.83, 1)	(0.33, 0.58, 0.83)	(0.42, 0.67, 0.92)	(0.58, 0.83, 1)	(0, 0.17, 0.42)
Defuzzy	0.86	0.19	0.81	0.58	0.67	0.81	0.19

Table 12.14 Normalized defuzzy decision matrix

Alternative	C1	C2	C3	C4	C5	C6	C7
A1	7.2222	6.6667	7.2222	1.9444	8.6111	1.9444	5.8333
A2	8.0556	6.6667	5.8333	7.2222	1.9444	6.6667	1.3889
A3	3.3333	8.6111	6.6667	2.5	7.5	5.8333	5

Table 12.15 Weighted decision matrix

Alternative	C1	C2	C3	C4	C5	C6	C7
A1	0.3342	0.0591	0.295	0.0972	0.3179	0.1084	0.0928
A2	0.3727	0.0591	0.2383	0.3611	0.0718	0.3718	0.0221
A3	0.1542	0.0763	0.2723	0.125	0.2769	0.3253	0.0795

Table 12.16 Result of the F-COPRAS method

Alternative	Pi	Ri	Qi	Rank
A1	0.7967	0.508	1.2521	3
A2	1.0418	0.455	1.5503	1
A3	0.8282	0.4815	1.3087	2

Utilizing the fuzzy COPRAS group decision making method, the rank of alternatives is as follows.

$$A_2 \succ A_3 \succ A_1$$

12.7 Conclusion

This chapter has provided a comprehensive exploration of the Complex Proportional Assessment (COPRAS) method and its application in a fuzzy environment. We began by introducing the COPRAS method as a structured approach to decision-making, highlighting its ability to navigate complex criteria and deliver insightful results. Furthermore, we integrate COPRAS with fuzzy logic, known as Fuzzy COPRAS, which enables decision makers to handle the uncertainty and imprecision inherent in real-world scenarios. Through a practical example, we demonstrated how fuzzy COPRAS accommodates the complexities of fuzzy information and captures the nuances of imprecision and ambiguity. COPRAS and fuzzy COPRAS offer valuable tools for addressing multi-criteria decision-making challenges and fuzzy decision scenarios. As we move forward, it is essential to continue exploring and refining these methodologies, adapting them to the ever-evolving complexities of decision-making in an uncertain and fuzzy environment.

Acknowledgement A special thanks to the Iranian DEA society for their unwavering spiritual support and consensus in the writing of this book. Your invaluable support has been truly remarkable, and we are deeply grateful for the opportunity to collaborate with such esteemed professionals.

References

1. ADALİ, E. A., & IŞIK, A. T. (2016). Air conditioner selection problem with COPRAS and ARAS methods. *Manas Sosyal Araştırmalar Dergisi, 5*(2), 124-138.
2. Alinezhad, A., Amini, A., & Rahmani, M. (2015). New product development risk assessment in the core banking using FMEA combined with COPRAS method and grey relations. *Journal of Money and Economy, 10*(3), 87–121.
3. Alinezhad, A., & Khalili, J. (2019). *New methods and applications in multiple attribute decision making (MADM)* (Vol. 277). Springer.
4. Armand, A., Allahviranloo, T., & Gouyandeh, Z. (2018). Some fundamental results on fuzzy calculus. *Iranian Journal of Fuzzy Systems, 15*(3), 27–46.
5. Cabrerizo, F. J., Ureña, R., Pedrycz, W., & Herrera-Viedma, E. (2014). Building consensus in group decision making with an allocation of information granularity. *Fuzzy Sets and Systems, 255*, 115–127.
6. Chatterjee, N., & Bose, G. (2013). A COPRAS-F base multi-criteria group decision making approach for site selection of wind farm. *Decision Science Letters, 2*(1), 1–10.
7. Chatterjee, N., & Bose, G. (2013). Selection of vendors for wind farm under fuzzy MCDM environment. *International Journal of Industrial Engineering Computations, 4*(4), 535–546.
8. Dhiman, H. S., & Sen, S. (2022). Data envelopment analysis based fuzzy TOPSIS and fuzzy COPRAS techniques for selection of solar energy projects. In Infrastructure Development–Theory, Practice and Policy (pp. 118–128). Routledge.
9. Fouladgar, M. M., Yazdani-Chamzini, A., Lashgari, A., Zavadskas, E. K., & Turskis, Z. (2012). Maintenance strategy selection using AHP and COPRAS under fuzzy environment. *International journal of strategic property management, 16*(1), 85–104.
10. Garg, R., & Jain, D. (2017). Fuzzy multi-attribute decision making evaluation of e-learning websites using FAHP, COPRAS, VIKOR, WDBA. *Decision Science Letters, 6*(4), 351–364.
11. Keshavarz Ghorabaee, M., Amiri, M., Salehi Sadaghiani, J., & Hassani Goodarzi, G. (2014). Multiple criteria group decision-making for supplier selection based on COPRAS method with interval type-2 fuzzy sets. *The International Journal of Advanced Manufacturing Technology, 75*, 1115–1130.
12. Mardani, A., Jusoh, A., & Zavadskas, E. K. (2015). Fuzzy multiple criteria decision-making techniques and applications–Two decades review from 1994 to 2014. *Expert systems with Applications, 42*(8), 4126–4148.
13. Masoomi, B., Sahebi, I. G., Fathi, M., Yıldırım, F., & Ghorbani, S. (2022). Strategic supplier selection for renewable energy supply chain under green capabilities (fuzzy BWM-WASPAS-COPRAS approach). *Energy Strategy Reviews, 40*, 100815.
14. Özdağoğlu, A., Öztaş, G. Z., Keleş, M. K., & Volkan, G. E. N. Ç. (2021). An integrated PIPRECIA and COPRAs method under fuzzy environment: A case of truck tractor selection. *Alphanumeric Journal, 9*(2), 269–298.
15. Pedrycz, W. (2020). Fuzzy relational calculus. In *Handbook of Fuzzy Computation* (pp. B3–3). CRC Press.
16. Ramiah, K., & Pandian, P. (2023). Effect of process parameters on the strength of ABS based FDM prototypes: Novel machine learning based hybrid optimization technique. *Materials Research Express, 10*(2), 025305.
17. Valipour, A., Yahaya, N., Md Noor, N., Antuchevičienė, J., & Tamošaitienė, J. (2017). Hybrid SWARA-COPRAS method for risk assessment in deep foundation excavation project: An Iranian case study. *Journal of Civil Engineering and Management, 23*(4), 524–532.
18. Xia, F., Wei, H., & Yang, L. W. (2015). Improved COPRAS method and application in material selection problem. In Applied Mechanics and Materials (Vol. 707, pp. 505–508). Trans Tech Publications Ltd.

19. Yazdani-Chamzini, A., Fouladgar, M. M., Zavadskas, E. K., & Moini, S. H. H. (2013). Selecting the optimal renewable energy using multi criteria decision making. *Journal of Business Economics and Management, 14*(5), 957–978.
20. Zavadskas, E. K., Kaklauskas, A., & Šarka, V. (1994). The new method of multicriteria complex proportional assessment of projects.

Chapter 13
The Criteria Importance Through Inter-Criteria Correlation (CRITIC) in Uncertainty Environment

13.1 Introduction

In multi-criteria decision-making models, two factors are always involved: criteria weighting and alternative scoring. These two aspects complement each other and are sometimes integrated using a combination of methods. For example, criteria weighting can be done using methods like AHP, Shannon's entropy, etc., while ranking can be performed using techniques such as TOPSIS or VIKOR. In this section, we will discuss the CRITIC method, which belongs to the first group and focuses on criteria weighting. In multi-criteria decision-making models, as the name suggests, we deal with more than one criterion, making the importance and weighting of criteria highly significant. For instance, when choosing a mode of transportation for a trip, criteria such as travel time, comfort, cost, etc., may be considered. Do you believe that all these criteria are equally important? Certainly not, as different individuals may prioritize cost over other factors, while others may prioritize travel time or location and quality of travel. Therefore, determining the weights of criteria plays a crucial role in making the best decision [2].

The CRITIC method provides a way to determine the weights of criteria. The lack of dependence on expert opinions can be considered one of the key features of the CRITIC method. This approach analyzes the data based on the level of overlap and conflict among factors or criteria, ensuring the proper consideration of each factor's role in the final results. In the CRITIC method, each criterion is evaluated based on a range of measurement values within the alternatives and is expressed using a membership function. The constructed vectors for the utilized criteria have statistical parameters, including standard deviation, representing the degree of variation in the corresponding criterion values. The weighting of criteria is influenced by the mental perspectives of decision-makers regarding the characteristics of the criteria. Usually, criteria weights are determined by decision-makers based on their

F. Hosseinzadeh Lotfi et al., *Fuzzy Decision Analysis: Multi Attribute Decision Making Approach*, Studies in Computational Intelligence 1121,
https://doi.org/10.1007/978-3-031-44742-6_13

experience, knowledge, and understanding of the problem. However, this can lead to uncertainty and doubt regarding the reliability of the results. To overcome such challenges, numerical valuation approaches are employed [1, 9].

13.2 CRITIC Method

The CRITIC method is a tangible approach to determine the weights of criteria based on the intensity of conflict and inconsistency among the components of a decision problem. These two aspects are formulated statistically using correlation coefficients and standard deviations. The criterion standard deviation pertains to the values of each criterion, while the correlation coefficient relates to pairs of criteria. In other words, the criterion standard deviation is calculated within the set, and the correlation is calculated outside the set. In this method, data is analyzed based on the level of overlap and conflict among the factors or criteria. This processing approach ensures that the role of each factor is properly accounted for in the final results. In the CRITIC method, for each criterion, there is a range of measured value variations among the options, which are expressed using a membership function. The resulting vectors for the utilized criteria possess statistical parameters such as standard deviation, indicating the degree of dispersion within the respective criterion values. These parameters represent the degree of variation in the criterion values. The steps of the CRITIC method are as follows [4]:

Step 1: Construction of the decision matrix

According to the number of criteria, the number of alternatives and the evaluation of all alternative for different criteria, the decision matrix is formed as follows.

$$\begin{matrix} & C_1 \cdots C_n \\ X = & \begin{bmatrix} x_{11} & \cdots & x_{1n} \\ \vdots & \ddots & \vdots \\ x_{m1} & \cdots & x_{mn} \end{bmatrix} \end{matrix} \tag{13.1}$$

where x_{ij} the score of the alternative $i(i = 1, \ldots, m)$ is related to the criterion $j(j = 1, \ldots, n)$.

Step 2: Construction of the normalized decision matrix.

At this step, based on existing methods, the decision matrix is normalized. Naturally, all cost criterion are converted into benefit criterion and they will be treated the same.

$$\begin{cases} n_{ij} = \frac{x_{ij} - x_{Min,j}}{x_{Max,j} - x_{Min,j}}, & i = 1, ..., m;\ j \in Beneficial\ Criteria \\ n_{ij} = \frac{x_{Max,j} - x_{ij}}{x_{Max,j} - x_{Min,j}}, & i = 1, ..., m;\ j \in Cost\ Criteria \end{cases}$$

$$\begin{cases} x_{Max,j} = \underset{i=1,...,m}{Max} \{x_{ij}\},\ j = 1, ..., n; \\ x_{Min,j} = \underset{i=1,...,m}{Min} \{x_{ij}\},\ j = 1, ..., n. \end{cases} \quad (13.2)$$

Using Eq. (13.2), the normalized matrix will be as follows.

$$N = \begin{bmatrix} n_{11} & \cdots & n_{1n} \\ \vdots & \ddots & \vdots \\ n_{m1} & \cdots & n_{mn} \end{bmatrix} \quad (13.3)$$

Step 3: Calculation of the standard deviation for each Criteria.

$$\delta_j = \sqrt{\frac{\sum_{i=1}^{m} (\mathrm{n}_{ij} - \overline{\mathrm{n}}_j)^2}{m}}, \quad j = 1, \ldots, n. \quad (13.4)$$

In the given context, $\overline{\mathrm{n}}_j$ represents the mean or average.

Step 4: Creation of correlation matrix

The normalized values correlation is constructed using the Pearson correlation coefficient. This correlation matrix represents the pairwise comparisons between the criteria.

$$r_{jk} = \frac{\sum_{i=1}^{m} (\mathrm{n}_{ij} - \overline{\mathrm{n}}_j)(\mathrm{n}_{ik} - \overline{\mathrm{n}}_k)}{\sqrt{\sum_{i=1}^{m} (\mathrm{n}_{ij} - \overline{\mathrm{n}}_j)^2 \sum_{i=1}^{m} (\mathrm{n}_{ik} - \overline{\mathrm{n}}_k)^2}}, \quad j, k = 1, \ldots, n. \quad (13.5)$$

Step 5: Calculation of H index

$$H_j = \delta_j \sum_{k=1}^{n} (1 - r_{jk}), \quad j = 1, \ldots, n. \quad (13.6)$$

Step 6: Calculation of the final weight

$$w_j = \frac{H_j}{\sum\limits_{k=1}^{n} H_k}, \quad j = 1, \ldots, n. \quad (13.7)$$

From the above relationships, it can be inferred that as the variance of a criterion's values increases and its correlation with other criteria decreases, it will carry a higher weight.

13.3 Project Ranking: Evaluating and Prioritizing Projects Based on Criteria

The ranking of projects based on specific criteria is an essential process for decision-making and resource allocation. This introduction will discuss the key criteria used in project ranking, highlighting their significance in evaluating and prioritizing projects [10].

I. Production Cost (C1): One crucial criterion in project ranking is the production cost associated with each project. This cost represents the expenses incurred during the production process, including materials, labor, equipment, and overhead costs. By considering the production cost, decision-makers can assess the financial feasibility and efficiency of different projects.

II. Annual Sales (C2): Another vital criterion is the projected annual sales revenue generated by each project. This criterion reflects the expected revenue stream and market potential of a project. Projects with higher annual sales indicate greater market demand and potential profitability, making them more favorable in the ranking process.

III. Net Present Value (NPV) (C3): NPV is a financial metric that calculates the current value of future cash flows associated with a project. It considers the time value of money and helps decision-makers assess the profitability and financial viability of projects. Projects with higher NPV are generally considered more favorable, as they indicate higher returns on investment.

IV. Internal Rate of Return (IRR) (C4): IRR is another essential criterion used in project ranking. It represents the rate of return on the project's investment over its lifespan. Projects with higher IRR offer greater returns on investment, indicating their potential profitability and desirability in the ranking process.

V. Capital Return Period (C5): The capital return period refers to the time required to recover the initial investment in a project. This criterion is particularly relevant in assessing the liquidity and cash flow dynamics of projects. Projects with shorter capital return periods are often preferred, as they enable quicker recovery of invested capital.

VI. Average Profit Margin (C6): The average profit margin generated by a project annually is an important criterion in project ranking. It indicates the profitability and financial performance of the project over time. Projects with higher average profit margins demonstrate stronger financial performance and are generally prioritized in the ranking process.

VII. Labor Requirements (C7): The number of workers required for a project is a practical criterion considered in project ranking. Labor requirements impact the project's operational efficiency, resource allocation, and overall cost structure. Projects with lower labor requirements may be more favorable, as they can result in cost savings and streamlined operations.

Project ranking involves evaluating and prioritizing projects based on various criteria. The production cost, annual sales, net present value, internal rate of return, capital return period, average profit margin, and labor requirements are key factors considered in this process. By carefully assessing these criteria, decision-makers can make informed choices and allocate resources effectively to the most promising projects. In order to prioritize the criteria for project ranking, the CRITIC method can be employed. The CRITIC (Criteria Importance Through Inter-criteria Correlation) method allows for a systematic and structured approach to assess the relative importance of each criterion. By considering the interrelationships and correlations among the criteria, the CRITIC method provides a framework to determine the significance and weightage of each criterion in the project ranking process. Through this method, decision-makers can make informed judgments and effectively prioritize the criteria to ensure a comprehensive and objective evaluation of projects.

The decision matrix is presented as follows, and in this situation, the goal is to determine weights based on the decision matrix (Tables 13.1, 13.2 and 13.3).

Table 13.1 Decision matrix of the project ranking

Project	C1	C2	C3	C4	C5	C6	C7
P1	126.8	376	57.8	33.6	3.1	45.7	59
P2	190	265	49	30.7	3	61.2	72
P3	122.8	321	97.3	29.7	3	75.3	66
P4	110.4	276	67.4	26.2	3.2	49.1	59
P5	145.67	234	176	30.4	3.5	54.3	57

Table 13.2 Normalized decision matrix of the project ranking

Project	C1	C2	C3	C4	C5	C6	C7
P1	0.7940	1.0000	0.0693	1.0000	0.8000	0.0000	0.8667
P2	0.0000	0.2183	0.0000	0.6081	1.0000	0.5236	0.0000
P3	0.8442	0.6127	0.3803	0.4730	1.0000	1.0000	0.4000
P4	1.0000	0.2958	0.1449	0.0000	0.6000	0.1149	0.8667
P5	0.5569	0.0000	1.0000	0.5676	0.0000	0.2905	1.0000
Average	0.6390	0.4254	0.3189	0.5297	0.6800	0.3858	0.6267
Standard deviation	0.3497	0.3481	0.3639	0.3202	0.3709	0.3543	0.3738

Table 13.3 Correlation coefficients

	C1	C2	C3	C4	C5	C6	C7
C1	1.0000	0.4070	0.0930	−0.3270	−0.1440	−0.1800	0.6840
C2	0.4070	1.0000	−0.5260	0.5310	0.5610	−0.0970	0.0540
C3	0.0930	−0.5260	1.0000	−0.0480	−0.8320	0.1120	0.4980
C4	−0.3270	0.5310	−0.0480	1.0000	0.1350	−0.1160	−0.0610
C5	−0.1440	0.5610	−0.8320	0.1350	1.0000	0.4270	−0.7460
C6	−0.1800	−0.0970	0.1120	−0.1160	0.4270	1.0000	−0.6320
C7	0.6840	0.0540	0.4980	−0.0610	−0.7460	−0.6320	1.0000

Table 13.4 The H index and the final weights

	H	W
C1	1.9121	0.1269
C2	1.7645	0.1171
C3	2.439	0.1619
C4	1.8845	0.1251
C5	2.4479	0.1625
C6	2.2977	0.1525
C7	2.3189	0.1539

In this step, the index H for each criterion is obtained by multiplying the standard deviation of each criterion by the sum of the corresponding row of Table(13.3). H index and the final weights are presented in Table 13.4.

To calculate the H index and the final weights, you can follow these steps:

Calculate the index H: Multiply the standard deviation of each criterion by the sum of the corresponding row in the decision matrix. This will give you the index H for each criterion. Calculate the final weights: Divide each index H by the sum of all index H values. This normalization step will provide you with the final weights for each criterion. The index H represents the contribution of each criterion to the decision-making process, taking into account both the standard deviation and the relative importance of the criterion. The final weights reflect the relative significance of each criterion in the overall decision.

13.4 Fuzzy CRITIC Method

In the realm of decision-making, navigating through complex and uncertain environments often requires us to assign weights to various criteria. These weights play a crucial role in determining the importance or significance of each criterion within a given context. However, when faced with imprecise or vague information, traditional methods may fall short in providing accurate weight assignments. Enter the world of fuzzy environments, where the boundaries between criteria blur and uncertainties abound. In such settings, a powerful method known as CRITIC (Criteria Importance Through Inter-criteria Correlation) emerges as a guiding light. CRITIC offers a systematic approach to determining the weight of criteria, taking into account the inherent fuzziness and imprecision of the decision-making landscape. Unlike conventional decision-making techniques that rely on crisp and well-defined parameters, CRITIC embraces the uncertainty and ambiguity present in real-world scenarios. It acknowledges that criteria are not isolated entities but interconnected and interdependent, affecting each other in subtle yet significant ways.

By capturing these interrelationships, CRITIC enables a more nuanced understanding of the criteria's relative importance. The essence of CRITIC lies in leveraging past knowledge and experiences while embracing the potential for indefinite outcomes. It embraces the idea that decisions are not static, but rather evolve over time, shaped by the dynamic nature of our surroundings. By harnessing historical data and employing sophisticated algorithms, CRITIC empowers decision-makers to discern patterns, uncover hidden correlations, and extract meaningful insights from the vast sea of information available. Through a delicate interplay of past verbs and indefinite sentences, CRITIC delves into the depths of uncertainty and emerges with weighted criteria that reflect the true essence of the fuzzy environment. It recognizes that decisions are seldom black or white, but rather shades of gray, and offers a flexible framework for accommodating these intricacies. So, as we embark on this journey into the world of CRITIC, let us embrace the ambiguity and venture into the uncharted territory of fuzzy decision-making. With CRITIC as our compass, we have the opportunity to unlock new perspectives, overcome the challenges posed by uncertainty, and make informed choices that steer us toward success in the face of the unknown [3, 6, 8, 11].

Below are the steps of the fuzzy CRITIC method described, followed by its application in a decision-making problem.

Step 1. Identify the criteria: Determine the relevant criteria for the decision-making problem. These criteria should reflect the important factors that need to be considered.

Step 2. Define linguistic variables: For each criterion, define linguistic variables that represent different degrees or levels of that criterion. These linguistic variables should capture the uncertainty and fuzziness associated with the criteria.

Step 3. Establish membership functions: Assign membership functions to each linguistic variable. These functions describe the degree of membership or truthfulness for each linguistic term associated with the variable.

Step 4. Construct the decision matrix
According to the number of criteria, the number of alternatives and the evaluation of all alternative for different criteria, the decision matrix is formed as follows.

$$\tilde{x}_{ij} = (x_{ij}^{L}, x_{ij}^{M_1}, x_{ij}^{M_2}, x_{ij}^{U})$$

$$\tilde{X} = \begin{bmatrix} \tilde{x}_{11} & \dots & \tilde{x}_{1n} \\ \vdots & \ddots & \dots \\ \tilde{x}_{m1} & \dots & \tilde{x}_{mn} \end{bmatrix} \tag{13.8}$$

where x_{ij} is the score of the alternative $i(i = 1, \dots, m)$ related to the criterion $j(j = 1, \dots, n)$.

Step 5: Construct the normalized decision matrix

Normalizing the fuzzy decision matrix would result in the following representation, which is displayed using $\tilde{N}$:

$$\tilde{n}_{ij} = (n_{ij}^{L}, n_{ij}^{M_1}, n_{ij}^{M_2}, n_{ij}^{U}) = \left(\frac{x_{ij}^{L}}{x_j^{Max}}, \frac{x_{ij}^{M_1}}{x_j^{Max}}, \frac{x_{ij}^{M2}}{x_j^{Max}}, \frac{x_{ij}^{U}}{x_j^{Max}} \right),$$

$$i = 1, \dots, m;\ j \in Benfitial\ Criteria$$

$$\tilde{n}_{ij} = (n_{ij}^{L}, n_{ij}^{M_1}, n_{ij}^{M_2}, n_{ij}^{U}) = \left(\frac{x_j^{Min}}{x_{ij}^{U}}, \frac{x_j^{Min}}{x_{ij}^{M_2}}, \frac{x_j^{Min}}{x_{ij}^{M_1}}, \frac{x_j^{Min}}{x_{ij}^{L}} \right),$$

$$i = 1, \dots, m;\ j \in Cost\ Criteria$$

$$\tilde{N} = \begin{bmatrix} \tilde{n}_{11} & \dots & \tilde{n}_{1n} \\ \vdots & \ddots & \dots \\ \tilde{n}_{m1} & \dots & \tilde{n}_{mn} \end{bmatrix} \tag{13.9}$$

In which $x_j^{Max} = \underset{i}{Max}\left\{x_{ij}^{U}\right\}; (j = 1, \dots, n)$ and $x_j^{Min} = \underset{i}{Min}\left\{x_{ij}^{L}\right\}, (j = 1, \dots, n)$.

Step 6: Calculate fuzzy standard deviation for each Criteria

$$\begin{aligned} \tilde{\delta}_j &= \sqrt{\frac{\sum_{i=1}^{m} (\tilde{n}_{ij} - \tilde{\bar{n}}_j)^2}{m}} \\ &= \frac{1}{\sqrt{m}} \left(\sqrt{\sum_{i=1}^{m} (n_{ij}^{L} - \bar{n}_j^{L})^2}, \sqrt{\sum_{i=1}^{m} (n_{ij}^{M_1} - \bar{n}_j^{M_1})^2}, \sqrt{\sum_{i=1}^{m} (n_{ij}^{M_2} - \bar{n}_j^{M_2})^2}, \sqrt{\sum_{i=1}^{m} (n_{ij}^{U} - \bar{n}_j^{U})^2} \right) \\ &= \left(\frac{1}{\sqrt{m}} \cdot \sqrt{\sum_{i=1}^{m} (n_{ij}^{L} - \bar{n}_j^{L})^2}, \frac{1}{\sqrt{m}} \cdot \sqrt{\sum_{i=1}^{m} (n_{ij}^{M_1} - \bar{n}_j^{M_1})^2}, \frac{1}{\sqrt{m}} \cdot \sqrt{\sum_{i=1}^{m} (n_{ij}^{M_2} - \bar{n}_j^{M_2})^2}, \frac{1}{\sqrt{m}} \cdot \sqrt{\sum_{i=1}^{m} (n_{ij}^{U} - \bar{n}_j^{U})^2} \right), \\ & j = 1, \dots, n. \end{aligned} \tag{13.10}$$

In the given context, $\tilde{\bar{n}}_j = (\bar{n}_{ij}^{L}, \bar{n}_{ij}^{M_1}, \bar{n}_{ij}^{M_2}, \bar{n}_{ij}^{U})$ represents the fuzzy mean for each criterion, and the square root of the fuzzy number is calculated accordingly. Specifically, the square root of the fuzzy number [7] and [5] would be computed.

Step 7: Construct the correlation matrix

Normalized values correlation is constructed using Pearson correlation coefficient

$$\tilde{r}_{jk} = \frac{\sum_{i=1}^{m}(\tilde{\mathrm{n}}_{ij}-\bar{\tilde{\mathrm{n}}}_{j})(\mathrm{n}_{ik}-\bar{\tilde{\mathrm{n}}}_{k})}{\sqrt{\sum_{i=1}^{m}(\tilde{\mathrm{n}}_{ij}-\bar{\tilde{\mathrm{n}}}_{j})^2(\tilde{\mathrm{n}}_{ik}-\bar{\tilde{\mathrm{n}}}_{k})^2}} = (r_{jk}^{L}, r_{jk}^{M_1}, r_{jk}^{M_2}, r_{jk}^{U});\ j,k=1,\ldots,n.$$

$$r_{jk}^{L} = \frac{\sum_{i=1}^{m}\left((\mathrm{n}_{ij}^{L}-\bar{\mathrm{n}}_{j}^{L})(\mathrm{n}_{ik}^{L}-\bar{\mathrm{n}}_{k}^{L})\right)}{\sqrt{\sum_{i=1}^{m}\left((\mathrm{n}_{ij}^{U}-\bar{\mathrm{n}}_{j}^{U})^2\right)\sum_{i=1}^{m}\left((\mathrm{n}_{ik}^{U}-\bar{\mathrm{n}}_{k}^{U})^2\right)}},$$

$$r_{jk}^{M_1} = \frac{\sum_{i=1}^{m}\left((\mathrm{n}_{ij}^{M_1}-\bar{\mathrm{n}}_{j}^{M_1})(\mathrm{n}_{ik}^{M_1}-\bar{\mathrm{n}}_{k}^{M_1})\right)}{\sqrt{\sum_{i=1}^{m}\left((\mathrm{n}_{ij}^{M_2}-\bar{\mathrm{n}}_{j}^{M_2})^2\right)\sum_{i=1}^{m}\left((\mathrm{n}_{ik}^{M_2}-\bar{\mathrm{n}}_{k}^{M_2})^2\right)}},$$

$$r_{jk}^{M_2} = \frac{\sum_{i=1}^{m}\left((\mathrm{n}_{ij}^{M_2}-\bar{\mathrm{n}}_{j}^{M_2})(\mathrm{n}_{ik}^{M_2}-\bar{\mathrm{n}}_{k}^{M_2})\right)}{\sqrt{\sum_{i=1}^{m}\left((\mathrm{n}_{ij}^{M_1}-\bar{\mathrm{n}}_{j}^{M_1})^2\right)\sum_{i=1}^{m}\left((\mathrm{n}_{ik}^{M_1}-\bar{\mathrm{n}}_{k}^{M_1})^2\right)}},$$

$$r_{jk}^{U} = \frac{\sum_{i=1}^{m}\left((\mathrm{n}_{ij}^{U}-\bar{\mathrm{n}}_{j}^{U})(\mathrm{n}_{ik}^{U}-\bar{\mathrm{n}}_{k}^{U})\right)}{\sqrt{\sum_{i=1}^{m}\left((\mathrm{n}_{ij}^{L}-\bar{\mathrm{n}}_{j}^{L})^2\right)\sum_{i=1}^{m}\left((\mathrm{n}_{ik}^{L}-\bar{\mathrm{n}}_{k}^{L})^2\right)}}, \tag{13.11}$$

Step 8: Calculate H index

$$\tilde{H}_j = (H_j^{L}, H_j^{M_1}, H_j^{M_2}, H_j^{U}) = \tilde{\delta}_j \sum_{k=1}^{n}(1-\tilde{r}_{jk})$$
$$= \left(\delta_j^{L}\sum_{k=1}^{n}(1-r_{jk}^{L}),\ \delta_j^{M_1}\sum_{k=1}^{n}(1-r_{jk}^{M_1}),\ \delta_j^{M_2}\sum_{k=1}^{n}(1-r_{jk}^{M_2}),\ \delta_j^{U}\sum_{k=1}^{n}(1-r_{jk}^{U})\right),$$
$$j=1,\ldots,n. \tag{13.12}$$

Therefore, after this step, the components of the trapezoidal number may not be in order. In this stage, we arrange them in ascending order.

Step 9: Calculate the final weight

$$\tilde{w}_j = \frac{\tilde{H}_j}{\sum_{k=1}^{n}\tilde{H}_k} = \left(\frac{H_j^{L}}{\sum_{k=1}^{n}H_k^{U}}, \frac{H_j^{M_1}}{\sum_{k=1}^{n}H_k^{U}}, \frac{H_j^{M_2}}{\sum_{k=1}^{n}H_k^{U}}, \frac{H_j^{U}}{\sum_{k=1}^{n}H_k^{U}}\right),\quad j=1,\ldots,n. \tag{13.13}$$

It is necessary to mention that to handle fuzzy numbers, one can calculate the square root of a fuzzy trapezoidal number based on [4, 6]. For a better understanding of the subject, readers can refer to Chapter 2 of this book.

Step 10: In this step, the obtained weights from Step 9 can be combined with the weights obtained from expert(s) opinions. For this purpose, the convex combination of these weights can be used to determine the final weights. The decision-maker specifies the coefficient of this convex combination.

$$\tilde{w}_j^{final} = \lambda \tilde{w}_j^{CRITIC} + (1 - \lambda)\tilde{w}_j^{expert}; \ (0 \le \lambda \le 1); \ j = 1, ..., n. \tag{13.14}$$

13.5 Finding the Perfect Spot: Criteria for Selecting Optimal Locations for Solar Farm

As society faced the pressing challenges of rapid energy consumption and the need for sustainable practices, the importance of prioritizing the production and utilization of renewable energy sources became increasingly evident. The limited availability of fossil fuels and the environmental repercussions associated with their consumption were being grappled with by the global community. In light of these concerns, solar energy emerged as a cost-effective and clean alternative among various renewable energy sources [9].

Solar energy, characterized by its temporal and spatial variability, held immense potential in addressing the existing challenges associated with fossil fuel and nuclear energy consumption. With an average of over 280 sunny days annually, Iran, for instance, exhibited a high potential for solar energy resources. The power of the sun had applications ranging from providing warmth and hot water to residential, commercial, and industrial buildings, to generating electricity on a larger scale.

When prioritizing regions for the installation of solar farms, several criteria play a crucial role in determining the suitability of a particular location. These criteria include:

I. Temperature: The temperature levels in a region affect the efficiency and performance of solar panels. Generally, higher temperatures can lead to a decrease in solar panel efficiency, making regions with moderate temperatures more favorable.
II. Humidity: Humidity levels can impact the overall performance and durability of solar panels. Excessive humidity may increase the risk of corrosion and reduce the lifespan of the panels.
III. Relative humidity: The relative humidity of a region influences the atmospheric conditions and can affect the performance of solar panels by impacting the transmission of sunlight.
IV. Wind speed: Wind speed is an important factor to consider as high wind speeds can potentially damage solar panels. It is essential to assess the wind conditions in a region to ensure the durability and stability of the solar farms.
V. Solar radiation: The amount of solar radiation received in a particular region directly affects the energy generation potential of solar farms. Areas with higher solar radiation levels are generally more favorable for solar energy production.
VI. Population: The population density of an area can impact the electricity demand and consumption patterns. Prioritizing solar farms in regions with

Table 13.5 Criteria for Selecting Optimal Locations for Solar Farm

Criterion 1	Temperature	(In degrees Celsius)
Criterion 2	Humidity	(In percentage)
Criterion 3	Wind speed	(In meters per second)
Criterion 4	Solar radiation	(In watts per square meter)
Criterion 5	Population	(Number of individuals)
Criterion 6	Rainfall	(In millimeters)
Criterion 7	Number of days	-
Criterion 8	Distance	(In meters)
Criterion 9	Slope	(In percentage)
Criterion 10	Distance from substations	(In kilometers)
Criterion 11	Distance From roads	(In kilometers)

higher population densities can help meet the energy requirements of a larger number of people.

VII. Rainfall: The amount and distribution of rainfall in a region can impact the overall solar energy generation. Regions with low rainfall levels and limited cloud cover are often more suitable for solar energy production.

VIII. Number of days with dust pollution: Dust particles in the air can reduce the efficiency of solar panels by obstructing the sunlight. Regions with fewer days of dust pollution are preferable for optimal solar energy production.

IX. Height above sea level: The altitude or height above sea level can affect the solar irradiance levels. Higher altitudes often receive more intense sunlight, making them potentially advantageous for solar energy generation.

X. Slope of the land: The slope or incline of the land can influence the positioning and alignment of solar panels. Flat or gently sloping terrains are usually more favorable for efficient solar panel installation.

XI. Distance from substations: The proximity of solar farms to existing electrical substations reduces transmission losses and makes it easier to connect the generated electricity to the grid.

XII. Distance from roads: Easy access to transportation infrastructure, such as roads, facilitates the installation and maintenance of solar farms. Proximity to roads simplifies logistics and reduces costs.

Table 13.5 shows all criteria and their unit. Considering these criteria can help decision-makers prioritize and select suitable locations for solar farms, maximizing their potential for electricity generation and contributing to a sustainable energy future.

Table 13.6 shows the decision matrix that the elements are trapezoidal fuzzy numbers. Applying Eq. (13.9) the normalized decision matrix is illustrated in Table 13.7.

Table 13.6 Decision matrix of Selecting Optimal Location for Solar Farm

	Location 1				Location 2				Location 3				Location 4			
	L	M1	M2	U	L	M1	M2	U	L	M1	M2	U	L	M1	M2	U
C1	22.05	22.09	23.58	24.88	17.62	17.67	18.09	18.49	25.02	25.11	26.83	27.01	23.24	23.36	25.23	26.69
C2	25.43	25.68	27.25	27.6	34.55	34.77	35.22	36.1	21.9	22.05	22.82	23.64	42.12	42.27	42.72	43.36
C3	4.81	4.84	5.16	5.28	5.88	5.92	6.29	6.48	6.76	6.77	7.04	7.16	5.71	5.76	5.89	6.05
C4	552.94	556.22	588.38	625.35	686.87	690.05	712.87	721.26	597.72	599.6	601.83	622.2	660.26	661.46	695.63	734.64
C5	1816.06	1830.05	1873.65	1997.51	8214.87	8276.96	8692.42	8939.17	4566.71	4570.66	4608.14	4903.64	5112.2	5140.08	5153.04	5561.35
C6	126.5	126.59	128.37	135.12	191.98	193.42	196.17	196.84	232.33	233.69	245.3	254.85	37.3	37.43	38.16	39.1
C7	126.21	126.38	135.82	140.98	119.1	120.04	122.92	124.46	10.46	10.54	10.6	10.82	61.76	61.76	63.09	66.25
C8	1436	1438.17	1458.21	1506.43	1288	1292.08	1325.37	1453.14	862	870.56	896.33	929.75	1683	1695.99	1749.81	1909.59
C9	2.1	2.1	2.23	2.24	0.2	0.2	0.2	0.2	1.74	1.74	1.82	1.93	2.54	2.55	2.63	2.7
C10	10	10.02	10.77	11.22	18	18.02	18.58	19.28	11	11.03	11.87	12.09	33	33.1	33.3	34.63
C11	0.88	0.88	0.91	0.94	0.79	0.79	0.8	0.85	0.11	0.11	0.11	0.12	0.6	0.61	0.63	0.67

Table 13.7 Normalized decision matrix of Selecting Optimal Location for Solar Farm

	Location 1				Location 2				Location 3				Location 4			
	L	M1	M2	U	L	M1	M2	U	L	M1	M2	U	L	M1	M2	U
C1	0.637	0.6721	0.7854	0.7989	0.8971	0.9172	0.9708	1	0.577	0.5809	0.6804	0.7042	0.5689	0.6018	0.7219	0.758
C2	0.6713	0.68	0.7837	0.8614	0.5633	0.5774	0.5958	0.6339	0.8323	0.8622	0.9373	1	0.4822	0.4894	0.5023	0.52
C3	0.629	0.666	0.7767	0.7946	0.7695	0.8207	0.9481	0.9769	0.8846	0.8944	0.9842	1	0.7473	0.8125	0.8556	0.8801
C4	0.6897	0.7307	0.8363	0.8889	0.8568	0.8965	0.9706	0.9821	0.7456	0.769	0.7762	0.8025	0.8236	0.8386	0.9469	1
C5	0.822	0.8763	0.9285	1	0.1776	0.1826	0.2055	0.2211	0.3631	0.3863	0.3943	0.3977	0.3102	0.3348	0.3369	0.3552
C6	0.2686	0.2828	0.2927	0.2948	0.1739	0.1745	0.1807	0.1943	0.1298	0.1349	0.1517	0.1605	0.8985	0.9206	0.9654	1
C7	0.0664	0.069	0.0818	0.0829	0.0759	0.0769	0.0815	0.0879	0.8995	0.9176	0.9303	1	0.1531	0.1607	0.1694	0.1695
C8	0.6725	0.6826	0.7064	0.7298	0.6032	0.6223	0.6624	0.7262	0.4037	0.4438	0.4766	0.4944	0.7881	0.849	0.9163	1
C9	0.0818	0.0822	0.0943	0.0954	0.9353	0.9359	0.9455	1	0.0945	0.1004	0.1109	0.1153	0.0696	0.0715	0.0771	0.0789
C10	0.796	0.8296	0.985	1	0.4905	0.5089	0.5482	0.5556	0.7304	0.7442	0.8853	0.9091	0.2788	0.29	0.2943	0.303
C11	0.11	0.1136	0.1223	0.1236	0.1247	0.1322	0.1351	0.1377	0.8744	0.9428	0.9513	1	0.1395	0.148	0.1656	0.1813

Table 13.8 presents the correlation of C_{11} obtained by Eq. (13.11). The fuzzy H index calculated by Eq. (13.12) is shown in Table 13.9. Ultimately, using Eq. (13.13) the final weight of all criteria is presented in Table 13.10.

Moreover, with respect to step 10, considering the weights obtained from expert(s) opinions, it is possible to cumulate the results by Eq. (13.14), which is a convex combination of obtained Wight by CRITIC method and the weights attained from expert(s) opinions.

Table 13.8 Coloration for C11

	L	M1	M2	U
C1	−0.38	−0.49	−0.48	−0.52
C2	0.52	0.6	0.66	0.72
C3	0.8	0.67	0.63	0.62
C4	−0.2	−0.26	−0.72	−0.82
C5	−0.11	−0.13	−0.15	−0.22
C6	−0.33	−0.37	−0.36	−0.37
C7	0.79	0.87	0.88	1
C8	−0.6	−0.62	−0.63	−0.75
C9	−0.26	−0.28	−0.28	−0.32
C10	0.27	0.27	0.37	0.4
C11	0.77	0.9	0.9	1

Table 13.9 The fuzzy H index of Selecting Optimal Location for Solar Farm

	L	M1	M2	U
C1	1.206	1.227	1.408	1.43
C2	1.351	1.441	1.749	1.968
C3	0.798	0.82	0.841	0.839
C4	0.664	0.683	0.903	0.911
C5	2.926	3.146	3.171	3.452
C6	3.742	3.789	3.987	4.062
C7	3.396	3.52	3.667	3.97
C8	1.72	1.755	1.876	2.131
C9	3.78	3.785	3.852	4.121
C10	2.216	2.308	2.902	2.976
C11	3.159	3.454	3.581	3.792

Table 13.10 Fuzzy weights of Selecting Optimal Location for Solar Farm

	L	M1	M2	U
w_1	0.0407	0.0414	0.0475	0.0482
w_2	0.0456	0.0486	0.0590	0.0664
w_3	0.0269	0.0277	0.0284	0.0283
w_4	0.0224	0.0230	0.0305	0.0307
w_5	0.0987	0.1061	0.1069	0.1164
w_6	0.1262	0.1278	0.1345	0.1370
w_7	0.1145	0.1187	0.1237	0.1339
w_8	0.0580	0.0592	0.0633	0.0719
w_9	0.1275	0.1276	0.1299	0.1390
w_{10}	0.0747	0.0778	0.0979	0.1004
w_{11}	0.1065	0.1165	0.1208	0.1279

13.6 Conclusion

In conclusion, this chapter emphasized the significance of weighting in decision-making processes and provided a practical implementation of the CRITIC (Criteria Importance Through Inter-criteria Correlation) method. Weighting criteria allows for a systematic evaluation of alternatives and enables decision-makers to prioritize factors based on their relative importance. By applying the CRITIC method to a real-world example, we witnessed how it aids in deriving objective weights for criteria. This approach considers the interrelationships and dependencies among criteria, resulting in more accurate and reliable weight assignments. The CRITIC method offers a robust framework that mitigates potential biases and subjectivity in the weighting process, leading to informed and well-balanced decisions. Furthermore, this chapter delved into the fuzzy CRITIC method, which extends the applicability of the CRITIC approach to scenarios where uncertainties and imprecisions exist. Fuzzy logic enables the representation of vague or uncertain information, ensuring a more comprehensive assessment of criteria weights in real-world decision problems. Step by step, we examined the implementation of the fuzzy CRITIC method using a practical example. By incorporating fuzzy sets and linguistic variables, the fuzzy CRITIC method accounted for the inherent uncertainty and ambiguity in decision-making. The combination of the CRITIC method and its fuzzy variant presented in this chapter provides a powerful toolkit for decision-makers to effectively handle multi-criteria decision-making problems. By incorporating both the inter-criteria correlations and fuzzy logic, decision-makers can gain valuable insights and make informed choices that consider various factors and their interdependencies. In summary, the chapter highlighted the importance of weighting in decision-making, showcased the application of the CRITIC method, and demonstrated the extended capabilities of the fuzzy CRITIC method. By employing these methods, decision-makers can enhance the objectivity, accuracy, and reliability of their decision processes, leading to more robust and effective outcomes in a wide range of real-world scenarios.

Acknowledgement A special thanks to the Iranian DEA society for their unwavering spiritual support and consensus in the writing of this book. Your invaluable support has been truly remarkable, and we are deeply grateful for the opportunity to collaborate with such esteemed professionals.

References

1. Asante, D., Ampah, J. D., Afrane, S., Adjei-Darko, P., Asante, B., Fosu, E., & Amoh, P. O. (2022). Prioritizing strategies to eliminate barriers to renewable energy adoption and development in Ghana: A CRITIC-fuzzy TOPSIS approach. *Renewable Energy, 195*, 47–65.
2. Diakoulaki, D., Mavrotas, G., & Papayannakis, L. (1995). Determining objective weights in multiple criteria problems: The critic method. *Computers & Operations Research, 22*(7), 763–770.
3. Mitrović Simić, J., Stević, Ž, Zavadskas, E. K., Bogdanović, V., Subotić, M., & Mardani, A. (2020). A novel CRITIC-Fuzzy FUCOM-DEA-Fuzzy MARCOS model for safety evaluation of road sections based on geometric parameters of road. *Symmetry, 12*(12), 2006.
4. Mohamadghasemi, A., Hadi-Vencheh, A., & Hosseinzadeh Lotfi, F. (2020). The multiobjective stochastic CRITIC–TOPSIS approach for solving the shipboard crane selection problem. *International Journal of Intelligent Systems, 35*(10), 1570–1598.
5. Mondal, S. P., & Roy, T. K. (2014). Non-linear arithmetic operation on generalized triangular intuitionistic fuzzy numbers. *Notes on Intuitionistic Fuzzy Sets, 20*(1), 9–19.
6. Mishra, A. R., Rani, P., & Pandey, K. (2022). Fermatean fuzzy CRITIC-EDAS approach for the selection of sustainable third-party reverse logistics providers using improved generalized score function. *Journal of Ambient Intelligence and Humanized Computing*, 1–17.
7. Puri, J., & Yadav, S. P. (2013). A concept of fuzzy input mix-efficiency in fuzzy DEA and its application in the banking sector. *Expert Systems with Applications, 40*(5), 1437–1450.
8. Rostamzadeh, R., Ghorabaee, M. K., Govindan, K., Esmaeili, A., & Nobar, H. B. K. (2018). Evaluation of sustainable supply chain risk management using an integrated fuzzy TOPSIS-CRITIC approach. *Journal of Cleaner Production, 175*, 651–669.
9. Wu, Y., Deng, Z., Tao, Y., Wang, L., Liu, F., & Zhou, J. (2021). Site selection decision framework for photovoltaic hydrogen production project using BWM-CRITIC-MABAC: A case study in Zhangjiakou. *Journal of Cleaner Production, 324*, 129233.
10. Xiao, L., Chen, Z. S., Zhang, X., Chang, J. P., Pedrycz, W., & Chin, K. S. (2020). Bid evaluation for major construction projects under large-scale group decision-making environment and characterized expertise levels. *International Journal of Computational Intelligence Systems, 13*(1), 1227–1242.
11. Yazdani, M., Torkayesh, A. E., Stević, Ž, Chatterjee, P., Ahari, S. A., & Hernandez, V. D. (2021). An interval valued neutrosophic decision-making structure for sustainable supplier selection. *Expert Systems with Applications, 183*, 115354.

Chapter 14
The Multi-Objective Optimization Ratio Analysis (MOORA) in Uncertainty Environment

14.1 Introduction

The MOORA method has emerged as a prominent multi-objective optimization technique that has garnered significant attention across diverse fields. In their seminal work [2], Brauers and Zavadskas presented the MOORA method as a powerful combination of three fundamental components: the ratio system, reference point, and the full multiplicative form. Together, these components form the MULTIMOORA framework, which has been widely recognized for its effectiveness in addressing multi-objective optimization problems.

At the heart of the MOORA method lies its ability to provide robust rankings for alternatives through the utilization of dominance analysis, as proposed by Brauers and Zavadskas [4]. By considering multiple ranking approaches, the MULTIMOORA technique ensures a comprehensive and reliable assessment, satisfying the essential conditions for robustness in multi-objective optimization. The versatility and applicability of the MOORA method have attracted researchers from various domains. Notable studies conducted by Jagadish and Ray [10], Kumar and Ray [13], Aytaç et al. [1], Brauers et al. [3, 5], Karande et al. [11], Kracka et al. [12], Kumar and Ray [14], and Tian et al. [18] have successfully applied the MOORA method to rank alternatives and explore its practical implications in real-world scenarios.

In this chapter, our focus deepens as we delve into the MOORA method within the fuzzy environment. We will explore the underlying principles of the method and shed light on its practical implications. Specifically, we will investigate the MOORA method's effectiveness in solving complex multi-objective optimization problems and examine various real-world applications where it has been successfully employed.

F. Hosseinzadeh Lotfi et al., *Fuzzy Decision Analysis: Multi Attribute Decision Making Approach*, Studies in Computational Intelligence 1121,
https://doi.org/10.1007/978-3-031-44742-6_14

By gaining a comprehensive understanding of the MOORA method's strengths and limitations, we can unlock valuable insights into its potential for decision-making and optimization in diverse contexts. This chapter aims to provide readers with a thorough exploration of the MOORA method within the fuzzy environment, enabling them to harness its capabilities and leverage its benefits in their own research and practical endeavors.

14.2 The MOORA and MOOSRA Methods

There are three approaches in MOORA method: the ratio system, reference point and full multiplicative form. In this section, first the ratio system is expressed by the following 4 steps, then the reference point and full multiplicative form is explained. Next, to integrate these three methods, it is suggested to use the preferential voting. Ultimately, the MOOSRA method is also declared. [9]

Ranking "m" alternatives $\{A_1, \ldots, A_m\}$ with respect to "n" criteria $\{C_1, \ldots, C_n\}$, the MOORA method is summarized as follows:

Step 1: Constructing the decision matrix.

If x_{ij} denotes the performance of ith alternative and jth criterion, then the decision matrix is equal to:

$$X = \begin{array}{c} \\ A_1 \\ A_2 \\ \vdots \\ A_m \end{array} \overset{\begin{array}{cccc} C_1 & C_2 & \ldots & C_n \end{array}}{\begin{bmatrix} x_{11} & x_{12} & \ldots & x_{1n} \\ x_{21} & x_{22} & \ldots & x_{2n} \\ \vdots & \vdots & \ddots & \vdots \\ x_{m1} & x_{m2} & \ldots & x_{mn} \end{bmatrix}} = \left[x_{ij}\right]_{m \times n} \tag{14.1}$$

Step 2: Normalizing the decision matrix.

Using Eq. (14.2) the normalized decision matrix is as follows:

$$R = \begin{bmatrix} r_{11} & r_{12} & \ldots & r_{1n} \\ r_{21} & r_{22} & \ldots & r_{2n} \\ \vdots & \vdots & \ddots & \vdots \\ r_{m1} & r_{m2} & \ldots & r_{mn} \end{bmatrix} = \left[r_{ij}\right]_{m \times n}$$

where

$$r_{ij} = \frac{x_{ij}}{\sqrt{\sum_{i=1}^{m} x_{ij}^2}}, \; j = 1, \ldots, n; i = 1, \ldots, m. \tag{14.2}$$

Step 3: Calculating the Overall performance score.

The overall performance score for each alternative is equal to the difference of the total performance of beneficial criteria and cost criteria. It obtains by Eq. (14.3)

$$RS_i = \sum_{j \in J^+} r_{ij} - \sum_{j \in J^-} r_{ij};\ i = 1, \ldots, m. \tag{14.3}$$

where J^+ shows the set of benefit criteria and J^- indicates the set of cost criteria.

Considering $w = (w_1, \ldots, w_j, \ldots, w_n)$ as the weight vector of criteria, Eq. (14.3) can be reformulated as bellow:

$$RS_i = \sum_{j \in J^+} w_j r_{ij} - \sum_{j \in J^-} w_j r_{ij};\ i = 1, \ldots, m. \tag{14.4}$$

Step 4: Ranking the alternatives.

The alternatives are ranked based on descending order of the Overall performance score $RS_i; i = 1, \ldots, m$.

The above four steps explained the **ratio system approach**.

Step 5: Calculating the reference point.

The reference point approach is as follows:

After attaining the normalized decision matrix in second step, Let r_j^* denotes the most desirable performance of jth criterion among all alternatives obtained by Eq. (14.5) that is called the reference point.

$$r_j^* = \begin{cases} \max_i \{r_{ij}\};\ j \in J^+ \\ \min_i \{r_{ij}\};\ j \in J^- \end{cases} \tag{14.5}$$

where J^+ shows the set of benefit criteria and J^- indicates the set of cost criteria.

Reference point score is the minimum total deviation of alternatives from the reference points. Applying the Tchebycheff norm, the reference point score is equal to:

$$RP_i = \max_j \left| r_j^* - r_{ij} \right|;\ i = 1, \ldots, m \tag{14.6}$$

The alternatives are ranked based on ascending order of the Reference point score $RP_i; i = 1, \ldots, m$.

Considering $w = \left(w_1, \ldots, w_j, \ldots, w_n\right)$ as the weight vector of criteria, Eq. (14.6) can be reformulated as bellow:

$$RP_i = \max_j \left| w_j r_j^* - w_j r_{ij} \right|;\ i = 1, \ldots, m \tag{14.7}$$

The alternative with the minimum total deviation from the reference points is ranked first. Hence, the less RP_i the better rank.

Step 6. Computing the full multiplicative form.

The full multiplicative form was first developed by Miller and Starr [16]. It is nonlinear and also, applicable when the decision makers consider both cost and profit criteria.

The Overall utility of alternatives is as bellow:

$$MF_i = \prod_{j=1}^{n} x_{ij};\ i = 1, \ldots, m. \tag{14.8}$$

To combine benefit and cost criteria, the overall utility equals to Eq. (14.9)

$$MF_i = \frac{\prod_{j \in J^+} \left(x_{ij}\right)^{w_j}}{\prod_{j \in J^-} \left(x_{ij}\right)^{w_j}};\ i = 1, \ldots, m. \tag{14.9}$$

where J^+ shows the set of benefit criteria and J^- indicates the set of cost criteria.

The higher the value of MF_i, the better the rank.

If $x_{ij} = 0$, withdrawal it from the decision matrix.

Step 7. Aggregating the results.

It should be noted that after calculating these three procedures, three ranking vectors are obtained. In many cases, the rank of each alternative in the three approaches is not the same. So, the issue of aggregating the ranks is raised here [11].

As an approach, using preferential voting system is suggested [18]. That, each of the three indicators can be considered as a voter who ranks the alternatives. Soltanifar et al. [16] investigated and developed preferential voting methods from the perspective of data envelopment analysis.

Assuming each of three procedures as voters who rank the alternatives, separately. Then applying on of the ranking methods of preferential voting, the final rank is obtained.

Equation (14.10) is one of the preferential voting methods known as Cook et al. [6] method. Which is applied to aggregate three above mentioned procedures to obtain the final ranking.

$$
\begin{aligned}
Z_p &= Max \sum_{r=1}^{k} w_r v_{rp} \\
s.t. &\sum_{r=1}^{k} w_r v_{rj} \le 1, \qquad j = 1, \ldots, n \\
&w_r - w_{r+1} \ge d(r, \varepsilon), \qquad r = 1, \ldots, k-1 \\
&w_k \ge d(k, \varepsilon)
\end{aligned}
\tag{14.10}
$$

The above model is a linear programming problem. Where $d(r, \varepsilon)$ is called the intensity discrimination function and it is predetermined by the decision maker. In this chapter, the intensity discrimination function is considered as $d(r, \varepsilon) = \frac{\varepsilon}{r}$. To read more about the intensity discrimination function and different models of preferential voting, refer to Soltanifar et al. [14].

The MOOSRA method:

The MOORA and MOOSRA methods are similar expect that the former uses the multiplication operator, which is a non-compensatory operator while the latter uses the sum to calculate the overall performance score, that is a compensatory operator. MOOSRA method is less sensitive to large variation in the values of the criteria. The MOOSRA was first presented by Das et al. [7]. This method uses ratio of the sum of beneficial criteria to sum of normalized cost criteria.

The overall performance score is as follows:

$$
MS_i = \frac{\sum_{j \in J^+} x_{ij}}{\sum_{j \in J^-} x_{ij}}; \; i = 1, \ldots, m. \tag{14.11}
$$

Considering $w = (w_1, \ldots, w_j, \ldots, w_n)$ as the weight vector of criteria, Eq. (14.11) can be reformulated to Eq. (14.12).

$$
MS_i = \frac{\sum_{j \in J^+} w_j x_{ij}}{\sum_{j \in J^-} w_j x_{ij}}; \; i = 1, \ldots, m. \tag{14.12}
$$

The alternative with the highest overall performance score MS_i is the best.

14.3 Numerical Example of the MOORA Method

The evaluation of university faculty members plays a vital role in maintaining and enhancing the quality of education and research within academic institutions. To ensure that faculty members meet the desired standards and contribute effectively to the development of their respective fields, evaluation methods are employed. In this section, we explore the utilization of the MULTIMOORA method to evaluate seven members of the academic faculty within the industrial engineering department of a university.

The evaluation process incorporates nine evaluation indicators, each providing valuable insights into different aspects of faculty members' qualifications and performance. Each of these indicators are expressed in brief to gain a comprehensive understanding of the evaluation criteria.

C1: Academic Teaching Experience—This indicator measures the number of years of academic teaching service for each faculty member. It takes into account their accumulated experience in teaching, reflecting their familiarity with pedagogical methods, classroom dynamics, and student engagement.

C2: Scientific Rank—The scientific rank of faculty members, ranging from instructor to assistant professor, associate professor, and full professor, is considered in this evaluation. This indicator acknowledges the progressive growth of faculty members in their academic careers, with higher ranks reflecting increased expertise and contributions to their fields.

C3: Years of Executive Responsibility—This indicator assesses the amount of experience faculty members have in non-scientific and academic executive roles. It takes into account the years of executive work, acknowledging the additional responsibilities and leadership roles undertaken by faculty members within the academic institution.

C4: Number of Research Articles—Research productivity is a crucial aspect of academic faculty evaluation. This indicator considers the number of research articles published by each faculty member, reflecting their contribution to the body of knowledge in their respective fields.

C5: Number of Participated Research Projects—In addition to individual research articles, faculty members often collaborate on research projects. This indicator measures the number of research projects in which each faculty member has actively participated, highlighting their involvement in collaborative research endeavors.

C6: Number of Books Authored, Inventions, and Discoveries—This indicator recognizes the scholarly achievements of faculty members beyond research articles. It takes into account the number of books authored, as well as any inventions or discoveries credited to each faculty member, showcasing their intellectual contributions and innovative endeavors.

C7: Number of Scientific Lectures or Educational Workshops—The dissemination of knowledge and expertise is an essential responsibility of faculty members. This indicator considers the number of scientific lectures or educational workshops conducted by each faculty member, demonstrating their commitment to sharing knowledge and nurturing the learning environment.

C8: Educational Performance Score—This indicator evaluates faculty members' educational performance based on feedback from students. At the end of each semester, students complete a questionnaire assessing various aspects, including the professor's punctuality, effective use of class time, attention to student learning and queries, mastery of the subject matter, utilization of scientific and up-to-date resources, and creativity in education. The scores obtained from this evaluation contribute to the overall assessment of faculty members' teaching abilities.

C9: Student Withdrawal Ratio—The final indicator assesses the ratio of students withdrawing from education under the supervision of each faculty member. It takes into account the number of students who discontinued their studies compared to the total number of students under the faculty member's guidance. This indicator provides insights into the faculty member's ability to effectively guide and support students throughout their academic journey.

By incorporating these nine indicators into the MULTIMOORA method, a comprehensive evaluation framework is established, enabling a holistic assessment of faculty members' qualifications, performance, and impact within the industrial engineering faculty. In the following sections, we will discuss the implementation of the MOORA method and analyze the outcomes to gain valuable insights into the faculty members' strengths and areas for improvement.

The decision matrix for evaluating the seven faculty members based on the nine mentioned criteria is displayed in Table 14.1.

Applying Eq. (14.2) to normalize the matrix and Considering the weight vector $W = (0.12, 0.08, 0.10, 0.18, 0.17, 0.15, 0.10, 0.05, 0.05)$ the weighted normalized decision matrix is illustrated in Table 14.2.

Using Eq. (14.4) the overall performance of ratio system approach is attained and the ranking is illustrated in the second column of Table 14.3. The ranking result of the reference point approach is shown in the third column. Also, the fourth column of Table 14.3 shows the ranking result of the full multiplicative form obtained by Eq. (14.9).

With respect to preferential voting method, the final ranking obtained by Eq. (14.10) is illustrated in the last column of Table 14.3.

Accordingly, A2 ranked the first and $A_2 \succ A_4 \succ A_5 \succ A_7 \succ A_3 \succ A_1 \succ A_6$.

Table 14.1 Decision matrix

	C1 (Benefit)	C2 (Benefit)	C3 (Benefit)	C4 (Benefit)	C5 (Benefit)	C6 (Benefit)	C7 (Benefit)	C8 (Benefit)	C9 (Cost)
A1	14	1	8	105	5	2	200	80	10 %
A2	25	4	10	302	25	14	410	90	9 %
A3	12	3	7	93	14	2	100	85	5 %
A4	10	4	15	245	23	10	90	85	7 %
A5	21	2	5	175	10	4	70	65	17 %
A6	15	1	3	64	5	2	50	75	24 %
A7	10	3	9	191	4	6	140	60	15 %

Table 14.2 Weighted normalized decision matrix of faculty members

	C1 (Benefit)	C2 (Benefit)	C3 (Benefit)	C4 (Benefit)	C5 (Benefit)	C6 (Benefit)	C7 (Benefit)	C8 (Benefit)	C9 (Cost)
A1	0.039261	0.01069	0.034019	0.038411	0.021831	0.015811	0.039746	0.019426	0.013634
A2	0.07011	0.042762	0.042524	0.110478	0.109154	0.11068	0.08148	0.021854	0.01227
A3	0.033653	0.032071	0.029767	0.034022	0.061126	0.015811	0.019873	0.02064	0.006817
A4	0.028044	0.042762	0.063786	0.089627	0.100422	0.079057	0.017886	0.02064	0.009543
A5	0.058892	0.021381	0.021262	0.064019	0.043662	0.031623	0.013911	0.015783	0.023177
A6	0.042066	0.01069	0.012757	0.023413	0.021831	0.015811	0.009937	0.018212	0.032721
A7	0.028044	0.032071	0.038272	0.069872	0.017465	0.047434	0.027823	0.014569	0.02045

Table 14.3 Ranking results

	RS_i	RP_i	MF_i	Final ranking
A1	6	5	6	6
A2	1	1	1	1
A3	5	6	5	5
A4	2	2	2	2
A5	4	3	3	3
A6	7	7	7	7
A7	3	4	4	4

14.4 Fuzzy MULTIMOORA Method Using Triangular Fuzzy Number

To rank the alternatives by Fuzzy MULTIMOORA method, first the decision matrix is formed according to the linguistic variables. Then using three approaches, fuzzy ratio system, fuzzy reference point method, and fuzzy full multiplicative form, three overall performance values and as a result three ratings are obtained.

Then applying on of the ranking methods of preferential voting, the final rank is obtained.

Step 1: Constructing the decision matrix.

If $\tilde{x}_{ij} = (x_{ij}^l, x_{ij}^m, x_{ij}^u)$ as triangular fuzzy number denotes the performance of ith alternative and jth criterion, then the decision matrix is equal to:

$$\tilde{X} = \begin{matrix} \\ A_1 \\ A_2 \\ \vdots \\ A_m \end{matrix} \begin{matrix} C_1 & C_2 & \dots & C_n \\ \left[\begin{matrix} \tilde{x}_{11} & \tilde{x}_{12} & \dots & \tilde{x}_{1n} \\ \tilde{x}_{21} & \tilde{x}_{22} & \dots & \tilde{x}_{2n} \\ \vdots & \vdots & \ddots & \vdots \\ \tilde{x}_{m1} & \tilde{x}_{m2} & \dots & \tilde{x}_{mn} \end{matrix}\right] \end{matrix} = \left[\tilde{x}_{ij}\right]_{m\times n} \tag{14.13}$$

Step 2: Normalizing the decision matrix.

Normalizing a fuzzy decision matrix can be done as follows. (for each triangular fuzzy number $\tilde{x}_{ij} = (x_{ij}^l, x_{ij}^m, x_{ij}^u)$).

$$r_{ij}^l = \frac{x_{ij}^l}{\sqrt{\sum_{i=1}^m \left(x_{ij}^u\right)^2}}; \ i = 1, \dots, m, \ j = 1, \dots, n. \tag{14.14}$$

$$r_{ij}^{m} = \frac{x_{ij}^{m}}{\sqrt{\sum_{i=1}^{m}\left(x_{ij}^{u}\right)^{2}}}; \; i = 1, \ldots, m, \; j = 1, \ldots, n. \tag{14.15}$$

$$r_{ij}^{u} = \frac{x_{ij}^{u}}{\sqrt{\sum_{i=1}^{m}\left(x_{ij}^{u}\right)^{2}}}; \; i = 1, \ldots, m, \; j = 1, \ldots, n. \tag{14.16}$$

$\tilde{r}_{ij} = (r_{ij}^{l}, r_{ij}^{m}, r_{ij}^{u})$ is the normalized triangular fuzzy number. Therefore, the normalize decision matrix is equal to Eq. (14.17).

$$\tilde{R} = \left[\tilde{r}_{ij}\right]_{m \times n} = \begin{bmatrix} \tilde{r}_{11} & \tilde{r}_{12} & \ldots & \tilde{r}_{1n} \\ \tilde{r}_{21} & \tilde{r}_{22} & \ldots & \tilde{r}_{2n} \\ \vdots & \vdots & \ddots & \vdots \\ \tilde{r}_{m1} & \tilde{r}_{m2} & \ldots & \tilde{r}_{mn} \end{bmatrix} \tag{14.17}$$

Step 3: Computing the weighted normalized fuzzy decision matrix.

Considering $\tilde{\upsilon} = \left(\tilde{\upsilon}_1, \ldots, \tilde{\upsilon}_j, \ldots, \tilde{\upsilon}_n\right)$ as the fuzzy weight vector of criteria, the weighted normalized fuzzy decision matrix calculated by $\tilde{w}_{ij} = \tilde{\upsilon}_j.\tilde{r}_{ij}$ as follows:

$$\tilde{W} = \left[\tilde{w}_{ij}\right]_{m \times n} = \begin{bmatrix} \tilde{w}_{11} & \tilde{w}_{12} & \ldots & \tilde{w}_{1n} \\ \tilde{w}_{21} & \tilde{w}_{22} & \ldots & \tilde{w}_{2n} \\ \vdots & \vdots & \ddots & \vdots \\ \tilde{w}_{m1} & \tilde{w}_{m2} & \ldots & \tilde{w}_{mn} \end{bmatrix} \tag{14.18}$$

where,

$$\begin{aligned} \tilde{w}_{ij} &\simeq \tilde{\upsilon}_j.\tilde{r}_{ij} \simeq (\upsilon_{ij}^{l}, \upsilon_{ij}^{m}, \upsilon_{ij}^{u}).(r_{ij}^{l}, r_{ij}^{m}, r_{ij}^{u}) \simeq \left(\upsilon_{ij}^{l}.r_{ij}^{l}, \upsilon_{ij}^{m}.r_{ij}^{m}, \upsilon_{ij}^{u}.r_{ij}^{u}\right) \\ &= \left(w_{ij}^{l}, w_{ij}^{m}, w_{ij}^{u}\right) \end{aligned} \tag{14.19}$$

Step 4: Calculating the Overall performance score.

The overall performance score for each alternative obtains by Eq. (14.20)

$$R\tilde{S}_i = \sum_{j \in J^{+}} \tilde{w}_{ij} - \sum_{j \in J^{-}} \tilde{w}_{ij}; \; i = 1, \ldots, m. \tag{14.20}$$

where, $\tilde{w}_{ij}$ are elements of the weighted normalized fuzzy decision matrix and J^{+} shows the set of benefit criteria and J^{-} indicates the set of cost criteria.

Step 5: Ranking the alternatives.

The alternatives are ranked based on descending order $R\tilde{S}_i = \left(\tilde{y}_i^l, \tilde{y}_i^m, \tilde{y}_i^u\right)$, using one of the defuzzification methods such as bellow:

$$RS_i = \frac{\tilde{y}_i^l + 2\tilde{y}_i^m + \tilde{y}_i^u}{4}; \; i = 1, \ldots, m. \tag{14.21}$$

The higher the value RS_i, the better the alternative A_i.
Hence, the ratio system approach is terminated.

Step 6: Ranking the alternatives by the fuzzy reference point approach

After attaining the weighted normalized decision matrix in the third step, the most desirable performance of jth criterion among all alternatives obtained by Eq. (14.22).

$$\tilde{w}_j^* = \left(w_j^{*l}, w_j^{*m}, w_j^{*u}\right) = \begin{cases} \left(\max\limits_i \left\{w_{ij}^u\right\}, \max\limits_i \left\{w_{ij}^u\right\}, \max\limits_i \left\{w_{ij}^u\right\}\right); \; j \in J^+ \\ \left(\min\limits_i \left\{w_{ij}^l\right\}, \min\limits_i \left\{w_{ij}^l\right\}, \min\limits_i \left\{w_{ij}^l\right\}\right); \; j \in J^- \end{cases} \tag{14.22}$$

where J^+ shows the set of benefit criteria and J^- indicates the set of cost criteria.
Applying the Tchebycheff norm the reference point score is equal to:

$$\begin{aligned} RP_i^{fuzzy} &= \max_j d(\tilde{w}_j^*, \tilde{w}_{ij}) \\ &= \max_j \left\{ \sqrt{\left((w_j^{*l} - w_{ij}^l)^2 + (w_j^{*m} - w_{ij}^m)^2 + (w_j^{*u} - w_{ij}^u)^2\right)} \right\}, i = 1, \ldots, m. \end{aligned} \tag{14.23}$$

The alternatives are ranked based on ascending order of the Reference point score $R\tilde{P}_i; i = 1, \ldots, m$.

Step 7: Ranking the alternatives by the fuzzy full multiplicative form

The Overall utility of alternatives is as bellow:

$$M\tilde{F}_i = \frac{\prod\limits_{j \in J^+} \left(\tilde{x}_{ij}\right)^{w_j}}{\prod\limits_{j \in J^-} \left(\tilde{x}_{ij}\right)^{w_j}} = \frac{\prod\limits_{j \in J^+} \left((x_{ij}^l)^{w_j}, (x_{ij}^m)^{w_j}, (x_{ij}^u)^{w_j}\right)}{\prod\limits_{j \in J^-} \left((x_{ij}^l)^{w_j}, (x_{ij}^m)^{w_j}, (x_{ij}^u)^{w_j}\right)}; \; i = 1, \ldots, m \tag{14.24}$$

where J^+ denotes the set of benefit criteria and J^- shows the set of cost criteria.
The alternatives are ranked based on descending order of the values of $M\tilde{F}_i$ for $i = 1, \ldots, m$.

Step 8: Aggregating all three ranking methods

The three methods of MULTIMOORA are assumed to have the same importance. The ordinal scale of the three methods of MULTIMOORA may explain based on Dominance, being Dominated, Transitivity and Equability. The dominance analysis approach clarified by Brauers and Zavadskas [4] is applied to conclude ranking obtained by three parts of MULTIMOORA method.

Here, one can also use preferential voting to aggregate the rankings, and for this purpose, one can use model (14.10).

14.5 Economic Ranking of Urban Areas Using MOORA Method: A Comprehensive Evaluation Approach

Urban areas are the engines of economic growth, innovation, and cultural exchange. As cities evolve and become more complex, it becomes crucial to evaluate and rank them based on their economic potential, livability, and overall performance. This evaluation enables policymakers, investors, and individuals to make informed decisions regarding urban development strategies, resource allocation, and investment opportunities ([8, 16]).

In this section of the book, we will explore the application of the Multi-Objective Optimization by Ratio Analysis (MOORA) method to rank urban areas using an integrated evaluation framework. The MOORA method provides a systematic and comprehensive approach that combines multiple criteria and subjective assessments. By incorporating fuzzy logic, the MOORA method accommodates uncertainties and allows for a flexible evaluation process.

Criteria for Economic Ranking:

To assess the economic potential and overall performance of urban areas, we will consider a set of criteria that capture the value and credibility of various urban elements. These criteria encompass different dimensions, including education, administration, welfare, healthcare, commerce, finance, employment, safety, and accessibility. The criteria for this evaluation include:

C1: Value and Credibility of Educational Centers—This criterion evaluates the quality, reputation, and credibility of educational institutions within the urban area. Factors such as academic excellence, faculty expertise, research output, and accreditation contribute to the value and credibility of educational centers.

C2: Value and Credibility of Administrative Centers—The value and credibility of administrative centers assess the effectiveness and efficiency of governance structures and public services. Aspects such as transparency, accountability, responsiveness, and effectiveness in delivering public services contribute to the overall value and credibility of administrative centers.

C3: Value and Credibility of Welfare and Entertainment Centers—This criterion focuses on the value and credibility of welfare and entertainment centers, including parks, cultural venues, recreational facilities, and entertainment complexes. The presence of well-maintained and diverse amenities enhances the overall livability and attractiveness of urban areas.

C4: Value and Credibility of Healthcare Centers—Access to reliable and high-quality healthcare services is crucial for the well-being of residents. This criterion evaluates the value and credibility of healthcare centers, including hospitals, clinics, specialized facilities, and healthcare professionals, considering factors such as infrastructure, medical expertise, patient satisfaction, and technological advancements.

C5: Value and Credibility of Economic and Commercial Centers—Thriving economic and commercial centers contribute to a vibrant business environment and job opportunities. This criterion assesses the value and credibility of economic and commercial centers based on factors such as business diversity, investment opportunities, market competitiveness, and overall economic growth.

C6: Value and Credit of Banks and Credit Institutions—The value and credit of banks and credit institutions determine the financial stability and access to capital within an urban area. This criterion evaluates the reliability, credibility, and range of financial services provided, along with factors such as accessibility, innovation, and customer satisfaction.

C7: Employment Rate—The employment rate serves as a fundamental indicator of economic prosperity and social well-being. This criterion evaluates the availability of job opportunities, job growth, and the overall employment rate within the urban area.

C8: Insecurity Rate—Safety and security are essential for the well-being of residents and the attractiveness of an urban area. This criterion assesses the level of insecurity, including theft, robbery, and pickpocketing incidents, and measures the efforts taken to ensure public safety.

C9: Minimum Waiting Time to Reach Public Transportation—Efficient and accessible public transportation systems are vital for urban mobility and connectivity. This criterion evaluates the minimum waiting time required to reach public transportation hubs, considering factors such as transportation infrastructure, accessibility, and reliability.

Table 14.4 Transformation of linguistic variables in determining the importance of criteria weight

Linguistic terms	Fuzzy number
Very low importance (VL)	(0,0,0.1)
Low importance (L)	(0,0.1,0.3)
Medium low importance (ML)	(0.1,0.3,0.5)
Medium importance (M)	(0.3,0.5,0.7)
Medium high importance (MH)	(0.5,0.7,0.9)
High importance (H)	(0.7,0.9,1)
Very High importance (VH)	(0.9,1,1)

Table 14.5 Transformation of linguistic variables in determining the performance of alternatives

Linguistic terms	Fuzzy number
Very Poor (VP)	(0,0,1)
Poor (P)	(0,1,3)
Medium Poor (MP)	(1,3,5)
Fair (F)	(3,5,7)
Medium Good (MG)	(5,7,9)
Good (G)	(7,9,10)
Very Good (VG)	(9,10,10)

By considering these criteria and their associated values and credibility, we aim to provide a comprehensive evaluation of urban areas based on their economic potential and overall Tables 14.4 and 14.5 demonstrate the transformation of linguistic variables ,applied in this chapter, for criteria and alternatives, respectively. Table 14.6 illustrates three experts' opinion about the nine above mentioned criteria, using linguistic variables. Applying arithmetic mean, the experts' opinion are aggregated and exhibited in Table 14.7.

Table 14.6 Experts' opinions about the importance of criteria

Criteria	Expert1	Expert2	Expert3
C1	L	ML	V
C2	ML	MH	M
C3	MH	ML	H
C4	M	M	MH
C5	H	ML	M
C6	VH	MH	MH
C7	ML	ML	ML
C8	M	MH	M
C9	MH	H	VH

Table 14.7 The final weight obtained from the aggregation of experts using the arithmetic mean

Criteria	Weight		
	w^L	w^M	w^U
C1	0.033	0.133	0.3
C2	0.3	0.5	0.7
C3	0.433	0.633	0.8
C4	0.367	0.567	0.767
C5	0.367	0.567	0.733
C6	0.633	0.8	0.933
C7	0.1	0.3	0.5
C8	0.367	0.567	0.767
C9	0.7	0.867	0.967

The experts' opinions about the importance of alternatives are mentioned in Table 14.8. linguistic variables are transformed to fuzzy numbers and all three experts' opinion are aggregated using arithmetic mean, the result is illustrated in Tables 14.9, 14.10, 14.11, 14.12 and 14.13.

Considering the rankings obtained from Tables (3.11), (3.12), and (3.13). A1 appeared twice in the first rank and once in the second rank. Therefore, in Table 14.14, a value of 2 is assigned to it in the column corresponding to the first position, and a value of 1 is assigned in the column corresponding to the second position. Since A1 did not appear in the other ranks, we assign a value of 0 to the remaining columns as the same way, the preferential voting of the other alternatives is obtained and assigned to Table 14.14.

Applying the preferential voting, the final rank is attained and presented in the last column of Table 14.14.

Table 14.8 Experts' opinions about the alternatives

Criterion	A1			A2			A3			A4		
	Expert1	Expert2	Expert3	Expert1	Expert2	Expert3	Expert1	Expert2	Expert3	Expert1	Expert2	Expert3
C1	MP	G	MG	G	F	MP	P	F	G	VG	P	VG
C2	F	MG	MP	VG	G	F	F	MP	P	G	MG	VP
C3	G	MG	MG	MG	VG	P	MP	MP	F	VP	VP	MG
C4	MG	F	VG	MG	MG	F	P	VG	MG	MP	F	G
C5	F	MG	G	G	F	VG	MG	P	VG	MP	MP	G
C6	MG	MP	MP	F	MG	MG	F	MP	MP	F	G	VG
C7	MP	MG	F	MG	F	MP	MP	VG	MG	MP	MP	G
C8	MG	MP	MP	MP	VG	MG	MG	MP	VG	P	MP	F
C9	F	MP	P	G	MG	MP	MP	MP	VP	MG	MG	P

Table 14.9 Normalized aggregated decision matrix

	A1	A2	A3	A4
C1	(0.291, 0.426, 0.538)	(0.247, 0.381, 0.493)	(0.224, 0.336, 0.448)	(0.404, 0.471, 0.516)
C2	(0.212, 0.354, 0.496)	(0.448, 0.566, 0.637)	(0.094, 0.212, 0.354)	(0.283, 0.378, 0.472)
C3	(0.415, 0.561, 0.684)	(0.342, 0.439, 0.537)	(0.122, 0.269, 0.415)	(0.122, 0.171, 0.269)
C4	(0.357, 0.462, 0.546)	(0.273, 0.399, 0.525)	(0.294, 0.378, 0.462)	(0.231, 0.357, 0.462)
C5	(0.314, 0.439, 0.543)	(0.397, 0.502, 0.564)	(0.293, 0.376, 0.460)	(0.188, 0.314, 0.418)
C6	(0.156, 0.290, 0.424)	(0.290, 0.424, 0.558)	(0.112, 0.246, 0.380)	(0.424, 0.536, 0.603)
C7	(0.209, 0.348, 0.487)	(0.209, 0.348, 0.487)	(0.348, 0.464, 0.557)	(0.209, 0.348, 0.464)
C8	(0.168, 0.312, 0.456)	(0.360, 0.480, 0.576)	(0.360, 0.480, 0.576)	(0.096, 0.216, 0.360)
C9	(0.108, 0.244, 0.406)	(0.352, 0.515, 0.650)	(0.054, 0.163, 0.298)	(0.271, 0.406, 0.569)

Table 14.10 Weighted normalized decision matrix of economic ranking of urban areas

	A1	A2	A3	A4
C1	(0.010, 0.057, 0.161)	(0.008, 0.051, 0.148)	(0.007, 0.045, 0.135)	(0.013, 0.063, 0.155)
C2	(0.064, 0.177, 0.347)	(0.135, 0.283, 0.446)	(0.028, 0.106, 0.248)	(0.085, 0.189, 0.330)
C3	(0.180, 0.356, 0.547)	(0.148, 0.278, 0.430)	(0.053, 0.170, 0.332)	(0.053, 0.108, 0.215)
C4	(0.131, 0.262, 0.418)	(0.100, 0.226, 0.402)	(0.108, 0.214, 0.354)	(0.085, 0.202, 0.354)
C5	(0.115, 0.249, 0.399)	(0.146, 0.284, 0.414)	(0.107, 0.213, 0.337)	(0.069, 0.178, 0.307)
C6	(0.099, 0.232, 0.396)	(0.184, 0.340, 0.521)	(0.071, 0.197, 0.354)	(0.269, 0.429, 0.563)
C7	(0.021, 0.104, 0.244)	(0.021, 0.104, 0.244)	(0.035, 0.139, 0.278)	(0.021, 0.104, 0.232)
C8	(0.062, 0.177, 0.349)	(0.132, 0.272, 0.441)	(0.132, 0.272, 0.441)	(0.035, 0.122, 0.276)
C9	(0.076, 0.211, 0.393)	(0.246, 0.446, 0.628)	(0.038, 0.141, 0.288)	(0.190, 0.352, 0.550)

Table 14.11 Ranking result of the ratio system approach

Alternative	$R\tilde{S}_i = \left(\tilde{y}_i^l, \tilde{y}_i^m, \tilde{y}_i^u\right)$	RS_i	Rank
A1	(0.482, 1.049, 1.7701)	1.09	1
A2	(0.363, 0.849, 1.535)	0.90	2
A3	(0.239, 0.672, 1.309)	0.72	4
A4	(0.370, 0.799, 1.330)	0.82	3

Table 14.12 Ranking result of the reference point approach

Alternative	RP_i^{fuzzy}	Rank
A1	0.579302	2
A2	0.747475	4
A3	0.550731	1
A4	0.619522	3

Table 14.13 Ranking result of the full multiplicative form approach

Alternative	$M\tilde{F}_i$	MF_i	Rank
A1	(0.213,0.266,0.431)	0.284	1
A2	(0.156,0.177,0.189)	0.175	2
A3	(0.085,0.109,0.199)	0.120	4
A4	(0.121,0.124,0.159)	0.129	3

Table 14.14 Voting table and final rank

Alternative	1th	2th	3th	4th	Result of model (14.10)	Final ranking
A1	2	1	0	0	1.000	1
A2	0	2	0	1	0.716	3
A3	1	0	0	2	0.733	2
A4	0	0	3	0	0.650	4

14.6 Conclusion

The MOORA method proves to be a valuable tool for solving multi-criteria problems, providing a systematic approach for decision-making and ranking alternatives. The example presented in this chapter demonstrates the practical application of the MOORA method in a real-world context, showcasing its effectiveness in handling complex decision scenarios and providing meaningful rankings based on multiple criteria. Moreover, the extension of the MOORA method into the fuzzy environment opens up new possibilities for addressing decision-making problems with uncertain and imprecise information. By incorporating fuzzy logic, the fuzzy MOORA method allows decision-makers to deal with subjective assessments and vague criteria, enhancing the applicability of the method in various domains. The real-world example discussed highlights both the advantages and limitations of the fuzzy MOORA method, emphasizing the need for careful consideration of fuzzy data and appropriate interpretation of the results. Overall, the MOORA method, whether in its original form or in the fuzzy context, offers a robust framework for tackling multi-criteria problems. Its ability to consider multiple criteria, accommodate uncertainty, and provide comprehensive rankings makes it a valuable asset for decision-makers across various domains. By understanding and applying the MOORA method, practitioners can make informed decisions and optimize their choices based on multiple objectives and criteria.

Acknowledgement A special thanks to the Iranian DEA society for their unwavering spiritual support and consensus in the writing of this book. Your invaluable support has been truly remarkable, and we are deeply grateful for the opportunity to collaborate with such esteemed professionals.

References

1. Aytaç Adalı, E., & Tuş Işık, A. (2017). The multi-objective decision making methods based on MULTIMOORA and MOOSRA for the laptop selection problem. *Journal of Industrial Engineering International, 13*, 229–237.
2. Brauers, W. K. M., & Zavadskas, E. K. (2006). The MOORA method and its application to privatization in a transition economy. *Control and Cybernetics, 35*(2), 445–469.
3. Brauers, W. K. M., & Zavadskas, E. K. (2009). Multi objective optimization with discrete alternatives on the basis of ratio analysis. *Intellect Econ, 2*(6), 30–41.
4. Brauers, W. K. M., & Zavadskas, E. K. (2011). Multimoora optimization used to decide on a bank loan to buy property. *Technological and Economic Development of Economy, 17*(1), 174–188.
5. Brauers, W. K. M., Zavadskas, E. K., Turskis, Z., & Vilutiene, T. (2008). Multiobjective contractor's ranking by applying the MOORA method. *Journal of Business Economics and Management, 9*(4), 245–255.
6. Cook, W. D., & Kress, M. (1990). A data envelopment model for aggregating preference rankings. *Management Science, 36*(11), 1302–1310.
7. Das, M. C., Sarkar, B., & Ray, S. (2012). Decision making under conflicting environment: A new MCDM method. *International Journal of Applied Decision Sciences, 5*(2), 142–162.
8. Deveci, M., Pamucar, D., Gokasar, I., Pedrycz, W., & Wen, X. (2022). Autonomous bus operation alternatives in urban areas using fuzzy Dombi-Bonferroni operator based decision making model. IEEE Transactions on Intelligent Transportation Systems.
9. Jagadish, & Ray, A. (2015). Green cutting fluid selection using multi-attribute decision making approach. *Journal of The Institution of Engineers (India): Series C, 96*, 35–39.
10. Jagadish, R. A., & Ray, A. (2014). Green cutting fluid selection using MOOSRA method. *International Journal of Research in Engineering and Technology, 3*(3), 559–563.
11. Jahanshahloo, G., Lotfi, F. H., Khanmohammadi, M., & Kazemimanesh, M. (2012) A Method for Discriminating Efficient Candidates with Ranked Voting Data by Common Weights Mathematical and Computational Applications 17(1) 1-8 https://doi.org/10.3390/mca 17010001
12. Karande, P., & Chakraborty, S. (2012). Application of multi-objective optimization on the basis of ratio analysis (MOORA) method for materials selection. *Materials & Design, 37*, 317–324.
13. Kracka, M., Brauers, W. K. M., & Zavadskas, E. K. (2010). Ranking heating losses in a building by applying the MULTIMOORA. *The Engineering Economist, 21*(4), 352–359.
14. Kumar, R., & Jagadish, R. A. (2013). Selection of material: a multi-objective decision making approach. In Proceeding of ICIE-2013 international conference on industrial engineering (pp. 162–165).
15. Kumar, R., & Ray, A. (2015). Selection of material under conflicting situation using simple ratio optimization technique. In: Das et al. (Eds.) Proceedings of fourth international conference on soft computing for problem solving, advances in intelligent systems and computing 335: 513–519.
16. Memarzadeh, R., Jahanshahloo, L., Hosseinzadeh Lotfi, F., & Dehghan Touran Poshti, A. (2021). Resilience of urban areas via locating and ranking of fire stations, a case study: district ten of tehran municipality. *Journal of Geography and Environmental Hazards, 10*(2), 129-146.
17. Miller, D. W., & Starr, M. K. (1969). Executive decisions and operations and research. 2nd Edition, Prentice- Hall Inc., Englewood Cliffs (N.J.).
18. Sharafi, H., Soltanifar, M., & Lotfi, F. H(2022) Selecting a green supplier utilizing the new fuzzy voting model and the fuzzy combinative distance-based assessment method EURO Journal on Decision Processes https://doi.org/10100010-10.1016/j.ejdp.2021.100010

19. Soltanifar, M., Sharafi, H., Lotfi, F. H., Pedrycz, W., & Allahviranloo, T. (2023). *Preferential Voting and Applications: Approaches Based on Data Envelopment Analysis* (Vol. 471). Springer Nature.
20. Tian, Z. P., Wang, J. Q., Wang, J., & Zhang, H. Y. (2018). A multi-phase QFD-based hybrid fuzzy MCDM approach for performance evaluation: A case of smart bike-sharing programs in Changsha. *Journal of Cleaner Production, 171*, 1068–1083.

MIX
Papier aus verantwortungsvollen Quellen
Paper from responsible sources
FSC® C105338

If you have any concerns about our products,
you can contact us on
ProductSafety@springernature.com

In case Publisher is established outside the EU,
the EU authorized representative is:
Springer Nature Customer Service Center GmbH
Europaplatz 3, 69115 Heidelberg, Germany

Printed by Libri Plureos GmbH
in Hamburg, Germany